Insight and Innovation in International Development

Edited by International Development Research Centre
Ottawa, Ontario, Canada

Titles in this series:
Göransson, B., Brundenius, C., eds., Universities in Transition
Charron, D.F., ed., Ecohealth Research in Practice

For further volumes:
http://www.springer.com/series/8850

Dominique F. Charron
Editor

Ecohealth Research in Practice

Innovative Applications of an Ecosystem Approach to Health

International Development Research Centre
Ottawa • Cairo • Dakar • Montevideo • Nairobi • New Delhi • Singapore

Editor
Dominique F. Charron
International Development Research Centre
PO Box 8500
Ottawa, ON, Canada K1G 3H9
dcharron@idrc.ca; ecohealth@idrc.ca

A co-publication with the
International Development Research Centre
PO Box 8500
Ottawa, ON, Canada K1G 3H9
info@idrc.ca/ www.idrc.ca
ISBN (e-book) 978-1-55250-529-8

Farmers in Ecuador discuss potato production methods that use less pesticide.
Photo: Xavier Mera.

ISBN 978-1-4614-0516-0 e-ISBN 978-1-4614-0517-7
DOI 10.1007/978-1-4614-0517-7
Springer New York Dordrecht Heidelberg London

Library of Congress Control Number: 2011937450

© International Development Research Centre, 2012
All rights reserved. This work may not be translated or copied in whole or in part without the written permission of the publisher (Springer Science+Business Media, LLC, 233 Spring Street, New York, NY 10013, USA), except for brief excerpts in connection with reviews or scholarly analysis. Use in connection with any form of information storage and retrieval, electronic adaptation, computer software, or by similar or dissimilar methodology now known or hereafter developed is forbidden.
The use in this publication of trade names, trademarks, service marks, and similar terms, even if they are not identified as such, is not to be taken as an expression of opinion as to whether or not they are subject to proprietary rights.
While the advice and information in this book are believed to be true and accurate at the date of going to press, neither the authors nor the editors nor the publisher can accept any legal responsibility for any errors or omissions that may be made. The publisher makes no warranty, express or implied, with respect to the material contained herein.

Printed on acid-free paper

Springer is part of Springer Science+Business Media (www.springer.com)

Foreword

Advancing the field of ecosystem approaches to human health (ecohealth) has been a major contribution of IDRC in its efforts to improve the health of communities in the poorest regions of the world. Research, education, and practice in ecohealth have seen almost logarithmic development since 1996, and the number of scientists who have adopted this approach has steadily increased. This book examines the fundamentals of ecohealth from the perspective of the achievements and shortfalls of 15 research projects that are presented as study cases.

In the first chapter of this book, the main principles and intrinsic goals are discussed: transdisciplinarity, systems thinking, multistakeholder participation, equity, environmental sustainability, and evidence for community-based interventions. The excellent discussion of these principles, along with a description of how to put them in practice in ecohealth projects, adds new understanding to the various research strategies that have been used in the field. The book builds on earlier work by Gilles Forget (1997) and Gilles Forget and Jean Lebel (2001) (both with IDRC at the time).

These foundational publications have guided ecohealth researchers and been beacons for navigation and progress in the field of ecohealth. These efforts by IDRC emerged within the context of a broader international effort to better link human activities with ecosystems to solve health problems. In this sense, it was part of a global effort to restore our planet's ecological equilibrium. By considering humans as integral parts of ecosystems, ecohealth's main contribution is its recognition of the interdependence of human and environmental health.

Ecohealth continues to demonstrate its relevance and effectiveness. Globalization and heightened social and economic interactions, overexploitation of the earth's resources, climate change, and an increase in the number, severity, and extension of natural disasters have all contributed to our awareness of the interdependency of the fate of human societies and the well-being of our planet. This awareness is more often influencing the scientific frameworks of health-research projects. Many investigators and their students are now more familiar with the advantages of conducting transdisciplinary research in partnership with stakeholders, including the affected communities. The active participation of the ecohealth community has contributed to this awareness and familiarity. The field continues to communicate its approach to students and other researchers through project-development workshops and, more recently, through regional communities of practice. The role of voluntary organizations of ecohealth practitioners is well described in Chap. 21.

The active participation of the ever-growing ecohealth community is probably the most effective driver of further development and dissemination of the approach. This participation is characteristically based on the ecohealth principles of transdisciplinary partnership among equals for the strategic development of the field. Its importance was evident in major international meetings, notably the International Forum on Ecosystem Approaches to Human Health in Montreal, Canada, in 2003 and the Second EcoHealth Forum and Biennial Conference of the International Association of Ecology and Health held in Mérida, Mexico, in 2008. During these events, intellectual and empirical wisdom combined to advance our understanding of what ecohealth meant.

Since the first Forum in 2003, practitioners in ecohealth have shared a similar philosophy based on the right of all individuals and social groups to access health; the fundamental need for community development and environmental sustainability; the pertinence and advantages of transdisciplinary research; the need for an open-minded approach to understand how others see reality; the importance of social and gender considerations to the creation of equal-opportunity societies; and that consensus was attainable.

The second Forum in Mérida helped position the field of ecohealth as a key international advocate for the idea that the preservation of human health is indissolubly linked to the health of the environment. This was achieved in no small part through the collaboration and participation of the organizing partners in the second Forum (International Association for Ecology and Health; the Oswaldo Cruz Foundation, Brazil; Institute of Ecological Research, Brazil; University of São Paulo, Brazil; Pan American Health Organization; Instituto Nacional de Salud Publica, Mexico; and IDRC). A call for action (Soskolne and Westra 2010) from the Forum's deliberations advocated global adoption of the ecohealth movement.

This book provides examples of how ecohealth projects can be used to develop environmentally friendly interventions. It also shows that there are many scientifically sound strategies to conduct research with an ecosystem approach to achieve successful outcomes. However, there remain conspicuous challenges to the implementation of an ecohealth approach, and these are discussed in the last chapter. The normal tensions that arise from the interactions and different interests of the diversity of participating stakeholders reflect the complex nature of human interactions. No recipes can be offered. Navigating through these hardships requires visionary leadership and constructive imagination to build trusting partnerships. Readers can imagine the magnitude of these difficulties by considering the complexity of the processes involved in attaining the many achievements described in the projects. Nevertheless, what permeates the case studies is that, *malgré tout,* conducting this type of research is not only effective but also much fun.

Still, other challenges must be overcome before the field (or practitioners) will fully exploit the potentials of ecohealth. Although achieving social equity is a main driving force in the design of ecosystem interventions, current practice does not go far enough in recognizing the disparities in the health conditions of men and women and does not yet lead to specific interventions that result in gender equity. The inclusion of social scientists in ecohealth research teams is necessary for achieving this, but not sufficient. Dialogue and education among communities of practice must have a tighter focus on the issue of gender equity. Building on the gains of the worldwide feminist movement would contribute a cultural substrate to help address this persistent (and neglected) issue. The projects described in this book offer good opportunities to examine gender perspectives by identifying gaps and opportunities within the interventions that are described.

The ultimate objective of ecohealth research and practice is to develop environmentally sustainable, community-based interventions to improve the health of affected communities. Much success has resulted from the incorporation of community transformation and empowerment as key project objectives. In some cases, the participation of government health services has guided the design of interventions suitable for adoption by health programs. However, besides the relatively successful examples of six projects (in Ecuador, Cuba, Guatemala, Mexico, Nepal, and Tanzania) presented in this book, project outcomes often have limited direct influence on transforming health programs and even more limited influence on health policy. In this book, the inclusion of projects with variable influence on health policy offers the opportunity to examine both the suitability and scope of the proposed interventions and the nature of the external factors that influence their adoption by health services.

Ecohealth research and practice have all the attributes of public health and should be identified as such. Both ecohealth and public health aspire at social equity through healthy societies, and share strategies for community participation and empowerment for the solution of health problems. But, there is currently an important difference in the emphasis put on the main actors and promoters of health care and promotion activities by these two fields. Public health is traditionally considered the responsibility of the state; whereas, ecohealth stresses the involvement of communities and seems to have (maybe out of frustration) relegated the role of the state to second place in the solution of the problems.

Ecohealth's influence on policymaking depends on how suitable the interventions are for application in communities outside those where they were developed and tested, and to some extent on the involvement of the state. It also depends on the pragmatic sustainability of the activities, processes, and social organization that are needed for their application. In these respects, ecohealth projects are no different from successful public-health interventions. As discussed in the last chapter of this book, strategies for up-scaling will facilitate the adoption by wider populations of project results. However, possible difficulties are likely to arise from both the specificity and the high level of community involvement in ecohealth interventions. These challenges require further analysis and discussion by the ecohealth community. Consideration and inclusion of interactions of regional and country health services with the communities they serve would clearly facilitate their inclusion in public health programs. Strategies that link the state's responsibility for public health with the participation of communities in decision making about their own health and their environments could provide opportunities for the up-scaling of sustainable ecohealth interventions.

This book, besides being informative, is inspirational. The introductory chapters, the description of the research projects, and the closing chapters that analyze the scope and challenges of ecohealth all together present evidence that this is a lively, evolving field with a clear mission, driven by an ever-growing scientific community of practice.

Cuernavaca, Morelos, Mexico   Mario-Henry Rodriguez

References

Forget, G. (1997). From Environmental Health to Health and the Environment: Research that Focuses on People. In: Shahi, G.S., Levy, B.S., Binger, A., Kjellström, T., and Lawrence, R. (Editors). International Perspectives on Environment, Development and Health: Towards a Sustainable World, Springer, New York, NY, USA.

Forget, G., and Lebel, J. (2001). An Ecosystem Approach to Human Health. International Journal of Occupational and Environmental Health, 7(2)(Suppl), S3–38.

Soskolne, C.L., and Westra, L. (2010). Public Health in the Face of Global Ecological and Climate Change. In: Engel, J.R., Westra, L., and Bosselmann, K. (Editors). Democracy, Ecological Integrity and International Law [Editors:]. Cambridge Scholars Publishing, Newcastle upon Tyne, UK. (Draft call for action from Mérida appears as an appendix on pages 261–263.)

Preface

Ecohealth Research in Practice: Innovative Applications of an Ecosystem Approach to Health represents a stocktaking by Canada's International Development Research Centre (IDRC). It builds on the results from nearly 15 years of support for ecohealth research, capacity-building activities, and networks. Created by the Parliament of Canada in 1970, IDRC helps developing countries use science and technology to find practical, long-term solutions to their social, economic, and environmental problems. IDRC does this by supporting developing country researchers, building research capacity, and fostering the uptake and use of research findings in policy and decision-making processes in developing countries.

In 1997, building on decades of experiences in health and environment research, IDRC initiated the Ecosystem Approaches to Human Health (or Ecohealth) research program. Its initial aim was to support innovative research on how improved human health could be attained from better natural resource and environmental management. This approach went beyond the then-dominant "environmental determinants of health" paradigm; it drew on developments in public and international health and emphasized three pillars: transdisciplinarity, multistakeholder participation, and gender and social analysis. This thinking has continued to inform IDRC ecohealth programming and provided the foundation for the practices, experiences, and achievements presented in this book.

Ecosystem approaches to health were initially widely disseminated in *Health: An Ecosystem Approach* as part of IDRC's *In Focus* series, which is openly accessible online (http://www.idrc.ca/in_focus_health/). The booklet was launched during the International Forum on Ecosystem Approaches to Human Health held in Montreal in 2003. Originally intended for an audience of nonspecialist users of research results, this small book written in layman's terms was immediately taken up by researchers to develop research projects and by academics to inform new curricula (Communications Division, IDRC, Ottawa, Canada. Evaluation of the In Focus Series, 2009 (unpublished data)). An unofficial textbook for implementing IDRC projects, it has been translated into French, Spanish, Arabic, and Chinese.

Around the same time, IDRC joined with organizations and researchers pursuing related ideas to launch a scientific journal called *EcoHealth*. An International Association for Ecology and Health was formed in 2006 to support the journal. Meanwhile, as greater numbers of researchers became interested in this approach, IDRC funded communities of practice and other networks to further develop expert peer groups to advance ecohealth theory, practice, and curriculum development.

What followed was a rapid expansion and development of the new field of ecohealth. Other donors joined IDRC in supporting research using an ecosystem approach to health. Communities of practice co-opted existing research and adapted it to ecosystem approaches. Journal publications contributed to an expanding knowledge base. New generations of researchers were trained and undertook ecohealth research of their own in the developing south as well as in northern universities. This work, often supported by IDRC, both informed and was influenced by wider scholarly debate around ecohealth and ecosystem approaches, and the global community's coming-to-terms with the impacts on health of wide-scale global environmental change. Ideas of systemic links between ecosystems and human health and well-being, mediated through social, economic, and cultural processes and practices, are now well established, as is the need for research to take greater account of a complex context.

In December 2008, IDRC, in partnership with Mexico's National Institute for Public Health (INSP), the Pan-American Health Organization and several other organizations, convened a second International EcoHealth Forum, in conjunction with the second biennial conference of the International Association for Ecology and Health. It was a watershed moment in the evolution of the field. The Forum revealed how much the field had grown (700 delegates from 70 countries), highlighted the impact of past IDRC investments in ecohealth research for development, and identified existing challenges. In his provocative Foreword to this book, Dr. Mario-Henry Rodriguez, Director General of INSP, discusses many of these challenges, including the need for greater emphasis on achieving gender equity, more persuasive arguments for decision makers to make policy changes based on research, and the need to develop strategies to translate local success into wider-scale impacts.

For more than a decade, researchers applying ecohealth ideas have successfully published peer-reviewed scientific results. However, the sum of the rich experience of ecohealth research has not yet been published in a book that articulates the evolution of ecosystem approaches to health. *Ecohealth Research in Practice* strives to help fill this gap in publication and to transmit the successes and strengths of an ecosystem approach to health. In doing so, it exposes many remaining challenges for researchers, educators, and practitioners. It also sets out some ideas on what might be needed to move the field forward.

The book is structured around 15 case studies of ecohealth research grouped into four sections, each focusing on international development themes of particular relevance to environmental health and represented in IDRC's programming history: agricultural transformations, environmental pollution, vector-borne diseases, and degraded urban ecosystems. The case studies are narratives that describe the process of implementing an ecosystem approach, from design to outcome. They represent a wide range of experiences, each specific to the nature of the problem at hand and the local context. Each case study represents a success in terms of positive and lasting changes that might not otherwise have been achievable. In many cases, the results were achieved after years of striving to develop and apply research findings to a problem through the engagement of communities and decision makers. They represent long-term investments by IDRC and by the researchers and stakeholders involved.

Three additional chapters help frame the case studies. The first chapter gives an overview of the history and conceptual underpinnings of ecohealth. Informed by the case studies that follow, it also describes a new, expanded understanding of an ecosystem approach to health. Chapter 21 explores the contributions of networks and communities of practice to the development of the field of ecohealth. The final chapter summarizes the key results of ecohealth research and describes the current practice, including gaps and challenges to be addressed in the future.

The experiences and insights presented are necessarily retrospective and selective. IDRC's portfolio includes far more excellent projects than could be described in this book. Unfortunately, entire thematic areas could not be included. For example, we are supporting a growing body of research on the social and environmental change dimensions of emerging diseases like avian influenza, and on the implications of climate change for health, including nutrition and food security. Other vehicles will be found to disseminate these and other achievements.

Who should read this book? It should appeal to anyone interested in concepts and experiences on how to study health issues that result from interacting social, economic, and ecological processes, or in strategies to address these issues in the real world, with real communities, and in real time. Because its contributors are primarily academically trained researchers (mostly from developing countries), we think it will appeal to academicians and students in both developed and developing countries interested in ecohealth theory and its application. But it will also be useful to practitioners in public health, development, agriculture, and environment who are interested in learning about a participatory form of inquiry that leads to positive and lasting change. For anyone curious about ecosystem approaches to health, the book provides background and context, as well as examples of its application to understand and better manage different kinds of health problems, like those from agroecosystems or slums, mining pollution, or vector-borne diseases.

Ecohealth Research in Practice presents a field of research that is responding to the need for evidence-based strategies to improve health through practical, equitable, and sustainable changes in practices, policies, and programs. We hope it will inspire and better equip people to continue to work with the ideas of ecohealth and to make a difference for people struggling to live healthy and productive lives in developing countries.

Lastly and on a more personal note, the *ecosystem approaches to health* experience at IDRC went far beyond any of our original expectations. Back in 1997, only a few of us appeared to recognize the scientific potential of such approaches. The building of this field of knowledge remains a demanding task. It is a collective endeavor of researchers – many supported by IDRC but also a growing number of others – requiring regular debate. We would like to acknowledge the efforts of IDRC staff, past and present, who have worked tirelessly and in sometimes very challenging situations to advance the field of ecohealth. This book stands as part of their legacy to a global community striving for healthier, more equitable, and more environmentally sustainable lives and livelihoods all over the world.

Ottawa, ON

Dominique F. Charron
Jean Lebel

Acknowledgments

First thanks are due to the contributing authors for their enthusiastic response to this opportunity to collect and present ecohealth research supported by IDRC. I am also grateful for the participation and engagement of IDRC's research partners, as well as community members and other stakeholders around the world who enabled and contributed to the work presented in this book. Special thanks also to Dr Mario-Henry Rodriguez for his insightful foreword.

I am very grateful to Craig Stephen who provided invaluable critique and suggestion at various stages of writing and whose input and debate improved the book immeasurably. Several others similarly challenged me, including David Waltner-Toews, Jean Lebel, and Margot Parkes, and I thank them for engaging in this essential role.

Some of the ideas in Chaps. 1 and 22 were influenced by discussions with students and colleagues at IDRC, CoPEH Canada, Faculté de médecine vétérinaire de l'Université de Montréal, the 2008 International EcoHealth Forum in Mérida, Mexico, and the 2010 EcoHealth Conference in London, UK.

This book could not have been produced without the tireless efforts of the IDRC editing team. I am particularly grateful to Andrés Sánchez and Zsófia Orosz, who took on particularly large shares of the work, as well as Roberto Bazzani, Ana Boischio, Alicia Iglesias, and Martin Wiese. Other IDRC colleagues Hein Mallee, François Gasengayire, and Ernest Dabiré contributed elements and challenged ideas, leading to a better final product. Lamia El-Fattal is acknowledged for contributions to earlier drafts of parts of the book. I thank Francine Sinzinkayo, Jayne Bergeron, Michéle Lafleur, and Vilma Gamero for administrative support, Bill Carman and IDRC's Communications Division for publishing assistance, and the rest of Team IDRC. A special thank you is due to IDRC's Innovation and Impact Editorial Committee for their decision to include this book in IDRC's publication series of the same name. Many thanks also to Springer for co-publishing the book with IDRC.

The work in this book reflects years of IDRC efforts, including the contributions of former program leaders and staff too numerous to list here. For their efforts and contributions, I am in their debt.

Finally, I am particularly grateful to Michael Graham (www.mgedit.com) for editing support throughout.

Dominique F. Charron

Contents

1 Ecohealth: Origins and Approach 1
Dominique F. Charron

Part I Linking Human Health and Well-Being to Changing Rural Agro-Ecosystems

2 Introduction 33
Lamia El-Fattal and Andrés Sánchez

3 Growing Healthy Communities: Farmer Participatory Research to Improve Child Nutrition, Food Security, and Soils in Ekwendeni, Malawi 37
Rachel Bezner Kerr, Rodgers Msachi, Laifolo Dakishoni,
Lizzie Shumba, Zachariah Nkhonya, Peter Berti,
Christine Bonatsos, Enoch Chione, Malumbo Mithi,
Anita Chitaya, Esther Maona, and Sheila Pachanya

4 Tackling Challenges to Farmers' Health and Agro-Ecosystem Sustainability in Highland Ecuador 47
Fadya A. Orozco and Donald C. Cole

5 Coping with Environmental and Health Impacts in a Floricultural Region of Ecuador 59
Jaime Breilh

6 Dietary Diversity in Lebanon and Yemen: A Tale of Two Countries 69
Malek Batal, Amin Al-Hakimi, and Frédéric Pelat

Part II Natural Resources, Ecosystems, Pollution, and Health

7 Introduction 83
   Ana Boischio and Zsófia Orosz

8 An Ecosystem Study of Manganese Mining
   in Molango, Mexico 87
   Horacio Riojas-Rodríguez and Sandra Rodríguez-Dozal

9 Ecohealth Research for Mitigating Health Risks
   of Stone Crushing and Quarrying, India 99
   Raghwesh Ranjan, K. Vijaya Lakshmi, and Kalpana Balakrishnan

10 A Virtuous Cycle in the Amazon: Reducing Mercury Exposure
    from Fish Consumption Requires Sustainable Agriculture 109
    Jean Remy Davée Guimarães and Donna Mergler

11 Impacts on Environmental Health of Small-Scale Gold
    Mining in Ecuador 119
    Óscar Betancourt, Ramiro Barriga, Jean Remy Davée Guimarães,
    Edwin Cueva, and Sebastián Betancourt

Part III Poverty, Ecosystems, and Vector-Borne Diseases

12 Introduction 133
    Roberto Bazzani and Martin Wiese

13 Malaria Research and Management Need Rethinking:
    Uganda and Tanzania Case Studies 139
    Joseph Okello-Onen, Leonard E.G. Mboera, and Samuel Mugisha

14 An Ecosystem Approach for the Prevention of Chagas Disease
    in Rural Guatemala 153
    Carlota Monroy, Xochitl Castro, Dulce Maria Bustamante,
    Sandy Steffany Pineda, Antonieta Rodas, Barbara Moguel,
    Virgilio Ayala, and Javier Quiñonez

15 Preventing Dengue at the Local Level in Havana City 163
    Cristina Díaz

16 Eco-Bio-Social Research on Dengue in Asia:
    General Principles and a Case Study from Indonesia 173
    S. Tana, W. Abeyewickreme, N. Arunachalam, F. Espino,
    P. Kittayapong, K.T. Wai, O. Horstick, and J. Sommerfeld

Part IV Building Community Health into City Living

17 Introduction 187
    Andrés Sánchez

18 Rebuilding Urban Ecosystems for Better Community Health in Kathmandu 191
D.D. Joshi, Minu Sharma, and David Waltner-Toews

19 Understanding Water, Understanding Health: The Case of Bebnine, Lebanon 203
Rima R. Habib

20 Water, Wastes, and Children's Health in Low-Income Neighbourhoods of Yaoundé 215
Emmanuel Ngnikam, Benoît Mougoué, Roger Feumba, Isidore Noumba, Ghislain Tabue, and Jean Meli

Part V Building a New Field

21 Better Together: Field-Building Networks at the Frontiers of Ecohealth Research 231
Margot W. Parkes, Dominique F. Charron, and Andrés Sánchez

22 Ecohealth Research in Practice 255
Dominique F. Charron

Index 273

Contributors

Wimaladharma Abeyewickreme University of Kelaniya, Kelaniya, Sri Lanka

Amin Al-Hakimi Yemeni Genetic Resources Center, Sana'a University, Sana'a, Yemen

N. Arunachalam Centre for Research in Medical Entomology, Indian Council of Medical Research, Madurai, India

Virgilio Ayala Universidad de San Carlos de Guatemala (USAC), Guatemala City, Guatemala

Kalpana Balakrishnan Department of Environmental Health Engineering, Sri Ramachandra University, Chennai, India

Ramiro Barriga National Polytechnical School (Escuela Politécnica Nacional, EPN), Quito, Ecuador

Malek Batal Nutrition Program, University of Ottawa, Ottawa, ON, Canada

Peter Berti HealthBridge Canada, Ottawa, ON, Canada

Óscar Betancourt Health, Environment, and Development Foundation (Fundación Salud Ambiente y Desarrollo, FUNSAD), Quito, Ecuador

Sebastián Betancourt Health, Environment, and Development Foundation (Fundación Salud Ambiente y Desarrollo, FUNSAD), Quito, Ecuador

Christine Bonatsos Department of Geography, University of Western Ontario, London, ON, Canada

Jaime Breilh Universidad Andina Simón Bolívar, Quito, Ecuador

Dulce Maria Bustamante Laboratorio de Entomología Aplicada y Parasitología (LENAP), Universidad de San Carlos de Guatemala (USAC), Guatemala City, Guatemala

Xochitl Castro Laboratorio de Entomología Aplicada y Parasitología (LENAP), Universidad de San Carlos de Guatemala (USAC), Guatemala City, Guatemala

Enoch Chione Soils, Food and Healthy Communities Project, Ekwendeni Hospital, Ekwendeni, Malawi

Anita Chitaya Soils, Food and Healthy Communities Project, Ekwendeni Hospital, Ekwendeni, Malawi

Donald C. Cole Dalla Lana School of Public Health, University of Toronto, Toronto, ON, Canada and International Potato Center, Lima, Peru

Edwin Cueva Health, Environment, and Development Foundation (Fundación Salud Ambiente y Desarrollo, FUNSAD), Quito, Ecuador

Laifolo Dakishoni Soils, Food and Healthy Communities Project, Ekwendeni Hospital, Ekwendeni, Malawi

Cristina Díaz Formerly of Pedro Kourí Tropical Medicine Institute, La Habana, Cuba

Fe Esperanza Espino Research Institute for Tropical Medicine, Alabang, Muntinlupa City, Philippines

Roger Feumba Environment and Water Sciences Laboratory, Ecole Nationale Supérieure Polytechnique, Yaoundé, Cameroon

Jean Remy Davée Guimarães Biophysics Institute, Federal University of Rio de Janeiro, Rio de Janeiro, Brazil

Rima R. Habib Department of Environmental Health, American University of Beirut, Beirut, Lebanon

Olaf Horstick Formerly of Special Programme for Research and Training in Tropical Diseases (TDR), World Health Organization, Geneva, Switzerland, currently of Deutsche Gesellschaft für Internationale Zusammenarbeit (GIZ) GmbH

Durga Dat Joshi National Zoonoses and Food Hygiene Research Centre (NZFHRC), Chagal, Kathmandu, Nepal

Rachel Bezner Kerr Department of Geography, University of Western Ontario, London, ON, Canada

Pattamaporn Kittayapong Center of Excellence for Vectors and Vector-Borne Diseases, Mahidol University at Salaya, Nakhon Pathom, Thailand

K. Vijaya Lakshmi Development Alternatives, New Delhi, India

Esther Maona Soils, Food and Healthy Communities Project, Ekwendeni Hospital, Ekwendeni, Malawi

Contributors

Leonard E.G. Mboera National Institute for Medical Research, Dar es Salaam, Tanzania

Jean Meli Université de Yaoundé I, Yaoundé, Cameroon

Donna Mergler Centre de recherche interdisciplinaire sur la biologie, la santé, la société et l'environnement (CINBIOSE), Université du Québec à Montréal, Montréal, QC, Canada

Malumbo Mithi Soils, Food and Healthy Communities Project, Ekwendeni Hospital, Ekwendeni, Malawi

Barbara Moguel Laboratorio de Entomología Aplicada y Parasitología (LENAP), Universidad de San Carlos de Guatemala (USAC), Guatemala City, Guatemala

Carlota Monroy Laboratorio de Entomología Aplicada y Parasitología (LENAP), Universidad de San Carlos de Guatemala (USAC), Guatemala City, Guatemala

Benoît Mougoué Department of Geography, Université de Yaoundé I, Yaoundé, Cameroon

Rodgers Msachi Soils, Food and Healthy Communities Project, Ekwendeni Hospital, Ekwendeni, Malawi

Samuel Mugisha Department of Zoology, Makerere University, Kampala, Uganda

Emmanuel Ngnikam Environment and Water Sciences Laboratory, Ecole Nationale Supérieure Polytechnique, Yaoundé, Cameroon

Zachariah Nkhonya Soils, Food and Healthy Communities Project, Ekwendeni Hospital, Ekwendeni, Malawi

Isidore Noumba Université de Yaoundé II, Yaoundé, Cameroon

Joseph Okello-Onen Gulu University, Gulu, Uganda

Fadya A. Orozco Instituto de Saúde Coletiva, Universidad Federal da Bahia, Rua Basílio da Gama, Salvador, Brazil

Sheila Pachanya Soils, Food and Healthy Communities Project, Ekwendeni Hospital, Ekwendeni, Malawi

Margot W. Parkes Ecosystems and Society, Health Sciences Programs, University of Northern British Columbia, Prince George, BC, Canada

Frédéric Pelat Initiatives de Développement Durable et Equitable sur la base d'Actions Locales et d'Echanges de Savoirs (IDDEALES), Yemen Branch, Sana'a, Yemen

Sandy Steffany Pineda Institute for Molecular Biosciences, The University of Queensland, St Lucia, QLD, Australia

Javier Quiñonez Universidad de San Carlos de Guatemala (USAC), Guatemala City, Guatemala

Raghwesh Ranjan Development Alternatives, New Delhi, India

Horacio Riojas-Rodríguez Dirección de Salud Ambiental, Instituto Nacional de Salud Pública, Cuernavaca, Morelos, México

Antonieta Rodas Laboratorio de Entomología Aplicada y Parasitología (LENAP), Universidad de San Carlos de Guatemala (USAC), Guatemala City, Guatemala

Sandra Rodríguez-Dozal Dirección de Salud Ambiental, Instituto Nacional de Salud Pública, Cuernavaca, Morelos, México

Minu Sharma National Zoonoses and Food Hygiene Research Centre (NZFHRC), Chagal, Kathmandu, Nepal

Lizzie Shumba Soils, Food and Healthy Communities Project, Ekwendeni Hospital, Ekwendeni, Malawi

Johannes Sommerfeld Special Programme for Research and Training in Tropical Diseases (TDR), World Health Organization, Geneva, Switzerland

Ghislain Tabue Environment and Water Sciences Laboratory, Ecole Nationale Supérieure Polytechnique, Yaoundé, Cameroon

Susilowati Tana Centre for Health Policy and Social Change, Yogyakarta, Indonesia

Khin Thet Wai Department of Medical Research (Lower Myanmar), Yangon, Myanmar

David Waltner-Toews Ontario Veterinary College, University of Guelph, Guelph, ON, Canada

Contributors from IDRC's Ecosystems and Human Health program

Roberto Bazzani, Senior Program Specialist

Ana Boischio, Senior Program Specialist

Dominique Charron, Program Leader

Lamia El-Fattal, Senior Program Specialist

Alicia Iglesias, Program Management Officer

Zsófia Orosz, Program Management Officer

Andrés Sánchez, Senior Program Specialist

Martin Wiese, Senior Program Specialist

Chapter 1
Ecohealth: Origins and Approach

Dominique F. Charron

Improving people's health, while promoting thriving, resilient communities and environmental sustainability, is one of the great development challenges for the twenty-first century. This book is about how a growing international field of research, education, and practice called ecohealth is tackling this challenge, and using innovative ideas to build healthier communities and environments in developing countries.

Almost two decades since the 1992 Earth Summit in Rio de Janeiro, and its declaration of Agenda 21 for sustainable development (United Nations 1992), the world still faces a contradiction: economic and social development are needed to alleviate poverty and improve human lives, but globally, ecosystems are still deteriorating because of past and present patterns of development, with major implications for human health. Some suggest that the overall carrying capacity of the planet is being exceeded (Hassan et al. 2005; Wackernagel et al. 2002), preventing human beings from living healthy and productive lives now, and threatening similar conditions for future generations. Changes are needed in how people interact with ecosystems to resolve this contradiction, and to achieve better health and ecologically, socially, and economically sustainable development.

This book is about doing innovative research to achieve sustainable and equitable change in people's health and well-being through improved interactions with the environment. It presents experiences from the field of ecosystem approaches to health (or ecohealth research) and some insights and lessons learned. It builds on previous literature, notably Forget (1997), Forget and Lebel (2001), Lebel (2003), and Waltner-Toews et al. (2008). Through case studies and other contributions by researchers supported by Canada's International Development Research Centre (IDRC), the book presents evidence of real changes in conditions of people, their

---

D.F. Charron (✉)
International Development Research Centre, Ottawa, ON, Canada
e-mail: ecohealth@idrc.ca

health, and the ecosystems that support them. These changes were derived from applications of an ecosystem approach to health in developing regions of the world. The book also illustrates the resulting body of applied, participatory, and action research that improved health and environmental management in developing countries and, in many cases, influenced policies and practices.

To date, no publication has effectively captured the full range of outcomes of ecohealth research, including the socio-economic and ecological context for achieving the results, their implications, and impacts. This book addresses this gap with a series of 15 case studies from developing countries. Through the case studies, it demonstrates the added value of ecosystem approaches to health applied to problems in developing regions, and presents comprehensive results of research and its contributions to development and to the field of ecohealth.

The case studies represent different kinds of success stories. Achieving change through an applied, integrated, and participatory action-research endeavour such as ecohealth is not easy. Given that the outcomes of every project are unique and unpredictable, how is success defined? The case studies illustrate successes in terms of positive and lasting changes that might not otherwise have been achievable. There are many dimensions to these changes. Health is improved by changing the way people interact with their environment based on research findings. In many cases, local environments are also improved. Social and economic conditions are changed for the better. New scientific findings and innovations, community empowerment and initiative, and policy changes represent just some aspects of successful ecohealth projects.

Innovation refers to new ways of doing things. More than science and technology, innovation encompasses new ideas, institutions, practices, behaviours, and social relations that affect how science and technology are developed, for what purpose and for whom, and how results are applied (STEPS Centre 2010). This chapter sets the context for the case studies by outlining the challenges that face the field of ecohealth, tracing the history of ecosystem approaches to human health, discussing a set of principles that informed the work, and providing examples of frameworks and a map of common processes involved in the application of an ecosystem approach to health.

Why Do We Need Ecohealth Research?

The world's population is heading toward nine billion by 2050. Almost all population growth, now and over the next 40 years, is projected to occur in developing countries (United Nations 2008). These same developing regions bear the majority share of the global burden of illness and death. Despite progress toward Millennium Development Goals (MDGs) (United Nations 2000) of reducing child-death rates and improving the control of major diseases like tuberculosis and malaria, the lives of people in the poorest countries are still up to 30% shorter and less healthy than those of people in the richest countries (WHO 2008). Environmental and health problems are escalating in scale and are increasingly occurring simultaneously, for example climate extremes, natural disasters, and global pandemic threats. They are made worse by other global crises, for example crises in financial markets, and affect the world's poor the most. Ecosystems are showing signs of being unable to provide the services people require of them (Hassan et al. 2005). The needs and lifestyles of only a few of the world's seven billion people drive the continued pressure on ecosystems and reinforce the gap between the rich and poor. These escalating health and environmental problems are interdependent. Globally, the largest drivers of environmental change – climate change, globalization, urbanization, deforestation, and agricultural intensification – are affecting human health and are compounding social and economic disparities between rich and poor around the world.

Human health problems from degraded ecosystems occur locally too. Environmental hazards such as unsafe drinking water, inadequate sanitation, poor air quality, occupational hazards, pollution, and poorly managed environments contribute to the majority of all diseases (Prüss-Üstün and Corvalán 2006) and pose an economic burden ranging between 1.5 and 4% of GDP (World Bank 2009) in many developing countries.

In many parts of the world, as shown by the case studies in this book, poverty traps people in degraded environments and occupations that are harmful to their health, like the workers in Ecuador's flower farms and gold mines, India's stone quarries, or the slum dwellers in Kathmandu and Yaoundé. Poor or otherwise marginalized people often feel powerless and have the least capacity to adapt to environmental, economic, and social changes or to protect themselves from environmental hazards. They may resort to using ecosystems in ways that put their health at risk by increasing their exposure to infectious organisms and toxic substances, and accentuating vulnerabilities to physical threats like flooding. Impoverished families generally have limited access to health information and care. Poor people often need to leave their homes for work, making both the migrants and their families left behind more vulnerable to health hazards.

The wide gap between rich and poor means that the very poor may not benefit from economic development as much as others do. Development activities may change ecosystems in ways that threaten people's ability to obtain food, water, and fuel. Over-exploited ecosystems cannot sustain healthy livelihoods and are hazardous to human health. In many of the world's developing regions, people going about daily subsistence may have no alternative to activities that further degrade environments, and further endanger their health. Poverty is indeed a trap, deterring investment and growth, and degrading ecosystems and causing ill health (Lopez and Serven 2009).

Changes to ecosystems around the world are resulting in less reliable weather patterns and reduced productivity. The poor state of many of the world's ecosystems is affecting the likelihood of reaching economic and human development goals, including better health for the world's poor. It is an enormous challenge to halt degradation and restore ecosystems while using them to meet increasing demands for their services without compromising human health. Current efforts around the world may not be advancing quickly enough to meet this challenge.

Hope for the future can be found in the growing level of attention being paid to more sustainable development, the quality of the environment, and the efforts needed to address human health globally. As an indication of this attention, there are now several international targets and frameworks for improving human health and the environment – including the MDGs. The World Health Organization (WHO) has contributed to better global health with the revised International Health Regulations (WHO 2005) and reports like Preventing Disease through Healthy Environments (Prüss-Üstün and Corvalán 2006) and the report on the Social Determinants of Health (CSDH 2008). The Intergovernmental Panel on Climate Change identifies human health and well-being as a key vulnerability under climate change scenarios, and stresses the wide-ranging direct health impacts of climate change and the impacts mediated through the environment. The potential health risks and benefits of greenhouse gas mitigation strategies are also increasingly addressed (Parry et al. 2007). More recently, WHO coordinated a process to establish policy-research priorities to protect health from the impacts of climate change (WHO 2009a).

The Millennium Ecosystem Assessment (MEA) represents a landmark attempt to link human health and well-being with conservation and more sustainable use of ecosystems. The assessment describes how ecosystems provide for human well-being, inclusively defined as ecosystem services. These are the benefits people derive from ecosystems, including: provision of food, water, timber, and fibre; regulation of climate, floods, disease, wastes, and water quality; recreational, aesthetic, and spiritual benefits; and fundamental biophysical processes such as soil formation, photosynthesis, and nutrient cycling. The MEA conceptual framework articulates the relationships between human health and well-being in relation to ecosystems (Hassan et al. 2005). The many reports of the MEA series make substantial strides forward in integrating human well-being and ecosystems, particularly the Health Synthesis (Corvalán et al. 2005) published by WHO. However, WHO's International Health Regulations (WHO 2005) and Commission on the Social Determinants of Health (CSDH 2008) only peripherally address the contributions of ecosystems to health, despite their inclusion as part of the MDGs and the links made in the MEA. The compartmentalization of health, environmental, and various other policy agendas persistently contradicts a fundamental truth: human health depends on healthy environments, and human prosperity depends on both healthy people and ecosystems in good condition.

The field of ecohealth is striving to overcome this compartmentalization. Emphasizing integration, the field draws researchers from many disciplines who are seeking to overcome silos in their own domains. There are successes. Some countries, like Ecuador (Republic of Ecuador 2008) and Thailand (Bhumibhol Adulyadej (King of Thailand) 2007), make links between health and environment in law and policy. More jurisdictions are following these examples. As many of the case studies in this book illustrate, ecohealth research can generate strong evidence of contributions to social and environmental changes that improve people's health. The research is informed by local context and priorities, and is connected to individuals and processes that can take up this knowledge and apply it for bettering health and ecosystems over the long term.

Without sufficient attention to the state of ecosystems and the social and economic inequities between people who depend on these same ecosystems, efforts to improve global health and human development will falter. Climate change, pollution, disruptions in animal-disease ecology that lead to new human diseases, and degraded ecosystems that fail to produce nutritious food, clean water, and air will limit the success of public health initiatives all over the world. For our own collective sakes, and those of future generations, human health and the state of ecosystems need to be improved together. To do so, stronger evidence is needed from research that integrates different types of knowledge, and generates better strategies to help improve human health and ecosystems in communities around the world.

Definitions

Many of the terms used in this book have different meanings to different audiences. The constitution of the WHO describes human health as physical, mental, and social well-being, and not merely the absence of disease or infirmity (WHO 1948). It also states that the enjoyment of the highest attainable standard of health is one of the fundamental rights of every human being without distinction of race, religion, political belief, or economic or social condition. A later notion, adopted at a conference on health promotion at Ottawa in 1986, and now widely known as the Ottawa Charter, defined health in terms of an ability to achieve goals and purpose: *an individual or group must be able to identify and to realize aspirations, to satisfy needs, and to change or cope with the environment* (WHO 1986). Health can also be considered a dynamic and relative condition, a capacity or resource rather than a state (PHAC 1996).

Health determinants refer to the panoply of external conditions that affect health. In common public health terminology, determinants refer to environmental and socio-economic factors that have been associated with health outcomes, and are somewhat removed from the immediate causes of disease. The population health approach was developed to apply a multideterminant understanding of health and to set out criteria for measuring it (PHAC 1996). But the concept of external determinants of health (e.g. the physical environment, pathogenic agents, income, and education) belies the dynamic nature and interconnectedness of underlying processes that link these determinants, and predisposes toward separate assessments of non-independent determinants.

The MEA presents a multifaceted conception of human well-being:

> ... *including the basic material for a good life, such as secure and adequate livelihoods, enough food, shelter, clothing, and access to goods; health, including ... having a healthy physical environment ...; good social relations ...; security, including secure access to natural and other resources, personal safety, and security from natural and human-made disasters; and freedom of choice and action, including the opportunity to achieve what an individual values doing and being* (excerpts from Hassan et al. 2005, p. v).

In this book, health is defined inclusively as more than just the absence of disease and infirmity, and as a relative condition depending on context, expectations, and so forth. Ecosystem approaches to health are consistent with population health, but consider the dynamic interplay among determinants, and between them and health outcomes. In ecohealth, health is mostly inferred and assessed at a community or sub-group level. In accordance with common usage, health is also used in a metaphorical sense, for example healthy environments. The term well-being is used to refer to the broad conception of human health, aspiration, and capacity to achieve goals, as defined earlier.

The terms environment and ecosystem have already appeared several times in this chapter. Environment is used in a general sense to refer to the surroundings of a given person or object, household, or community. Ecosystem refers to a functional unit that encompasses the dynamics among plants, animals (including humans), microorganisms, and their physical surroundings. The research in this book defines boundaries of ecosystems according to the context of the problem under attention, such as an urban slum, or a rice-irrigation scheme, or an Amazonian riparian zone. The behaviour of ecosystems is complex, and efforts to understand them draw on elements of systems thinking. The use of "ecosystem approach" in this book is consistent with contemporary literature that addresses complexity and systems thinking as part of ecosystem approaches (e.g. Allen et al. 1993; Kay et al. 1999; and several chapters by Kay and others in Waltner-Toews et al. 2008).

People use ecosystems in direct and indirect ways, and derive services from them. Ecosystems also have intrinsic value. Around the world, people change and shape ecosystems – managing aspects of them for specific uses and benefits, such as agriculture, urban settlements, aquaculture, and energy and natural resource projects.

Ecosystem Approaches to Health

Ecosystem approaches to health (or ecohealth research) formally connect ideas of environmental and social determinants of health with those of ecology and systems thinking in an action-research framework applied mostly within a context of social and economic development. Ecosystem approaches to health focus on the interactions between the ecological and socio-economic dimensions of a given situation, and their influence on human health, as well as how people use or impact ecosystems, the implications for the quality of ecosystems, the provision of ecosystem services, and sustainability. Ecohealth also refers to a growing international field of research, education, and practice that encompasses several different schools of thought. The ideas presented in this book stem primarily from one school, initially developed and promoted by IDRC, but now influenced by a much wider and growing field. There is no single best or even (as yet) dominant approach, and this is reflected in the number of different frameworks and approaches put forth under the banner of ecohealth, and the gamut of research presented in the journal EcoHealth.[1] Indeed, it is evident in the different interpretations of an ecosystem approach presented in the case studies in this book, despite their common origin in the same school of thought. This diversity is an asset to the growing field of ecohealth and is consistent with its inclusive and transdisciplinary principles.

Elements of a research approach to coupled problems of environment and human health in developing countries were put forward as an *ecosystem approach to health* in Forget and Lebel (2001) and further developed in Lebel (2003). In essence, this approach linked better environmental management to better health within a transdisciplinary and participatory research framework. The consideration of health or illness as more than just the result of the (cumulative) effects of proximate and independent social or environmental determinants was a cornerstone of the approach, and different from the predominant view in environmental health research at the time. An ecosystem approach recognizes that health and well-being are the result of complex and dynamic interactions between determinants, and between people, social and economic conditions, and ecosystems. The conditions of ecosystems are also affected by a dynamic process of interactions, often determined by the social and economic activities of people. Thus, an ecosystem approach to health, although focused on improving human health, goes beyond prevailing biomedical or epidemiological approaches to health research (see Rapport et al. 1999 for more discussion of the limitations of a biomedical or clinical approach).

From the outset, this work addressed a growing need for practical guidance expressed by researchers seeking to respond to the challenge of understanding health within ecosystems. Ecosystem approaches to health arose partly in response to the frustrations of some researchers with the apparent limitations of their own and other disciplines when faced with the increasingly complicated problems of health and ecosystems, and the struggle to make a difference in local communities (DePlaen and Kilelu 2004). It was also influenced by the tendency of some environmental movements in the 1980s to ignore or externalize people from ecosystems (Forget 1997). The development of ecosystem approaches to health (and likely that of many intellectual cousins) has been influenced by several other movements and ideas including the ecosystem approach developed by the International Joint Commission for transboundary water management in Canada (Allen et al. 1993), sustainable development (Brundtland 1987), health promotion (WHO 1986), ecosystem health (Rapport et al. 1979, 1999), eco-epidemiology (Susser and Susser 1996; March and Susser 2006), Latin American social medicine (Waitzkin et al. 2001; Iriart et al. 2002), and the wider public health movement in Europe (see Krieger and Birn 1998 for a summary). Much of this early history and heritage in public health is presented in Forget and Lebel (2001). Parkes et al. (2003) expand this discussion to human ecology. Bunch et al. (2008) present an astute exploration of the intellectual parentage of the ecosystem approach. Parkes et al. (2005) trace the evolution of ecosystem approaches from a public-health perspective, and focus on the control of infectious diseases.

---

[1] http://www.ecohealth.net/.

The complex relationships and interactions between societies and ecosystems can be considered as coupled social–ecological systems (Berkes and Folke 1998). Social, cultural, and economic activities of people – from local household decisions to national policies and international treaties and systems – are occurring within ecosystems while determining human health and shaping the relationships people have with ecosystems. In their application to research, ecosystem approaches to health draw on both natural and social sciences, and their consideration of system behaviours, as well as methodologies and interpretation of results. The application of systems thinking to a broad set of ecosystem approaches is treated in greater depth in several chapters of Waltner-Toews et al. (2008), and in Kay and Regier (2000) and Regier and Kay (2001). Their work draws substantially on Checkland (2000) for systems thinking and on Allen and Hoekstra (1992) for complex systems ecology.

To better understand health in the context of coupled social–ecological systems, different scientific perspectives need to be integrated to fully describe the behaviour of the system. But academic perspectives alone are not sufficient. The knowledge and perspectives of people immersed in the situation and living in the ecosystem are also relevant, as are the perspectives of people holding decision-making power over the situation. By engaging these different people in an action-research process, research results can lead to changes in decisions, policies, and practices that lead to lasting improvements. In short, ecohealth research requires academics to work not only with experts from other specialties, but with civil-society stakeholders throughout the process of inquiry to incorporate different perspectives and forms of knowledge. In addition to differences between sometimes strongly held views, understanding and managing power dynamics come into play in this type of research. Success is not easily achieved, and the case studies in this book represent both achievements and challenges still to overcome.

This predominantly "ecologistic" view of health and well-being is balanced by contributions from social science philosophy. The predominant view in natural and medical sciences considers science to provide objective and independently validated information as evidence for making informed decisions. An ecosystem approach to health adopts the strengths of these views, while recognizing that scientists (and research) are part of ongoing political and social processes, and that because of this, science cannot claim to be fully empirical or objective (Kuhn 1970). Ethical dimensions are emphasized in ecohealth research and practice through integration of stakeholder perspectives and participation, and through the research intent to make positive changes in the world. Ecohealth research generates new knowledge that serves as evidence to help achieve these changes.

However, evidence is not automatically transformed into policies that change peoples' lives in villages, farms, and cities around the world. The research presented in this book is applied research – one important application being "sustainable development." It is not only about the production of knowledge that can be generalized beyond a specific context. Instead, this research deals with the production of knowledge that can be immediately applicable to change a given problematic situation. The knowledge produced is both rigorously tested through science and locally relevant, which leads to more effective interventions with greater initiative and leadership from communities.

Why is this kind of approach important? Today's human health and environmental challenges are interlinked and symptomatic of many problems facing the world: they are urgent and threatening; they are large-scale and multifaceted; they appear complicated and full of uncertainties; and yet, they require immediate action to reverse, resolve, or otherwise address the problems. There is international consensus that: *[p]roblems facing humanity are closely intertwined, and that each tends to complicate the solution of one or more others* (United Nations 2001). Despite this understanding, responses and interventions tend toward single sector, non-participatory practices and technological quick fixes that are inadequate.

Today, the field of ecohealth includes a heterogeneous global community of researchers and practitioners who are working on many different aspects linking environment, society, and health. The field of ecohealth includes groups of researchers focused on environment and health in developing countries (like in this book), global climate change and health, and links between diseases of many species that are associated with environmental change, conservation biology, and conservation medicine. More than an association of like-minded individuals from different disciplinary backgrounds, ecohealth is a growing field of research, education, and practice with distinct epistemological and historical roots (Forget and Lebel 2001; Bunch et al. 2008; Waltner-Toews et al. 2008).

The case studies in this book address the conditions of people living in poverty in many different types of degraded ecosystems around the world, and the implications of such conditions for a wide set of health problems. They have all applied ecosystem approaches to health influenced by IDRC (mostly Lebel 2003). Approach, in this context, refers not to a framework or methodology, but rather to a mindset that orients a process of inquiry that is meant to lead to some action or change in the conditions of these same people and their environment. The process of inquiry (or the research) is unconventional because it is investigating the reasons for the situation while also being a part of a change process – that of healthier and more environmentally sustainable development.

Principles of an Ecosystem Approach to Health

Ecohealth research is difficult to do because it relies on both empirical approaches and flexible, context-specific methodological protocols. However, practical experiences from ecohealth research suggest that a set of principles inform the application of ecosystem approaches to health. These principles are guide posts to the implementation of ecohealth research and inform the kinds of outcomes that can be expected. The principles are not a methodological checklist, and their consideration is no guarantee of success in resolving the problems that link health and ecosystems. They are, however, elements of an effective process of inquiry to generate knowledge and apply it to resolve such problems. All six principles can inform "how to" conduct ecohealth research, but the first three more strongly emphasize process; whereas, the last three focus on the intrinsic goals of ecohealth research.

Principle 1: Systems Thinking

Understanding how people and their health relate to ecosystems is a tough puzzle. Researchers tackling this puzzle, while also considering multiple other academic and expert points of view, may become overwhelmed by the many possible links, relationships, and components. Systems thinking helps apply some order to the complex reality of health in the context of social–ecological systems.

Framing a problem in coupled social–ecological systems terms can include consideration of several dimensions (ecological, social-cultural, economic, and governance). Systems thinking considers the relationships among these elements. The involvement of people and how they interact are key to the modelling of complex systems, and a good understanding of this dimension requires expertise from the fields of social sciences. Systems thinking can lead to a better understanding of the limits of the problem, its scale, and its dynamics. Ultimately, it leads to a richer, more effective research process.

From a systems perspective, scale is important because different parts of social–ecological systems operate on different time horizons – from the very short (e.g. daily routines to gather wood or water), to longer seasonal cycles, to much longer trends (e.g. climate change). At the same time, the interactions between the ecosystem and the regional, national, and global context can help researchers understand the drivers of a particular context or problem. As well, interactions between individuals in a household, and the links at this level to interactions within groups in a community, which may be different between men and women, are just as important. Many of the case studies succeed in exploring and addressing these local multiscale links. They also illustrate the challenges of linking the community to broader levels of the system (sub-nationally and nationally) (Freitas et al. 2007).

Research developed from systems thinking can lead to changes in policies and practices. The Guatemala Chagas disease case study in this book explains how the project succeeded both in convincing the national Ministry of Health to do things differently, and in influencing regional health initiatives. The research team was able to situate the ecology of this disease in a regional context, develop locally applicable interventions, and make links to existing policy mechanisms to broaden the scale and uptake of these innovations. But this success can also be attributed to good timing. The project capitalized on concurrent national and international trends in health and development policy.

The case studies also illustrate that applying systems thinking in ecohealth research can present challenges in practice. It can be difficult to set the boundaries for the study, refine the study design and conduct the analysis, and interpret the results. Choices and trade-offs need to be made between inclusiveness and feasibility, based on the time, skills, and resources available.

Principle 2: Transdisciplinary Research

Transdisciplinary research helps achieve an improved understanding of health in the context of coupled social–ecological systems, and the real world that such systems approximate. It also enhances the resulting innovations and the design of strategies to improve health and environmental conditions in a sustainable, contextually appropriate way. Community representatives and other stakeholders possess knowledge about the problem that is informed by their experience. They will necessarily have a role and a stake in achieving a better understanding of a problem and in designing strategies to resolve it. A transdisciplinary approach integrates different scientific perspectives (Parkes et al 2005; Wilcox and Kueffer 2008) and provides a formal platform for stakeholder participation in the research and in the development of new information, ideas, and strategies, their testing, and eventual application.

Transdisciplinary research involves the integration of research methodologies and tools across disciplines and includes non-academic perspectives and knowledge. The research team includes these different perspectives, and ideally functions as a cohesive collaborative unit from design, to data collection, to development of strategies for change, to implementation. As highlighted by Bopp and Bopp (2004): *health and natural resource management professionals, and the technical solutions they create, cannot, by themselves, solve many of the problems communities face. To be effective, solutions have to address a complex set of variables that may be largely invisible to professionals from outside the communities.* When scientists working in interdisciplinary teams involve community members, decision makers, and other non-scientific stakeholders, they are engaging in a transdisciplinary research process. They are creating new knowledge and theory around a set of common questions (Pohl and Hirsch Hadorn 2008).

Transdisciplinarity takes time to build into an ecohealth research process. The core research team almost never initiates a project with the ideal mix of scientists and key stakeholders. Even the most experienced teams evolve over time as relationships develop, contributors are added or changed, and new understandings, theories, and methods are applied. With time and effort, transdisciplinary practice develops as team members get to know one another, understand each other's perspectives, better understand the problem under study, and develop working relationships. In practice, the priorities of stakeholders tend to evolve with the emergence of this new understanding, which helps build trust and respect. Thus, transdisciplinarity is less about addressing the priorities of all and more about establishing an acceptable process for discussion and negotiation among the actors who are in joint pursuit of a new understanding of a given problem or situation.

To achieve transdisciplinarity, researchers must draw on a wide range of skill sets that are not usually part of academic training, including consensus building, negotiation, facilitation, communication, and strategic planning. There are a variety of methods for achieving effective multistakeholder processes, a necessary component of transdisciplinary research. For example, Social Analysis Systems (Chevalier and Buckles 2008) provide a framework for group dialogue and social inquiry for development; multicriteria evaluation has proven useful for conflict management (Paruccini 1994); and outcome mapping (Earl et al. 2001) has been used to engage stakeholders, identify the changes sought, and facilitate effective participation of all stakeholders.

Most of the projects presented as case studies in this book emphasize the process by which transdisciplinarity was achieved. As just one example, the Malawi case study on improving soils and nutrition achieved an integrated research design that linked improvements in soil quality with improved yields, changes in dietary habits, and improved child health, while integrating local knowledge and capacities.

Principle 3: Participation

Transdisciplinarity and participation go hand in hand as part of an ecosystem approach to health. Stakeholder participation adds to the knowledge generated by the research and enhances the action that can result from, or be integrated into, the research. Participation, a principle of ecohealth, reflects broader current trends in research for development. Community initiatives and wider social movements are recognized to lead to innovation strategies not otherwise discernable to experts or consultants. Such participation leads to locally rooted forms of innovation and increases the impact of the benefits of innovation (STEPS Centre 2010).

In theory, appropriate stakeholder engagement enhances the likelihood of finding and using new knowledge. More important than the number of stakeholders, however, is the process of engagement. There are strategically and ethically charged decisions to be made that also affect the research process: Who convenes? How is participation decided? What will be done if participation is not constructive or is used to advance particular interests that obstruct or derail the change sought by the majority? Some of these aspects of participative process in ecohealth are discussed in Mertens et al. (2005) as part of the project presented in the Amazon case study.

Change (in various forms) is the intent of ecohealth research. The participation of actors living with the consequences of a problem or issue, those contributing to it, and those influencing and bringing about change are vital to a research process that strives for change. Community participation came into public health practice in 1978 following the Alma Ata Declaration (WHO 1978), but it has waxed and waned in practical implementation ever since (Draper et al. 2010). Although participation is central to ecosystem approaches to health, not all communities are equally able or prepared to engage in such processes. A variety of tools are available to assess such capacities (Bopp and Bopp 2004). Action research focuses on the effects of researchers' actions in a community, with the usual goal of improving conditions or changing behaviours of the community (Reason and Bradbury 2007; MacIntyre 2008). Participatory action research seeks effective stakeholder participation in research and beyond, and is rooted in the notion that those who are affected should be part of the process of defining not only the problems, but also their solutions. Participatory Rural Appraisal (PRA) methodology (Chambers 1994) provides one effective framework for incorporating the community into a research process. The Ecosalud Ecuador and Malawi soils, food and health case studies both illustrate the effective use of PRA in ecohealth research.

Participation can lead to cooperation, collaboration, and eventually, to breakthroughs in resolving long-standing differences that impede progress. The case study of open-air slaughtering practices in urban Kathmandu illustrates the power of engaging multiple actors in community change. A participatory process helps develop a shared understanding of the socio-political environment that surrounds a particular development problem and its associated social and health problems. In this case, full participation of various groups led to their political empowerment and later to effective negotiation between community groups, community leadership, and the government (Neudoerffer et al. 2005). Morrison et al. (2008) present a detailed literature review and methodology for their participatory study of local social–ecological resilience to Ciguatera fish poisoning in Cuba. Goy and Waltner-Toews (2005) present a multilevel stakeholder analysis pertaining to environment and health in Peru.

Participatory processes also help identify barriers to change, clarify information and knowledge gaps, and provide means to negotiate concrete steps for moving forward. In Mexico's manganese mining case study, research demonstrated the toxic effects of manganese dust in the affected communities. Community representatives, government, and mining company officials were brought together to overcome decades of inaction on this pollution problem. Backed by scientific evidence, they developed a risk-management plan to reduce impacts from mining. Some changes in policy and practice to reduce different exposure routes are underway, while others remain under negotiation.

There are practical limits to a participatory research process. Community priorities may not be aligned with the problem that is motivating the researchers. The interests and intentions of different actors (researchers included) sometimes clash with scientific relevance or methodological requirements. Every stakeholder has particular concerns and interests, which are sometimes compatible with those of other stakeholders, but frequently conflict, and need to be resolved. Some participating stakeholders may also have motives that are incompatible with the research agenda or the change being sought by other stakeholders. Therefore, levels and terms of participation vary and may need to be renegotiated repeatedly, which can interrupt data collection or project timelines. Finally, community expectations for change may not be attained by the research, leading communities to experience disappointment or to feel exploited – an undesirable outcome of any project.

The iterative nature of ecohealth research can accommodate many of these issues, but not without transactions. Transaction costs are high in participatory research, requiring time, and sometimes presenting unforeseen and frustrating delays. The contribution of community participation to project outcomes is challenging to assess, although tools used in public health program evaluation may be useful in this regard (Draper et al. 2010).

Principle 4: Sustainability

An ecosystem approach to health is predicated on an understanding that protecting ecosystems and improving degraded environments are fundamental requirements for human health and well-being now and for future generations. Sustainability (ecological and social) is part of the change sought through ecohealth research and action, and seeking such change motivates many in the field of ecohealth (Soskolne et al. 2007; Waltner-Toews et al. 2008).

As research for development, ecohealth research aims to make ethical, positive, and lasting changes. Sustainability implies that these changes be environmentally sound and socially sustainable (socially and culturally responsible and appropriate, as well as easily systematized). The case studies on Chagas disease in Guatemala, sanitation in Yaoundé, Cameroon, and improved soil and nutrition in Malawi illustrate how this can be achieved.

Another sustainability issue relevant to research for development pertains to the uptake and use of research results for achieving change. In Kathmandu, ecohealth research led to the apparently permanent transformation of an entire neighbourhood and catalysed substantial modernization of the meat-processing sector across Nepal. Despite social and environmental problems in riverside slums in Nepal, there appears to be no going back to the original situation, at least in the two municipal wards where the project was initiated. However, change may not always be clear or clearly attributable to ecohealth research or any research-for-development process. Change processes are not linear. There may be slippage back into previous, negative patterns or relationships; setbacks as well as leaps forward; and new problems may arise. Ecohealth researchers should anticipate these dynamics and be prepared to learn from them.

A fundamental challenge arises when seeking sustainability. Environmentally and socially sustainable development is a very lofty goal. The dynamics of social–ecological systems are entrenched and not easily changed where people live in extreme poverty with very limited access to resources and even fewer choices in livelihoods. Researchers will face ethical quandaries when the short-term needs and priorities of people are not consistent with a longer term process for improving health and environment. Ecohealth research can provide some insights by addressing both local concerns and the wider forces that maintain cycles of poverty, environmental degradation, and ill health. A focus on achieving change locally can sometimes help shift perceptions and motivate people to tackle wider issues. There is a need for even more strategies that reach beyond the community to levels of the system where a wider difference can be made over the longer term.

Improving livelihoods and economic conditions with inadequate attention to environment and social inequities can imperil health and become unsustainable over time. Ignoring the drivers of environmental and social conditions when trying to improve health can be ineffective. But, there are trade-offs to be made in achieving sustainable development; gains in one dimension (health or economic) can come at a cost in other dimensions (ecological or social). There are setbacks caused by unforeseen events (natural disasters) or otherwise unpreventable ones (crisis in the world financial markets). Flexible and adaptive governance mechanisms are required. Ecosystem approaches to health contribute to an evidence base to better inform communities and decision makers, and to foster conditions for improved sustainability and health.

Principle 5: Gender and Social Equity

An ecosystem approach to health explicitly addresses unequal and unfair conditions impinging on the health and well-being of women and other disadvantaged groups in society. The differences between members of different social, economic, class, age, or gender groups in all societies are reflected in their relationships with ecosystems, their exposure to different health risks, their health status, and their well-being goals. In implementing this principle, research not only documents social and gender differences in causal pathways, outcomes, and proposed interventions and actions but it takes on ethical dimensions by becoming oriented toward reducing inequities.

Inequity underlies many of the world's current ills from environmental degradation. Together, inequity and degraded environments contribute to ill health and poor life expectancy, economic woes, and conflict. The ongoing high levels of ill health among poor people, and the pronounced health gradient according to wealth within and between countries, are due to inequity. The WHO Commission on the Social Determinants of Health (CSDH 2008) calls for renewed attention to inequity as a key limiting factor in achieving health for all. Power, income, goods, and services are unfairly distributed and this affects ecosystems and how they are used. There are inequities in access to health care, education, and work and living environments. The combined impacts of various inequities affect the power people have to improve their circumstances. The CSDH report urges action for health by reducing inequities through better social policies and programs, improved economic arrangements, and political will. Greater attention to environmental quality must also be a priority for equitable development, and ecosystems need to be restored and protected so that they may continue to support flourishing societies and improved health. The point is that the CSDH does not address environmental drivers of inequity or of poor health.

In most developing countries, women represent the largest segment of society facing persistently unequal and unfair access to opportunity and to health. Women hold most of the responsibility for the health of their families, the social fabric of their communities, and the socialization and education of children. They have stewardship of a substantial portion of the world's managed ecosystems and natural resources. Women's health is not as good as that of men in poor countries. They suffer greater levels of violence, malnutrition, and with their children, represent almost all the deaths from malaria worldwide (WHO 2008). Nearly all of the more than half a million maternal deaths every year occur in developing regions (WHO 2009b). Women are at greater risk of bearing the negative health impacts of climate change (WHO 2009c). Women have less land, wealth, and property in almost all societies, yet in all countries almost always carry a double burden – they are responsible for caring for the household, for children and elders, and for earning wages or income. Women are penalized in the workplace when family demands or their own health prevent them from working. The resulting pressures and fatigue put their health at further risk (WGEKN 2007). However, women are also agents of change and a resource that is often underutilized to achieve development goals (UNEP 2004).

Better environmental management can improve the lives and health of women (Prüss-Üstün and Corvalán 2006) – for example by improved drinking water and providing cleaner household energy. Healthier and productive ecosystems provide more livelihood opportunities for women and more resources for the household. Reductions in the burden of childhood diseases (the majority of which can be linked to poor ecosystem conditions) relieve women of the burden of care, improve survival of children, and reduce pressure on women to have large families. The time saved can be spent in economic activities that generate income and promote gender equality. It also reduces pressure on ecosystems.

In practice, changes in social and gender inequalities and resulting unfair health status are not easily achieved. The reasons for these inequities are often deeply rooted in multifaceted economic and cultural patterns that stubbornly resist change. Although most ecohealth research demonstrates awareness of the possible importance of gender or social difference in a particular context, it struggles to address inequity beyond the consideration of gender or social group in risk-factor analyses. Some of the case studies demonstrate how ecohealth research has addressed gender inequity in a more integrated manner. For example, in Malawi's rural north, much is expected from women in their child-bearing years. They must raise children, they are responsible for family health care, and also provide most agricultural labour. While elder women hold considerable power over household decision-making (notably in child rearing), women in general have little say in how meagre household income may be used. This dynamic contributed to chronic malnutrition and stunting of children. The Malawi case study describes how both women and men were engaged in developing strategies to introduce high-protein legume crops. Over time, relationships between men and women changed and led to better conditions for women.

There is a need for more gender and social analysis in ecohealth research; however, methodologies and tools remain few (WHO 2002a). Although not focused on health, the World Bank and partners have developed an extensive agriculture and gender sourcebook that includes training modules for many different contexts relevant to ecohealth research (gender and food security, livelihoods, and crises) (World Bank, FAO, IFAD 2009). Others have presented the case for integrating social and gender analysis in development research (Vernooy 2006). Beyond the development and adaptation of analytical tools, there is a need for further conceptual development of the implications of gender and social inequities in ecohealth research, education, and practice. This appears to be an area ripe for new contributions from experiences in ecohealth research.

Principle 6: Knowledge to Action

The notion that knowledge from research is used to improve health and well-being through an improved environment is fundamental to an ecosystem approach to health. A variety of terms are used to describe this process, but in the context of ecohealth, knowledge-to-action is preferred to the commonly-used *knowledge translation*. The point is not to achieve some near-prefect level of knowledge before making a change (the translation). In transdisciplinary and participatory research involving decision-makers, the situation may be changing while new knowledge is being produced over time through a series of research–action cycles. Other authors have defined knowledge-to-action in health research to include trailoring of knowledge and a defined number of steps for sustained use of knowledge (Graham et al. 2006), but these steps are not always applicable in ecohealth research.

That research conditions change at the onset of participatory research is widely understood in social science research (e.g. Bernard 2000), but is not as commonly considered in public health. Many of the case studies in this book describe in detail how the research became an ongoing intervention process, and yet some of them also invoke epidemiological designs like case-control studies, wherein conditions are assumed to remain the same or where changes are measured and controlled in the analysis. This tension between scientific endeavour and action to improve sometimes terrible conditions of people and their environment is characteristic of ecohealth research. Researchers have a responsibility to be aware of this tension and to document both pre-existing conditions and changes wrought by the research process. Outcome mapping (Earl et al. 2001) can be a useful tool for this.

The innovations, actions, and changes that result from ecohealth research involve multiple sectors, agencies, and stakeholders. In addition, ecohealth research may generate unintended positive (and sometimes negative) outcomes that can be difficult to link to the results of research or to the original research question. Ethical dilemmas are expected to arise – and researchers who anticipate and consider these beforehand are sometimes better equipped to navigate them (Funtowicz and Ravetz 2008; Lambert et al. 2003).

There is already a strong emphasis on uptake and application of knowledge in public health (Pablos-Mendez et al. 2005; WHO 2004) as reflected in the concepts of implementation, technology transfer, and knowledge translation. Knowledge translation is gaining particular prominence as a rubric for operational research and for developing and assessing the effectiveness of interventions (Tugwell et al. 2006). Knowledge translation has been described as the practice, science, and art of bridging the know–do gap between knowledge accumulation and use (WHO cited in Ottoson 2009) or as "*a dynamic and iterative process that includes the synthesis, dissemination, exchange and ethically sound application of knowledge*" (Tetroe 2007), and these are relevant to knowledge-to-action in ecohealth. Understanding how knowledge is used is an active field of study that contains many different fields of thought (Ottoson 2009). Sudsawad (2007) and Tetroe (2007) explore a variety of knowledge translation models in public health. Health sciences are generating knowledge translation tools that could potentially be adapted to an ecohealth research context (e.g. Campbell et al. 2010; Tugwell et al. 2006).

Lavis et al. (2006) point out that knowledge does not move in a single direction (from research process to results to action). Three different drivers of knowledge translation are reflected in examples from ecohealth research: (1) research pushing new knowledge forward into policy and action (many examples in this book); (2) policy requesting new knowledge (alternatives to DDT sought after NAFTA[2] – not in this book – Chanon et al. 2003); and (3) collaborative exchanges and platforms between these audiences (the manganese mining pollution case study from Mexico).

Formal policy processes and legal frameworks need to be engaged to achieve the far-reaching and permanent changes required for a more sustainable future (Soskolne 2007). Policy influence and change are important elements to successfully move research results into action, but knowledge can move without the help of policy. The Malawi case study illustrates how innovations from ecohealth research can be disseminated by socializing and institutionalizing in the community. The case study illustrates how the use of legumes to improve soil fertility and local diets was spread by word of mouth to several neighbouring communities. In the case of manganese mining pollution in Mexico, the project worked over several years to develop both the evidence base that indicated harm from manganese-laden aerosols in the mining region, and a platform for policy and private-sector engagement with the community. The process of achieving change in policy is ongoing, but has encountered setbacks due to political change and jurisdictional disputes.

Researchers necessarily represent only a fraction of the inputs and shaping forces of policy processes. These include the broader economy and societal context, the qualities of political leaders, and their responses to political pressure. The case studies describe many conditions that hinder the uptake of research by policy including: lack of demand for research; lack of knowledge translation processes or inadequate capacity in governmental institutions to act on research; lack of recognition by government of the importance or relevance of research findings; and the research community being perceived as hostile to government (Carden 2009).

---

[2] North American Free-Trade Agreement. Available at: www.nafta-sec-alena.org.

There are several strategies to overcome these pitfalls, and these too are exposed in the case studies. It helps if the project formulates an intent to influence policy and action. Outcome mapping (Earl et al. 2001) can help research teams understand both how knowledge moves and how processes of translation and uptake of that knowledge work. Researchers that employ networks for research collaboration can use these for policy advocacy (see Chap. 21). A communication and dissemination strategy that systematically introduces research into the policy process is also important (Carden 2009).

Putting Ecohealth Principles into Practice

The preceding principles form the basis for implementing research using ecosystem approaches to health. They are couched in an understanding that humans, and our social and economic systems, are embedded within ecosystems, and that these coupled social–ecological systems behave as complex systems. To achieve positive and sustainable changes in people's health through better interaction with ecosystems, a variety of actors and processes are needed in research. The knowledge generated by ecohealth research is intended for use by local communities and policymakers at local and wider scales. This set of principles is useful to understand the intent and process of ecohealth research, and to inform practice. However, researchers seeking to design and implement ecohealth research also need some understanding of process.

Since the 1980s, many different frameworks have been developed to apply ecosystems theory to other problems, particularly those of environmental management. A framework facilitates the structuring of a research process so that findings may be ordered and systematically recorded. Frameworks also inform the learning process because they provide a common language and set of methods and tools to allow different participants to share insights and learn together. Several research frameworks are consistent with an ecosystem approach to health. The process of choosing and adapting a framework can help develop or refine lines of inquiry and expose assumptions, omissions, and other potentially important information before the research gets underway.

Some frameworks address health, environment, and development issues (e.g. WHO 2002b), but do not meet the needs of researchers trying to make links between interacting social, economic, and ecological processes, and their influence on human health. Many researchers have since contributed important and practical lessons that informed the development of more explicit guidelines for implementing ecosystem approaches to health. But only a few practical manuals with a how-to approach have been developed (e.g. Waltner-Toews 2004).

Parkes et al. (2010) provide a useful analysis of several conceptual constructs, including the driving force, pressure, state, exposure, effect, and action (DPSEEA) model (WHO 2002b), butterfly model (VanLeeuwen et al. 1999), Prism framework (Parkes et al. 2010), and the MEA (Hassan et al. 2005). Many researchers have been influenced by Mergler's (2003) target framework (which appeared in Forget and Lebel 2001). A more explicit how-to approach can be found in adaptive methodologies for ecosystem sustainability and health (AMESH), which guides investigators through an iterative process of understanding the interlinked social and ecological dynamics of the problem (Waltner-Toews et al. 2004; Waltner-Toews and Kay 2005).

Many of the case studies in this book reflect research that was initiated before the publication of most of these frameworks. The research underpinning the case studies was influenced by IDRC's approach (Forget and Lebel 2001; Lebel 2003). Many of them included an initial extensive conceptual phase of research that aimed to gather data to more fully describe the coupled social–ecological system context of the problem at hand. The case studies on malaria in East Africa, mercury in the Amazon, and urban ecosystem health in Kathmandu provide some salient examples, although this aspect receives limited treatment in the text due to space limitations. The eco-bio-social study on dengue in Asia is based on a framework that emphasizes integration of methods to address the large and complex data sets generated in ecohealth research.

It is not the intent of this book to identify a single preferred framework for implementing an ecosystem approach to health. Few of the case studies in this book refer to any formal conceptual research framework beyond citing an ecosystem approach to health. However, some common patterns in the process of ecohealth research can provide insights on how positive outcomes can be achieved. This can be useful for researchers new to this approach as well as for those more experienced and interested in further reflecting on and refining their approach.

What follows is a description of the apparent process of ecohealth research, gleaned from IDRC's dozen years of experience. It is not meant to replace other frameworks. To do so would be a disservice to ecohealth researchers who seek to ensure that their research is appropriate to the problem and its context, and to others diligently striving to develop better frameworks and methodologies in this young field. Rather, the purpose is to map a common process of ecohealth research to assist in understanding how the case studies unfolded. It may also inform further development of ecohealth research and practice, but this is not the goal.

In IDRC's experience, and as illustrated by the case studies in this book, ecohealth research advances through iterative cycles of knowledge generation, action, and reflection. Many of the case studies point to this in their description of research that occurred in phases over several years. The studies on gold mining and pollution in Ecuador, drinking water quality in Lebanon, and mercury toxicity in the Amazon are cases in point. Each cycle consists of four overlapping phases: participatory research design; knowledge development; intervention strategy development and testing; and systematization of knowledge. Within each phase there may be several iterations and adaptations of the process. Theoretically, an ecohealth research process can begin in any phase (Fig. 1.1). The process is not particularly unidirectional, and the demarcation between phases is not very distinct. The research process tends to move back and forth between these phases, for example jumping ahead to seize an opportunity for piloting interventions while still pursuing the collection and analysis of data. The more transdisciplinary and participatory the project, the more

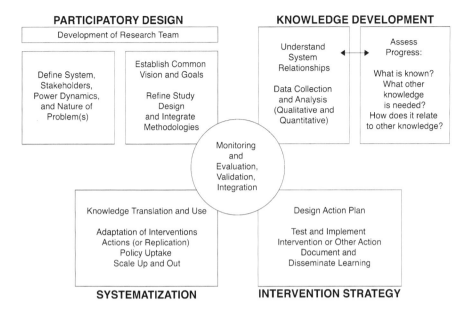

Fig. 1.1 Research process using an ecosystem approach to health – the case studies in this book illustrate how research generally proceeds through four main phases, allowing for back and forth among them, and over a number of iterations. Ecohealth research could be initiated in any quadrant, but tends to start in the *top left* with a participatory design phase

likely will be the integration of data collection and analysis, the consensus around findings, and even the strategies for acting on the findings. This is because effective mechanisms for addressing different points of view and trade-offs around possible actions will have been developed over time through an ecohealth research process. The transitions between phases seem to offer opportunities to reflect on what has happened and what has been learnt (monitoring and evaluation) and to validate the findings (through peer-review, ground truthing, expert and stakeholder discussion), ideally coming to some consensus with relevant stakeholders on relevance and the way forward. But the case studies reveal that in practice it is hard to achieve this with any predictable rhythm or regularity.

Participatory Design Phase

All of the case studies in this book represent research that included a participatory design phase. However, in quite a few cases, this came after a more conventionally academic initial design. The participatory design phase captures a series of activities common to most ecohealth research of this nature. During this part of the process, the investigation team becomes organized; ecohealth principles and frameworks are discussed and applied; and stakeholders are identified and included. The team strives for consensus around goals of the research, possible methodological approaches, and consideration of the end-users of the research results. The research questions and methodologies are initially defined, with attention to integration across disciplines. The research team may consider a plan for transdisciplinary practice, integrated analysis, end-user engagement, and monitoring and evaluating progress toward project goals. Data collection occurring in this phase informs a full description of the system context for the problem at hand. Social and gender inequities and ethical issues are considered in the design phase. There may be a redefining of research priorities or a narrowing of the elements of the system that will be studied. Stakeholder power dynamics come into play and affect both team dynamics and the research design. A common vision for the way forward may be developed.

Knowledge Development Phase

As the research shifts into active discovery mode, this marks a new phase: knowledge development. This phase necessarily overlaps with its neighbouring phases because of the complex reality of field research. Although the knowledge development phase is familiar and exciting for most academic researchers, it is sometimes frustrating for community members awaiting some sort of change to occur in their problematic conditions. During this phase, research instruments are developed, integrated across disciplines, challenged, tested, and applied; samples taken and data collected; and gender and social analysis applied. Analyses are undertaken, and new knowledge begins to emerge that informs a new understanding of the behaviour of the system, and the reasons for the problem. Research may be redesigned or new components added. Findings and progress are tracked, and refinements to methodological protocols recorded. Research findings may generate new questions and trigger additional research. Through processes that were planned and evolved throughout the research, findings are systematically shared and validated among researchers, the community, and other stakeholders. As consensus emerges around various findings, strategies for action begin to emerge. Ideally, before any intervention begins, there is an opportunity to reflect on and disseminate knowledge gained and other progress made, and to revisit stakeholder analysis and the participation of stakeholders in developing a common vision for action.

Intervention Strategy Phase

The intervention phase describes a period of activities targeted at making a change. In participatory research, some aspects of intervention are likely to be underway at the same time as the discovery phase. In the Guatemala study, the research team spent much time in the community (determining their needs, otherwise engaging the community in the research, and understanding the community's development), and as a result, the intervention and discovery phases were almost fully blended. In other cases, an intervention phase begins once enough knowledge has been acquired and validated to develop an action plan, or at least to begin testing various intervention ideas (e.g. Yaoundé and Kathmandu). The implementation of this plan ideally occurs within an evaluative framework so that its effectiveness can be gauged. The observed effects of these interventions and other actions contribute more information. The system's behaviour (does the intervention work, or not, or only partly so?) may provoke another round of knowledge development. But if successful, the ecohealth research process can move into a broader systematization of new knowledge and lessons learned about what works and what does not work.

Systematization Phase

In the systematization phase, the knowledge gained from research and action is applied on a wider scale. Advocacy for policy change or new programs may be successful, resulting in new policies and programs based on research evidence. There may be opportunities to test and apply knowledge and replicate interventions in other contexts. Knowledge translation is not relegated to the very end of the project. In participatory processes like those described here, there appear to emerge fast cycles (generation of knowledge that requires immediate action and implementation) and slower cycles that result in broader ecological, health outcome, and policy changes (Gitau et al. 2009). To maintain long-term credibility as well as ethical conduct, it is often essential for the researchers to respond to the short-term implications of their findings, and sometimes to questions or needs not directly related to the main research question, as illustrated in the Bebnine, Lebanon case study. This asynchronous advancement of different threads of a participatory research process presents challenges, particularly around capturing results of all of these threads and weaving them in an overall story of the project's accomplishments. AMESH (Waltner-Toews et al. 2004) or other ecohealth-like research process descriptions do not necessarily cover this, but it is a phenomenon that is described in some of the case studies.

Why These Case Studies?

The case studies reveal many idiosyncrasies that belie their common application of an ecosystem approach. There is no single "right" way to do ecohealth research, although like in any research endeavour or field, there is an emerging common practice, based on lessons learned that lead to preferred ways of doing things, and ways that are known to be flawed or problematic. The case studies in this book, each in their own way, reflect many principles of ecohealth research, but they do not all necessarily share a common framework or methodology. All of them have involved communities and other stakeholders in executing the research. All have strived for effective transdisciplinarity. They have used different methodologies and tools to arrive at various kinds of results.

Hundreds of IDRC projects and many others around the world have implemented these, and other, complementary approaches to develop the field of ecohealth. The field has emerged at the intersection of several others, including conservation medicine (Aguirre 2002), ecosystem health (Rapport et al. 1999), ecosystem integrity (Sieswerda et al. 2001), and international development research. In keeping with the richness to be found among many different perspectives, there is more than one ecosystem approach to health. Likewise, there are many related schools of thought with a common goal to work across disciplines in pursuit of knowledge to help improve human health, ecosystems, and their sustainability. In addition, other recent ideas in public health, such as One Health (Conrad et al. 2009; Karesh and Cook 2005; Zinsstag et al. 2011) and global health (Koplan et al. 2009; Stephen and Daibes 2010) are converging with ecohealth.

The 15 case studies in this book represent similar (although far from identical) applications of an ecosystem approach to health, each strongly influenced by IDRC's school of thought. As such, they do not explicitly or equally emphasize each of the six principles of ecohealth research presented earlier – this framing of ecohealth principles is a new contribution of this book. The case studies differ in style, with some emphasizing quantitative results, and others providing reflection on process. They are illustrations of experience, and are not intended to provide the reader with a road map to replicate the research. Other platforms exist to cover those needs, such as peer-reviewed technical publications, and most of this research can also be found there. Rather, for the first time, this book captures the accumulated experience in ecohealth research of dozens of researchers, working in four continents, over more than a decade.

Arranged in four thematic sections, the case studies bring experiences from the field where efforts to improve the lives of the poor are ongoing in response to agricultural transformations, pollution and environmental change, infectious diseases, and urbanization. With experiences from developing regions around the world, they illustrate how scientists from different disciplines and different countries have collaborated with communities and with leaders from industry and government to address tough problems of health related to environmental pollution, degradation, or change. They also show how ecohealth research has led to lasting changes for the betterment of peoples' lives and the ecosystems that support them. The case studies are followed by a chapter exploring the added value of ecohealth networks and their significance as part of an evolving trend in development research. The book concludes with a discussion and synthesis of lessons from the case studies, and implications for the field of ecohealth.

References

Aguirre, A. (Editor) (2002). Conservation Medicine: ecological health in practice. Oxford University Press, Oxford, UK.

Allen, T.H.F., and Hoekstra, T.W. (Editors). 1992. Toward a Unified Ecology. Complexity in Ecological Sciences Series. Columbia, New York, NY, USA.

Allen, T.F.H., Bandursky, B.L., and King, A.W. (1993). The Ecosystem Approach: Theory and Ecosystem Integrity. A Report to the International Joint Commission of the Great Lakes. International Joint Commission, Washington, DC, USA.

Berkes, F., and Folke, C. (Editors). (1998). Linking Social and Ecological Systems. Cambridge University Press, Cambridge, UK.

Bernard, H.R. (2000). Social Research Methods. Qualitative and Quantitative Approaches. Sage Publications: Thousand Oaks, CA. USA.

Bhumibhol Adulyadej (King of Thailand) (2007). National Health Act, B.E. 2550 (A.D. 2007). Available at: http://whothailand.healthrepository.org/bitstream/123456789/590/1/National%20Health%20Act_2007.pdf

Bopp, M., and Bopp, J. (2004). Welcome to the Swamp: Addressing Community Capacity in Ecohealth Research and Intervention. EcoHealth, 1(2)(Suppl), 24–34.

Brundtland, G. (Editor). (1987). Our Common Future: The World Commission on Environment and Development. Oxford University Press, Oxford, UK.

Bunch, M., McCarthy, D., and Waltner-Toews, D. (2008). A Family of Origin for an Ecosystem Approach to Managing for Sustainability. In: Waltner-Toews, D., Kay, J.J., and Lister, N.M. (Editors). The Ecosystem Approach. Complexity, Uncertainty and Managing for Sustainability. Columbia University Press, New York, NY, USA.

Campbell, B. (2010). Applying Knowledge to Generate Action: A Community-Based Knowledge Translation Framework. Journal of Continuing Education in the Health Professions, 30(1), 65–71.

Carden, F. (2009). Knowledge to Policy: Making the Most of Development Research. SAGE Publications, New Delhi and Thousand Oaks, CA. USA. Available at: http://www.idrc.ca/en/ev-135779-201-1-DO_TOPIC.html.

Chambers, Robert. (1994). The origins and practice of participatory rural appraisal. World Development, 22(7), 953–969.

Chanon, K.E., Méndez-Galván, J.F., Galindo-Jaramillo, J.M., Olguín-Bernal, H., and Borja-Aburto, V.H. (2003). Cooperative Actions to Achieve Malaria Control Without the Use of DDT. International Journal of Hygiene and Environmental Health, 206(4-5), 387–394.

Checkland, P. (2000). Soft Systems Methodology: A Thirty Year Retrospective. Systems Research and Behavioural Science, 17, S11–S58.

Chevalier, J.M., and Buckles, D.J. (2008). SAS²: A Guide to Collaborative Inquiry and Social Engagement. Sage Publications, Thousand Oaks, CA, USA and IDRC, Ottawa, Canada. Available at: http://www.idrc.ca/en/ev-130303-201-1-DO_TOPIC.html and http://www.sas2.net/.

Conrad, P.A., Mazet, J.A., Clifford, D., Scott, C., and Wilkes, M. (2009). Evolution of a Transdisciplinary "One Medicine-One Health" Approach to Global Health Education at the University of California, Davis. Preventive Veterinary Medicine. 92(4), 268–274.

Corvalán, C., Hales, S., Anthony, J., and McMichael, A.J. (2005). Ecosystems and Human Well-Being: Health Synthesis. World Health Organization, Geneva, Switzerland.

CSDH (Commission on Social Determinants of Health). (2008). Closing the Gap in a Generation: Health Equity through Action on the Social Determinants of Health. Final Report of the Commission on Social Determinants of Health. World Health Organization, Geneva, Switzerland. Available at: http://www.who.int/social_determinants/thecommission/finalreport/en/index.html.

DePlaen, R., and Kilelu, C. (2004). From Multiple Voices to a Common Language: Ecosystem Approaches to Human Health as an Emerging Paradigm. EcoHealth, 1(2)(Suppl), 8–15.

Draper, A.K., Hewitt, G., and Rifkin, S. (2010). Chasing the dragon: Developing indicators for the assessment of community participation in health programmes. Social Science and Medicine, 71(6), 1102–1109.

Earl, S., Carden, F., and Smutylo, T. (2001). Outcome Mapping: Building Learning and Reflection into Development Programs. International Development Research Centre, Ottawa, Canada. Available at: http://www.idrc.ca/en/ev-9330-201-1-DO_TOPIC.html.

Forget, G. (1997). From Environmental Health to Health and the Environment: Research that Focuses on People. In: Shahi, G.S., Levy, B.S., Binger, A., Kjellström, T., and Lawrence, R. (Editors). International Perspectives on Environment, Development and Health: Towards a Sustainable World, Springer, New York, NY, USA.

Forget, G., and Lebel, J. (2001). An Ecosystem Approach to Human Health. International Journal of Occupational and Environmental Health, 7(2)(Suppl), S3–38.

Freitas, C.M. de, Oliveira, S.G. de, Schütz, G.E., Freitas, M.B., and Camponovo, M.P.G. (2007). Ecosystem Approaches and Health in Latin America. Cadernos de Saúde Pública, 23(2), 283–296. Available at: http://www.scielo.br/scielo.php?script=sci_arttext&pid=S0102-311X2007000200004&lng=en.%20doi:%2010.1590/S0102-311X2007000200004.

Funtowicz, S., and Ravetz, J. (2008). Beyond Complex Systems: Emergent Complexity and Social Solidarity. In: Waltner-Toews, D., Kay, J.J., and Lister, N.M. (Editors). (2008). The Ecosystem Approach. Complexity, Uncertainty and Managing for Sustainability. Columbia University Press, New York, NY, USA.

Gitau, T., Gitau, M.W., Waltner-Toews, D. (2009). Integrated Assessment of Health and Sustainability of Agroecosystems. CRC Press, Boca Raton, FL, USA, and Taylor and Francis, London, UK.

Goy, J., and Waltner-Toews, D. (2005). Improving Health in Ucayali, Peru: A Multisector and Multilevel Analysis. EcoHealth, 2(1), 47–57.

Graham, I.D., Logan, J., Harrison, M., Straus, S., Tetroe, J.M., Caswell, W., and Robinson, N. (2006). Lost in knowledge translation: Time for a map? Journal of Continuing Education in the Health Professions, 26(1), 13–24.

Hassan, R., Scholes, R., and Ash, N. (Editors). (2005). Ecosystems and Human Well-Being: Current State and Trends, Volume 1 of the Millennium Ecosystem Assessment Series. Island Press, Washington, DC, USA. Available at: http://www.millenniumassessment.org/documents/document.765.aspx.pdf.

Iriart, C., Waitzkin, H., Breilh, J., Estrada, A., and Merhy, E.E. (2002). Latin American Social Medicine: Contributions and Challenges (*in Spanish*). Revista Panamericana de Salud Pública, 12(2), 128–136.

Karesh, W.B., and Cook, R.A. (2005). The Human-Animal Link. Foreign Affairs, 84, 38–50.

Kay, J., and Regier, H. (2000). Uncertainty, Complexity, and Ecological Integrity: Insights from an Ecosystem Approach. In: Crabbe, P., Holland, A., Ryszkowski, L., and Westra, L. (Editors). Implementing Ecological Integrity: Restoring Regional and Global Environmental and Human Health. NATO Science Series, Environmental Security, Kluwer Academic Publishers, Dordrecht, The Netherlands.

Kay, J., Regier, H., Boyle, M., and Francis, G. (1999). An Ecosystem Approach for Sustainability: Addressing the Challenge of Complexity. Futures, 31(7), 721–742.

Koplan, J.P., Bond, T.C., Merson, M.H., Reddy, K.S., Rodriguez, M.H., Sewankambo, N.K., and Wasserheit, J.N. (2009). Consortium of Universities for Global Health Executive Board. Towards a Common Definition of Global Health. Lancet, 373(9679), 1993–1995.

Krieger, N., and Birn, A.E. (1998). A Vision of Social Justice as the Foundation of Public Health: Commemorating 150 Years of the Spirit of 1848. American Journal of Public Health, 88(11), 1603–1606.

Kuhn, T.S. (1970). The Structure of Scientific Revolutions. 2nd. Edition. University of Chicago Press, Chicago, USA.

Lambert, T.W., Soskolne, C.L., Bergum, V., Howell, J., and Dossetor, J.B. (2003). Ethical Perspectives for Public and Environmental Health: Fostering Autonomy and the Right to Know. Environmental Health Perspectives, 111(2), 133–137.

Lavis, J.N., Lomas, J., Hamid, M., and Sewankambo, N.K. (2006). Assessing Country-Level Efforts to Link Research to Action. Bulletin of the World Health Organization, 84(8), 620–628. Erratum in Bulletin of the World Health Organization, 2006 84(10), 840.

Lebel, J. (2003). Health: An Ecosystem Approach. International Development Research Centre, Ottawa Canada. Available at: http://www.idrc.ca/in_focus_health/.

Lopez, H., and Serven, L. (2009). Too Poor to Grow. Policy Research Working Paper 5012. Development Research Group, Macroeconomics and Growth Team, World Bank, Washington, DC, USA. http://econ.worldbank.org.

MacIntyre, A. (2008). Participatory Action Research. Sage Publications, Thousand Oaks, CA, USA.

March, D., and Susser, E. (2006). The Eco- in Eco-Epidemiology. International Journal of Epidemiology, 35(6), 1379–1383.

Mergler, D. (2003). Integrating Human Health into an Ecosystem Approach to Mining. In: Rapport, D.J., Lasley, W.L., Rolston, D.E., Nielsen, N.O., Qualset, C.O., and Damania, A.B. (Editors.) Managing for Healthy Ecosystems. CRC Press, Boca Raton, Florida, USA.

Mertens, F., Saint-Charles, J., Mergler, D., Passos, C. J., Lucotte, M. (2005). A network Approach for Analysing and Promoting Equity in Participatory Ecohealth Research. EcoHealth, 2, 113–126. http://www.unites.uqam.ca/gmf/caruso/doc/caruso/mertens/mertens_et_al_2005.pdf

Morrison, K., Aguiar Prieto, P., Castro Domínguez, A., Waltner-Toews, D., and Fitzgibbon, J. (2008). Ciguatera Fish Poisoning in la Habana, Cuba: A Study of Local Social-Ecological Resilience. EcoHealth, 5(3), 346–359.

Neudoerffer, R.C., Waltner-Toews, D., Kay, J.J., Joshi, D.D., and Tamang, M.S. (2005). A Diagrammatic Approach to Understanding Complex Eco-Social Interactions in Kathmandu, Nepal. Ecology and Society, 10(2), 12. Available at: http://www.ecologyandsociety.org/vol10/iss2/art12/.

Ottoson, J.M. (2009). Knowledge-for-Action Theories in Evaluation: Knowledge Utilization, Diffusion, Implementation, Transfer and Translation. In: JM Ottoson, J.M., and Hawe, P. (Editors), Knowledge Utilisation, Diffusion, Implementation Transfer and Translation: Implications for Evaluation. New Directions for Evaluation, 124, 7–20.

Pablos-Mendez, A., Chunharas, S., Lansang, M.A., Shademani, R., and Tugwell, P. (2005). Knowledge Translation in Global Health. Bulletin of the World Health Organization, 83(10), 723.

Parkes, M., Panelli, R, Weinstein, P. (2003). Converging Paradigms for Environmental Health Theory and Practice. Environmental Health Perspectives, 111, 669-675.

Parkes, M.W., Bienen, L., Breilh, J., Hsu, L-N., McDonald, M. Patz, J.A., Rosenthal, J.P., Sahani, M. Sleigh, A.,Waltner-Toews, D., and Yassi, A. (2005). All Hands on Deck: Transdisciplinary Approaches to Emerging Infectious Disease. EcoHealth, 2(4), 258–272.

Parkes, M.W., Morrison, K.E., Bunch, M.J., Hallstrom, L.K., Neudoerffer, R.C., Venema, H.D., and Waltner-Toews, D. (2010). Towards Integrated Governance for Water, Health and Social–Ecological Systems: The Watershed Governance Prism. Global Environmental Change 20:693–704.

Parry, M.L., Canziani, O.F., Palutikof, J.P., van der Linden, P.J., and Hanson, C.E. (Editors). (2007). Contribution of Working Group II to the Fourth Assessment Report of the Intergovernmental Panel on Climate Change, 2007. Cambridge University Press, Cambridge, United Kingdom and New York, NY, USA.

Paruccini, M. (1994). Applying Multiple Criteria Aid for Decision to Environmental Management. Kluwer Academic Publishers, Dordrecht, The Netherlands.

PHAC (Public Health Agency of Canada). (1996). Towards a Common Understanding: Clarifying the Core Concepts of Population Health: Core Concepts of the Population Health Approach. A Discussion Paper. Catalogue No. H39-391/1996E ISBN 0-662-25122-9. Public Health Agency of Canada, Ottawa, Canada. Available at: www.phac-aspc.gc.ca/ph-sp/docs/common-commune/index-eng.php.

Prüss-Üstün, A., and Corvalán, C. (2006). Preventing Disease Through Healthy Environments: Towards an Estimate of the Environmental Burden of Disease. World Health Organization, Geneva, Switzerland. Available at: http://www.who.int/entity/quantifying_ehimpacts/publications/preventingdiseasebegin.pdf.

Pohl, C., and Hirsch Hadorn, G. (2008). Methodological Challenges of Transdisciplinary Research. Natures Sciences Sociétés, 16, 111–121.

Rapport, D.J., Böhm, G., Buckingham, D., Cairns, J., Costanza, R., Karr, J.R., de Kruijf, H.A.M., Levins, R., McMichael, A.J., Nielsen, N.O., Whitford, W.G. (1999) Ecosystem health: the concept, the ISEH, and the important task ahead. Ecosystem Health 5, 82–90.

Rapport D.J., Thorpe, C., and Regier, H.A. (1979). Ecosystem Medicine. Bulletin of the Ecological Society of America, 60, 180–192.

Reason, P., and Bradbury, H. (Editors). (2007). Handbook of Action Research. Participative Inquiry and Practice. 2nd Edition. Sage, London, UK.

Regier, H.A., and Kay, J.J. (2001). Phase Shifts or Flip-Flops in Complex Systems. In: Munn, R. (Editor in Chief). Volume 5, Encyclopedia of Global Environmental Change. John Wiley & Sons, London, UK.

Republic of Ecuador. (2008). Articles 14 (Right to Healthy Environment) and 32 (Right to Health). Constitution of 2008. Political Database of the Americas. Available at: http://pdba.georgetown.edu/Constitutions/Ecuador/ecuador08.html#mozTocId735388.

Sieswerda, L.E., Soskolne, C.L., Newman, S.C., Schopflocher, D., and Smoyer, K.E. (2001). Toward Measuring the Impact of Ecological Disintegrity on Human Health. Epidemiology, 12(1), 28–32.

Soskolne, C.L. (Editor). (2007). Sustaining Life on Earth: Environmental and Human Health through Global Governance. Lexington Books, Lanham, MD, USA.

Soskolne, C.L., Butler, C.D., Ijsselmuiden, C., London, L., and von Schirnding, Y. (2007). Toward a global agenda for research in environmental epidemiology. Epidemiology, 18(1), 162–166.

STEPS Centre (Social, Technological and Environmental Pathways to Sustainability Centre). (2010). Innovation, Sustainability, Development: A New Manifesto. STEPS Centre: Brighton, UK. Available at: http://anewmanifesto.org/wp-content/uploads/steps-manifesto_small-file.pdf

Stephen, C., and Daibes, I. (2010). Defining Features of the Practice of Global Health Research: An Examination of 14 Global Health Research Teams. Global Health Action. Available at: http://www.globalhealthaction.net/index.php/gha/article/viewArticle/5188/5772.

Sudsawad, P. (2007). Knowledge translation: Introduction to models, strategies, and measures. Southwest Educational Development Laboratory, National Center for the Dissemination of Disability Research. Austin, TX, USA. Available at: http://www.ncddr.org/kt/products/ktintro/allinone.html.

Susser, M., and Susser, E. (1996). Choosing a Future for Epidemiology: II. From Black Box to Chinese Boxes and Eco-Epidemiology. American Journal of Public Health, 86(5), 674–677. Erratum in: 86(8), 1093.

Tetroe, J. (2007). Knowledge Translation at the Canadian Institutes of Health Research: A Primer. Austin, TX: National Center for the Dissemination of Disability Research. Focus Technical Brief No. 18. http://www.ncddr.org/kt/products/focus/focus18/.

Tugwell, P., Robinson, V., Grimshaw, J., and Santesso, N. (2006). Systematic Reviews and Knowledge Translation. Bulletin of the World Health Organization, 84(8), 643–651.

UNEP (United Nations Environment Program). (2004). Women and the Environment. United Nations Environment Program, New York, NY, USA.

United Nations. (1992). Report of the United Nations Conference on Environment and Development (Rio de Janeiro, 3–14 June 1992). Annex I: Rio Declaration on Environment and Development, Principle 3. United Nations Publication, A/CONF.151/26 (Vol. I).

United Nations. (2000). United Nations General Assembly 55th Session. Agenda Item 60b, Resolution 2, Session 55, United Nations Millennium Declaration, 18 September 2000. Available at: http://www.un.org/ga/55/.

United Nations. (2001). U.N. General Assembly, 56th Session. Road Map Towards the Implementation of the United Nations Millennium Declaration: Report of the Secretary-General (A/56/326). 6 September 2001.

United Nations. (2008). Highlights from World Population Prospects: The 2008 Revision.

VanLeeuwen, J., Waltner-Toews, D., Abernathy, T. and Smit, B. (1999). Evolving Models of Human Health toward an Ecosystem Context. Ecosystem Health, 5, 204–219.

Vernooy, R. (Editor). (2006). Social and Gender Analysis in Natural Resource Management: Learning Studies and Lessons from Asia.Sage India/CAP/IDRC Canada. Available at: http://www.idrc.ca/en/ev-91907-201-1-DO_TOPIC.html.

Wackernagel, M., Schulz, M.B., Deumling, D., Linares, A.C., Jenkins, M., Kapos, V., Monfreda, C., Loh, J., Myers, N., Norgaard, R., and Randers, J. (2002). Tracking the Ecological Overshoot of the Human Economy. Proceedings of the National Academy of Science, 99(14), 9266–9271. Available at: http://www.pnas.org/content/99/14/9266.full.pdf.

Waitzkin, H., Iriart, C., Estrada, A., and Lamadrid, S. (2001). Social Medicine Then and Now: Lessons from Latin America. American Journal of Public Health, 91(10), 1592–1601.

Waltner-Toews, D. (2004). Ecosystem Sustainability and Health: A Practical Approach. Cambridge University Press, Cambridge, UK.

Waltner-Toews, D., and Kay, J. (2005). The Evolution of an Ecosystem Approach: The Diamond Schematic and an Adaptive Methodology for Ecosystem Sustainability and Health. Ecology and Society, 10(1), 38. Available at: http://www.ecologyandsociety.org/vol10/iss1/art38/.

Waltner-Toews, D., Kay, J.J., and Lister, N.M. (Editors). (2008). The Ecosystem Approach. Complexity, Uncertainty and Managing for Sustainability. Columbia University Press, New York, NY, USA. 383p.

Waltner-Toews, D., Kay, J., Tamsyn, P.M., and Neudoerffer, C. (2004). Adaptive Methodology for Ecosystem Sustainability and Health (AMESH): An Introduction. In: Midgley, G., and Ochoa-Arias, A.E. (Editors). Community Operational Research: Systems Thinking for Community Development. Kluwer and Plenum Press, New York, NY, USA.

WGEKN (Women and Gender Equity Knowledge Network). (2007). Unequal, Unfair, Ineffective and Inefficient. Gender Inequity in Health: Why It Exists and How We Can Change It. Final Report of the Women and Gender Equity Knowledge Network. World Health Organization, Geneva, Switzerland. Available at: www.who.int/social_determinants/resources/csdh_media/wgekn_final_report_07.pdf.

WHO (World Health Organization). (1948). Constitution of the World Health Organization. WHO, Geneva, Switzerland. Available at: http://www.who.int/governance/eb/who_constitution_en.pdf.

WHO (World Health Organization). (1978). Declaration of Alma-Ata. WHO, Geneva, Switzerland. Available at: http://www.who.int/hpr/NPH/docs/declaration_almaata.pdf.

WHO (World Health Organization). (1986). Ottawa Charter for Health Promotion. WHO, Geneva, Switzerland. Available at: http://www.who.int/healthpromotion/conferences/previous/ottawa/en/

WHO (World Health Organization). (2002a). Gender Analysis in Health: A Review of Selected Tools. WHO, Geneva, Switzerland. Available at: http://www.who.int/gender/documents/en/Gender.analysis.pdf.

WHO (World Health Organization). (2002b). Framework for Linkages Between Health, Environment and Development. Chapter 7 in: Health in Sustainable Development Planning: The Role of Indicators. WHO, Geneva, Switzerland. Available at: http://www.who.int/wssd/resources/indicators/en/.

WHO (World Health Organization). (2004). World Report on Knowledge for Better Health. WHO, Geneva, Switzerland. Available at: http://www.who.int/rpc/meetings/pub1/en/

WHO (World Health Organization). (2005). International Health Regulations. WHO, Geneva, Switzerland. Available at: http://www.who.int/ihr/en/.

WHO (World Health Organization). (2008). The Global Burden of Disease: 2004 Update. WHO, Geneva, Switzerland. Available at: http://www.who.int/healthinfo/global_burden_disease/2004_report_update/en/index.html.

WHO (World Health Organization). (2009a). Protecting Health from Climate Change. Connecting Science, Policy and People. WHO, Geneva, Switzerland. Available at: http://www.who.int/globalchange/publications/reports/9789241598880/en/index.html.

WHO (World Health Organization). (2009b). Women and Health: Today's Evidence, Tomorrow's Agenda. WHO, Geneva, Switzerland. Available at: http://www.who.int/gender/documents/9789241563857/en/index.html.

WHO (World Health Organization). (2009c). Gender, Climate Change and Health. Draft Discussion Paper. WHO, Geneva, Switzerland. Available at: http://www.who.int/globalchange/publications/reports/gender_climate_change/en/index.html.

Wilcox, B., and Kueffer, C. (2008). Transdisciplinarity in EcoHealth: Status and Future Prospects. EcoHealth, 5, 1–3.

World Bank. (2009). Global Monitoring Report 2008: MDGs and the Environment. Figure 2.14: Economic Burden Associated with Poor Environmental Health (p. 82). World Bank, Washington, DC, USA.

World Bank, FAO (Food and Agriculture Organization), and IFAD (International Fund for Agricultural Development). 2009. Gender in Agriculture Sourcebook. World Bank, Washington, DC, USA.

Zinsstag, J., Schelling, E., Waltner-Toews, D., and Tanner, M. (2011). From "one medicine" to "one health" and systemic approaches to health and well-being. Preventive Veterinary Medicine, 101(3–4), 148–56.

Part I
Linking Human Health and Well-Being to Changing Rural Agro-Ecosystems

Chapter 2
Introduction

Lamia El-Fattal and Andrés Sánchez

Since the 1950s, farmers around the world have made remarkable achievements in food production, notably in Asia and Latin America, but at a significant cost to their environment and health. Spurred by the technological advances of the Green Revolution, crop yields increased and domestic food production expanded substantially. In countries that had often suffered in the past from famine, chronic food shortages were diminished. The 2010 Summit on Millennium Development Goals (MDGs) reported that the proportion of hungry people in the world had shrunk from 20% in 1990 to about 16% today – considerable progress toward the MDG goal of less than 10% by 2015. But the number of hungry people remains around one billion and is clearly unacceptable in a world that has the knowledge and resources needed to eradicate hunger (FAO 2009, 2010).

In terms of environmental sustainability and human health costs, this increase in agricultural productivity has been accompanied by an unsustainable dependence on agrochemicals, overexploitation of water resources, and widespread neglect of soil health. Land degradation now affects 1.9 billion hectares and 2.6 billion farmers, accounting for about a third of all arable land. Agricultural inputs increased at exponential rates (e.g. a threefold increase in global expenditures on pesticides for agriculture and a tenfold increase in fertilizer use over the last 50 years), while there was a concomitant decline of the natural resource base on which agriculture depends. Up to 75% of the genetic base of agricultural crops is thought to have been lost (IAASTD 2008).

There are other negative impacts on health and the environment from the intensification of agriculture. For example, the misuse of pesticides has led to the contamination of land and water, biodiversity decline of nontarget species, and the emergence of pesticide-resistant pests and human disease vectors. Exposure to pesticides is also one of the principal occupational hazards of agriculture. It is estimated

---

L. El-Fattal (✉) • A. Sánchez
International Development Research Centre, Ottawa, ON, Canada
e-mail: ecohealth@idrc.ca

that 355,000 people die every year from pesticide poisoning. The intensification of livestock production in the last decades has also been accompanied by problems, including antimicrobial resistance, new human pathogens, and waste-management issues. Climate and environmental changes are now creating new challenges for agriculture, in some cases increasing requirements for chemical inputs and compounding the problems of managing agricultural wastes and runoff. The resulting combined harm from poorly managed agricultural pollution and extreme weather is likely to disproportionately affect the poor (World Bank 2008b).

In sharp contrast to this stark background is the potential of agriculture for improving the livelihoods of a billion rural people in the developing world who live on less than US$ 1 a day (World Bank 2008a). But in order to benefit from this potential, more attention is required to the linkages between human and animal health, the sustainability of agro-ecosystems, and rural livelihoods. Ecohealth research can contribute to highlighting alternative healthier and more sustainable agricultural livelihood strategies.

This section presents four examples of ecohealth research on agricultural transformations (understood here as intensification of agricultural production and/or agricultural area expansion for subsistence and cash-crop production). In all four cases, agriculture is not only seen as the basis for people's sustenance, but also as a productive activity that creates economic value and sustains healthier rural livelihoods. Agriculture, through its social, economic, and environmental dimensions, affects the health of farmers, agricultural workers, their families, and villages. Their health, in turn, impacts agriculture. In this sense, health is both an outcome and a driving force of sustainable production. Agro-ecosystems are a particularly interesting focus for ecosystem approaches to health because of the many interlinked health benefits and hazards determined by how agroecosystems are managed. Well-managed agroecosystems are beneficial in terms of environmental sustainability, social equity, and health.

The case studies cover a range of sub-themes, including occupational health, water management, and the issues around the misuse of agrochemicals. The importance of economically productive livelihoods is particularly evident as a key factor mediating environmental and occupational exposure to agrochemicals. A case in point is the Granobles River basin of Ecuador, where the newly introduced floriculture industry provided important employment and income opportunities for traditional potato growers and under-employed rural people. The research produced evidence of pesticide pollution in waterways and negative health impacts of pesticide exposure in children and flower workers. An environmentally and socially responsible accreditation program was developed as an innovative strategy for reducing pesticide exposure and improving livelihoods.

A second example from Ecuador describes how the Peruvian-based International Potato Centre (CIP) worked with local farmers to develop cultivation practices suited to highland conditions while relying less on costly pesticides and fertilizers. Through this work, health hazards and production costs were both effectively reduced. The discoveries made by the Ecosalud team were used to improve national legislation on the sale and use of pesticides in Ecuador.

In northern Malawi, the low-input agriculture theme continues. Building on farmers' knowledge, the project used an iterative learning process to implement new legume options for smallholder farmers, improving soil fertility, and reaping significant nutritional and food-security benefits.

The last example is from Lebanon and Yemen. The projects explored how eroding local food systems in marginal regions of the highlands of Yemen and the semi-arid lands of Lebanon have led to shrinking dietary diversity, diminished agro-ecological sustainability, and weakened community health. The projects developed policy options and local actions to promote agro-biodiversity and a diverse nutritious diet.

The applied research in all four cases aimed to understand social and agro-ecological interactions, while developing practical strategies to change the situation for improved human health. The contributions of participatory methods is clear in the Malawi and Ecuador (Ecosalud) projects, where farmers and community members became change agents, adopting the practices informed by the research findings. Ecohealth provided a basis for addressing the interplay between occupational health hazards and other social and environmental determinants in both Ecuadorian projects. The examples provide evidence on how healthier agricultural practices and policies are possible while maintaining productivity and mitigating negative effects. The research studies also broadened and facilitated social engagement and mobilization for healthier and more sustainable rural livelihoods, especially among the more marginalized groups – Yemeni highlanders, women farmers in Lebanon, and indigenous communities in Ecuador.

References

FAO (Food and Agriculture Organization of the United Nations). (2009). 1.02 Billion People Hungry: One Sixth of Humanity Undernourished — More Than Ever Before. Available at: http://www.fao.org/news/story/en/item/20568/icode/, FAO Media Centre, FAO Rome.

FAO (Food and Agriculture Organization of the United Nations). (2010). Brief: Global Hunger Declining, But Still Unacceptably High. Economic and Social Development Department, September 2010. Available at: http://www.fao.org/docrep/012/al390e/al390e00.pdf

IAASTD (International Assessment of Agricultural Knowledge, Science and Technology for Development). (2008). Agriculture at a Crossroads. Global Report. Available at: http://www.agassessment.org/reports/IAASTD/EN/Agriculture%20at%20a%20Crossroads_Global%20Report%20(English).pdf

World Bank. (2008a). Agriculture and Environment Policy Brief. World Development Report 2008. Available at: http://go.worldbank.org/GYSIT22AU0.

World Bank. (2008b). World Development Report. Millennium Ecosystem Assessment. Volume 1: Current State and Trends. Island Press, Washington, D.C., USA.

Chapter 3
Growing Healthy Communities: Farmer Participatory Research to Improve Child Nutrition, Food Security, and Soils in Ekwendeni, Malawi

Rachel Bezner Kerr, Rodgers Msachi, Laifolo Dakishoni, Lizzie Shumba, Zachariah Nkhonya, Peter Berti, Christine Bonatsos, Enoch Chione, Malumbo Mithi, Anita Chitaya, Esther Maona, and Sheila Pachanya

In 1997, a Malawian community nurse and a Canadian soil-science student interviewed families whose children were admitted to the Nutrition Rehabilitation Centre at Ekwendeni Hospital. They wanted to learn what had led to the severe malnourishment of children in their region in northern Malawi. The stories they heard from the community had a similar refrain – families were no longer able to afford the rising prices of commercial fertilizers. Soil fertility had declined and without fertilizers, farmers were unable to get adequate yields of maize, the dominant food crop in smallholder agriculture in Malawi (Snapp et al. 1998). People reported that this food shortage was leading to high levels of child malnutrition. The women reported that their husbands had serious problems with alcohol and were sometimes violent toward them. Many families felt hopeless about their situation (Bezner Kerr 2005).

These stories were so compelling that the nurse and the student decided to investigate alternatives to commercial fertilizers to improve soil fertility. They were also keen to address the social dimensions that were affecting child health and nutrition. They learned about on-farm research, which had identified appropriate legume options for smallholder farmers to improve soil fertility in Central Malawi (Snapp et al. 1998). These legume options included intercropping two legume crops (e.g., pigeon pea and groundnut) that together improved soil fertility if the crop residues were buried after harvest. Previous research had indicated that farmers preferred the

R.B. Kerr (✉) • C. Bonatsos
Department of Geography, University of Western Ontario, London, ON, Canada
e-mail: rbeznerkerr@uwo.ca

R. Msachi • L. Dakishoni • L. Shumba • Z. Nkhonya • E. Chione • M. Mithi • A. Chitaya
• E. Maona • S. Pachanya
Soils, Food and Healthy Communities Project, Ekwendeni Hospital, Ekwendeni, Malawi

P. Berti
HealthBridge Canada, Ottawa, ON, Canada

"doubling up" of edible legumes because they improved soil fertility, provided food for consumption and possibly for the market, and reduced weeding requirements (Snapp et al. 1998). As edible food, the legumes added nutritional and food-security benefits because they added a better source of protein and micronutrients to the diet (FAO 1992).

The Approach

From this previous knowledge, the Soils, Food and Healthy Communities (SFHC) project in the Ekwendeni region of northern Malawi was initiated to explore the relationship between human health, social factors, and the ecosystem (Forget and Lebel 2001). The region is a mid-altitude, semi-humid ecosystem, with one rainy season during the months of December to April. The majority of people are smallholder farmers growing maize as their primary crop and other crops such as groundnuts, cassava, and beans. Participatory and interdisciplinary methods were used to test various legume options and to understand and address the links among degraded soils, food insecurity, and child malnutrition. The focus was to improve child nutrition *through* land management, which caught the attention of farming families struggling with malnourished children at home.

Accustomed to agricultural extension workers lecturing them about farming techniques, and nurses telling them how to care for their children, the community was surprised when hospital staff first approached them to talk about agriculture. Spearheaded by a multidisciplinary team that included a sociologist, nutritionist, and agronomist along with hospital staff and farmers, the project began in 2000 in seven villages near Ekwendeni and drew on farmer participatory research done elsewhere (Gubbels 1997; Humphries et al. 2000). Participating villages were asked to select representatives for a Farmer Research Team (FRT), which would learn about different agricultural options, test them in their own fields, and teach other farmers. This 30-member team traveled to central Malawi to learn about different legume options to improve soil fertility. After they returned, they held village-level meetings to present what they had learned and invite other interested farmers to join the FRT. In the first year of the project, 183 farmers joined, and each secured enough legume seeds to plant a 10 m by 10 m plot.

The case–control longitudinal study used multiple methods to assess change. The design type was appropriate for examining whether this ecohealth approach could make a significant impact on child nutrition (Bryman et al. 2009). To address ethical concerns about measurement of human subjects with no clear benefit, control communities were able to join the project after a set time. During 2000–2009, 200 in-depth interviews, 30 focus groups, and 8 surveys were conducted. In addition, the team made more than 3,000 anthropometric measurements of children (weight and height), weighed crop harvests, visited hundreds of farmers' fields to assess residue practices, and held many participatory workshops. The process was iterative – as new issues arose, research activities were adjusted accordingly. This iterative design is a necessary element of an ecohealth approach because it allows for new findings and community concerns to influence research activities.

The FRT was involved in all major project activities: training new participants; seed distribution; field visits; and organizing community meetings. The FRT explicitly linked agricultural innovations to the nutritional outcomes of the children through talks, songs, and drama. Initially, farmers were quite skeptical about whether these legume options could improve soil fertility enough to improve maize yields, but this changed as the evidence emerged. The FRT played a key role in testing the legumes, conducting research, and developing innovative solutions to many of the identified problems. The impact of its efforts was felt over time, as more and more farmers and villages joined the project.

Early in the project, the need to incorporate the legume residue into the soil soon after harvest arose as a key issue. Because the women were usually responsible for harvesting legumes, they now had an additional labor-intensive task to carry out during a busy time of the year (Bezner Kerr 2008). The FRT decided to organize "crop-residue incorporation days" at the village level. This activity involved a public demonstration of crop-residue incorporation at village plots, with men taking a lead role in encouraging other men to take on this responsibility.

Achievements

By 2005, more than 4,000 farmers had joined the experiment and there was geographic evidence of an expanded legumes cultivation area in this region (Bezner Kerr et al. 2007b). The FRT visited more than 100 farmers' fields annually to assess whether or not farmers were burying legume residue after harvest. The data consistently demonstrated that participating farmers were burying crop residue significantly more often than nonparticipating farmers (Fig. 3.1).

Increased legume production also led to improved soil fertility. Farmers reported improved maize yields and soil structure after several years of incorporating legume residue. Crop-yield data collected in 2008 indicated that farmers who grew legume crops the previous year had significantly higher maize yields compared with farmers who grew maize only (SFHC Project 2008). Knowledge about different ways to improve soil fertility has also been gained. During interviews conducted in 2007 and 2008, farmers had begun to refer to the crop residues as a type of fertilizer (Shumba and Bezner Kerr 2008). As one farmer noted:

> I have learnt how to grow mucuna, pigeon pea, and lots of soya, as we were not growing a lot of soya. I am amazed, because once you bury the residue the soil gets black quickly and so fertile, something which I never knew.[1]

---
[1] Semi-structured interview with 49-year-old male farmer, April 2009.

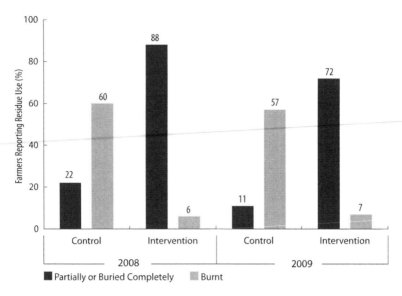

Fig. 3.1 Use of crop residue in 2008 and 2009 by control and intervention farmers (February survey, $n=177$ in 2008 and $n=231$ in 2009)

Not only did the project focus on improving soil fertility, the ecosystem approach to health also encouraged the team to consider the potential impact of increased productivity on gender relations, and intersections between child-feeding practices, household dynamics, and nutritional outcomes. Using qualitative methods, participatory workshops, a structured survey, and anthropometric measures, early child-feeding practices were examined (Bezner Kerr et al. 2007a, 2008a). The project team learned that while families were increasing their legume production, some legumes were being sold by the men who often spent the proceeds for other purposes such as on discretionary drinking (Bezner Kerr 2008). Another finding was that grandmothers were key decision makers about early child feeding, and at times fostered unhealthy feeding practices such as very early introduction of porridge to infants (Bezner Kerr et al. 2008a). These findings led to the initiation of Agriculture and Nutrition Discussion Groups (ANDGs), composed of small groups by gender and age. The groups discussed agriculture, nutrition, and social topics related to improving child nutrition. The goal was to solve key issues that affected child nutrition. These discussion groups provided an important arena to generate ideas and share knowledge in the communities (Satzinger et al. 2009). For example, farmers shared ideas about seed storage and child-feeding practices. Surveys conducted in 2008 and 2009 showed a significant difference in farmer practices between ANDG-participating households and other project households. In 2009, 81% of ANDG participants buried crop residue compared with 69% of participating farmers not in the ANDGs, out of a sample of 231 randomly selected households (Bonatsos et al. 2009).

3 Growing Healthy Communities: Farmer Participatory Research to Improve... 41

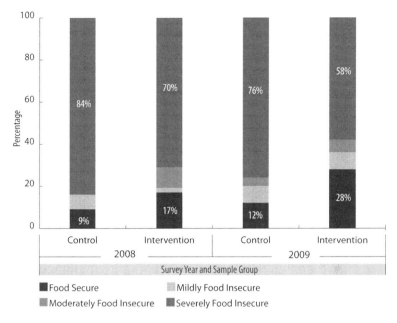

Fig. 3.2 Prevalence of food security in control and intervention groups in 2008 and 2009

The face of agriculture is changing around Ekwendeni – crops are more diverse now because of the project. In the past, mono-cropping was common, especially of maize, which not only depleted soil fertility, but limited the diversity of the community diet (Shaxson and Tauer 1992). With legume crops, dietary diversity has improved. For example, a survey conducted in 2008 indicated that control households grew an average of 4.4 crops compared with intervention households who grew an average of 6.16 crops, with the additional crops usually being legumes (Bonatsos et al. 2009).

There were also interesting results in the area of food security. Of a sample of 177 and 231 households in 2008 and 2009, respectively, higher levels of control households were severely food insecure compared with those households participating in the project (Fig. 3.2). For example, in 2009, 76% of control households were severely food insecure compared with 58% of intervention households.

There was also evidence of a significant increase in children's consumption of legumes as reported in the surveys conducted in 2002 and 2005 (Bezner Kerr et al. 2007b). In the past, people often only fed their children maize porridge, but the surveys showed a change toward supplementing porridge with legumes. These behavioral changes have had measurable impacts on child health, including their growth, a common and effective child-health indicator (de Onis et al. 2004). Children fed porridge too early in life grew more poorly than children who were not introduced to porridge until later (Bezner Kerr et al. 2007a). Porridge, which is easily contaminated by pathogens when prepared in unsanitary conditions, provides few of the nutrients a child normally needs like iron, and is much less nutrient-dense than breast milk (Kramer and Kakuma 2002; WHO 2000). From this initial research, intervention efforts focused on improving feeding practices for young children – eliminating porridge during infancy and reducing reliance on traditional herbal infusions (dawale and mzuwula), which are believed to have protective powers and are promoted by grandmothers (Bezner Kerr et al. 2007a; Sikstrom et al. 2011).

The height and weight of children in participating and control villages was measured 10 times over 6 years. A total of 3,838 measurements were made. Straight comparisons between participants and control-village children were not sufficiently informative because many "control" villages joined the intervention group (they received information and seeds from participants, through normal community interactions). As well, some households classified as being in the intervention did not fully participate, in that they stopped growing the legumes after one season, or did not attend any of the educational sessions. To avoid this possible misclassification issue between intervention and control villages, child-growth status was tested instead against two factors: duration of village participation in project and intensity of involvement. The results from these tests showed that the longer the involvement of a village, and the greater its intensity of involvement, the better the growth of their children. Children in villages involved in the project for longer and with greater intensity of participation were on average 1 kg heavier at 1 year of age, and 1.5 kg heavier at 3 years of age, than those in other villages (Bezner Kerr et al. 2010).

The emphasis on "family cooperation" in the ANDGs, in particular around a more equal division of labor and decision making about household resources, has also been an important change. Qualitative interviews conducted in 2009 found that 23 of 35 respondents, both men and women, reported a positive change in gender relations. Most of the changes were associated with the division of labor, such as men helping more with child-care and cooking. Others noted a change in household decision-making linked to agriculture and income.

After participating in the project, the majority of farmers interviewed in 2009 could link soil fertility, food, health, and gender relations when they talked about the effect of the project. They sometimes discussed broader changes in community relations and expressed a sense of well-being. In an interview, one 49-year-old male farmer described significant changes at multiple levels – his land, his household, and his community:

> I have good fertile soils because of burying residue in my garden. I have plenty of food for my family and then I have again enough for selling … at first I was growing maize on the same land with very little harvest. I was running short of food most of the time and we were having problems with different types of illnesses. We were very poor because each time I found money I was thinking of buying food more than anything, because of shortage of food. Therefore, we were never having money in the house. After harvest, we were having food only for 3–4 months. When the family has no food, there is no peace. My wife and I were having conflicts. But now we are happy. There is no reason to have conflicts … SFHC has made us have friends within the community, country, and outside Malawi. The things which we learn and see in the exchange visits have brought a lot of change in my family.

As the project expanded, the need arose for a system for the community to ensure long-term access to legume seeds. A Community Legume Seed Bank was built under the management of the FRT. Legume seeds are stored after harvest, and distributed by the project to new participants, who then replenish the seed bank with new seed after harvest while also retaining some seed for future plantings. The FRT has shown high competence in record-keeping and seed collection, with over 70% of farmers repaying their seed "loans."

An unexpected outcome of the project was the formation of the Ekwendeni Farmer Association, initiated by the farmers themselves. The aim of the association is to work collectively to improve prices received for their crops. It also promotes legume cultivation as a source of soil fertility, food security, and child health. Although the association is still young, it points to the extent to which farmers feel empowered.

Most Important Successes

The most important success of the project was the enhanced knowledge and use of affordable ways to improve soil fertility and food security. The use of legume residues to improve soil fertility allows farming families to address food security without relying on expensive inputs. Linked to this success is evidence of improved food security and dietary diversity of project households. The participatory nutrition-education effort has in turn helped improve child nutrition.

Farmers were empowered, especially those who worked in the FRT, and they have effectively reached out and involved others. They have also increased their knowledge in agriculture, nutrition, child-feeding practices, and gender relations, and been able to apply this knowledge in practice.

The Challenges Faced

One unexpected challenge was the number of times the project lost its "control" villages, either formally as they asked to be included in the intervention, or informally as farmers shared their seed and knowledge with family, neighbors, and friends. Although this was a positive outcome for the villages, it presented challenges to statistical analysis.

Farmers face many difficulties in Malawi, most of which are beyond the control of researchers and development organizations. AIDS-affected farmers struggle with labor shortages, ongoing illnesses, and increased medical costs (Bezner Kerr et al. 2008b). Rising prices for food and expensive school fees push farmers to seek higher prices for their crops, but these prices often fluctuate. The growing of legumes is accompanied by technical, social, and economic problems, which range from livestock that eat the crops, to legume pests, to low prices. There were also issues with community politics around competing roles and responsibilities, which led to misunderstandings and sometimes to uncompleted tasks (Bezner Kerr and Chirwa 2004). Farmers did not always feel supported in their work and it was difficult to maintain their enthusiasm, particularly for the FRT members who put in countless volunteer hours distributing seed, visiting farmers, collecting data, and conducting training:

> It is hard work. We work long hours as volunteers, and then in addition we have to farm our own land. It takes years for the legumes to have an effect on the soils. Now the [fertilizer] subsidy people are saying, why are you doing this? We can get discouraged.[2]

The current Malawian government's commercial fertilizer subsidies have undermined the work of the project to some extent and present a challenge for the future. Already, there has been a noted drop in residue incorporation in 2009 by about 16% compared to 2008 (Bonatsos et al. 2009). To counterbalance this development, the project held additional crop residue "celebration days" that included prizes for the best fields. Even if farmers use chemical fertilizer, they can still continue to incorporate residues. The project team was trying to help farmers look beyond the immediate quick-fix provided by chemical fertilizers to see the long-term impact and reliability of additions of organic matter to the soil to improve its structure, fertility, and water retention.

The Way Forward

At the heart of this project's success is linking more sustainable land-management methods to positive child-health outcomes, using farmer knowledge and practice. Ekwendeni farmers still face many challenges. AIDS-affected families are struggling to produce enough food and income. A changing and variable climate is requiring farmers to find more drought-resistant crops. In a changing economic and agricultural policy environment, the Farmer Association continues to need to negotiate better prices and support the communities.

Acknowledgments This paper is the culmination of long-term efforts by many farmers in Ekwendeni, and the work of volunteers, staff, and researchers of the Soils, Food and Healthy Communities (SFHC), Malawi project, including the late Marko Chirwa, Esther Lupafya, AIDS Coordinator, Ekwendeni Hospital, and Dr. Sieglinde Snapp, Michigan State University. IDRC support was provided through projects 101829 and 100670. Presbyterian World Service and Development and the Canadian Foodgrains Bank also provided financial support to SFHC.

---

[2]Comment of FRT member during a workshop, May 2009.

References

Bezner Kerr, R. (2005). Food Security in Northern Malawi: Historical Context and the Significance of Gender, Kinship Relations and Entitlements. Journal of Southern African Studies, 31(1), 53–74.

Bezner Kerr, R. (2008). Gender and Agrarian Inequality at the Local Scale. In: Snapp S. and Pound, B. (Editors). Agricultural Systems: Agroecology and Rural Innovation. Elsevier Inc., San Diego, California, USA. pp. 279–306.

Bezner Kerr, R., and Chirwa, M. (2004). Participatory Research Approaches and Social Dynamics that Influence Agricultural Practices to Improve Child Nutrition in Malawi. Ecohealth, 1(Suppl. 2), 109–119.

Bezner Kerr, R., Berti, P., and Chirwa, M. (2007a). Breastfeeding and Mixed Feeding Practices in Malawi: Timing, Reasons, Decision Makers, and Child Health Consequences. Food and Nutrition Bulletin, 28(1), 90–99.

Bezner Kerr, R., Snapp, S., Chirwa, M., Shumba, L., and Msachi, R. (2007b). Participatory Research on Legume Diversification with Malawian Smallholder Farmers for Improved Human Nutrition and Soil Fertility. Experimental Agriculture, 43(4), 1–17.

Bezner Kerr, R., Dakishoni, L., Chirwa, M., Shumba, L., Msachi, R. (2008a). We Grandmothers Know Plenty: Breastfeeding, Complementary Feeding and the Multifaceted Role of Grandmothers in Malawi. Social Science and Medicine, 66, 1095–1105.

Bezner Kerr, R., Shumba, L., Phiri, P., and Kanyimbo, P. (2008b). Resilience and Struggle: Agricultural Options for AIDS-Affected Farmers in Malawi. Paper presented at the American Association of Geographers Annual Meeting, April, Boston, USA, p. 10.

Bezner Kerr, R., Berti, P., and Shumba, L. (2010). Effects of a Participatory Agriculture and Nutrition Education Project on Child Growth in Northern Malawi. Public Health Nutrition, 9, 1–7.

Bonatsos, C., Bezner Kerr, R., and Shumba, L. (2009). SFHC Food Security Status, Crop Diversity and Dietary Diversity: 2007, 2008 and 2009 Results. (p. 41). SFHC Project, Ekwendeni Hospital, Ekwendeni, Malawi.

Bryman, A., Teevan, J.J., and Bell, E. (2009). Social Research Methods. 2nd Canadian Edition. Oxford University Press, Toronto, Canada.

de Onis, M., Garza, C., Victora, C.G., Onyango, A. W., Frongillo, E.A., and Martines, J. (2004). The WHO Multicentre Growth Reference Study: Planning, Study Design, and Methodology. Food and Nutrition Bulletin, 25(1), S15–26.

FAO (Food and Agriculture Organization) (1992). Maize in Human Nutrition. Food and Agriculture Organization, Rome, Italy.

Forget, G., and Lebel, J. (2001). An Ecosystem Approach to Human Health. International Journal of Occupational and Environmental Health, 7(2), S3–38.

Gubbels, P. (1997). Strengthening Community Capacity for Sustainable Agriculture. In: van Veldhuizen, L., Waters-Bayer, A., Ramirez, R., Johnson, D.A., and Thompson, J. (Editors), Farmers' Research in Practice: Lessons from the Field. IT Publications, London, UK. pp. 217–244.

Humphries, S., Gonzales, J., Jiminez, J., and Sierra, F. (2000). Searching for Sustainable Land Use Practices in Honduras: Lessons from a Programme of Participatory Research with Hillside Farmers. Agricultural Research and Extension Network Paper, 104.

Kramer, M.S., and Kakuma, R. (2002). The Optimal Duration of Exclusive Breastfeeding: A Systematic Review. World Health Organization, Geneva, Switzerland.

Satzinger, F., Bezner Kerr, R., and Shumba, L. (2009). Intergenerational Participatory Discussion Groups Foster Knowledge Exchange to Improve Child Nutrition and Food Security in Northern Malawi. Ecology of Food and Nutrition, 48(5), 369–382.

SFHC Project (2008). Annual Report to the Canadian Foodgrains Bank and Presbyterian World Service and Development. SFHC Project, Ekwendeni Hospital, Ekwendeni, Malawi.

Shaxson, L., and Tauer, L.W. (1992). Intercropping and Diversity: An Economic Analysis of Cropping Patterns on Smallholder Farms in Malawi. Experimental Agriculture, 28, 211–228.

Shumba, L., and Bezner Kerr, R. (2008). Food Security Qualitative Study (*Unpublished Report*). SFHC Project, Ekwendeni Hospital, Ekwendeni, Malawi.

Sikstrom, L., Bezner Kerr, R., and Dakishoni, L. (2011). Fluid Boundaries: Multiple Meanings of the Illness 'Moto' in Northern Malawi. In: *Infant Feeding Beliefs and Practices: A Cross-Cultural Perspective.* P. Liamputtong (Ed), Springer.

Snapp, S.S., Mafongoya, P.L., and Waddington, S. (1998). Organic Matter Technologies for Integrated Nutrient Management in Smallholder Cropping Systems of Southern Africa. Agriculture, Ecosystems and Environment, 71(1–3), 185–200.

WHO (World Health Organization). (2000). Effect of Breastfeeding on Infant and Child Mortality due to Infectious Diseases in Less Developed Countries: A Pooled Analysis. Lancet, 355, 451–455.

Chapter 4
Tackling Challenges to Farmers' Health and Agro-Ecosystem Sustainability in Highland Ecuador

Fadya A. Orozco and Donald C. Cole*

Over the last half century, agricultural "modernization" has transformed smallholder potato and horticultural production systems on the once fertile slopes of the Ecuadorian Andes (Sherwood 2009). Striving to maintain outputs, farms and fields dotted across the mountainous landscape have increasingly used external inputs (machinery, contract labour, fertilizers, and pesticides). Contrary to expectations, production has declined and pesticide neurotoxicity and poisonings have increased (Cole et al. 2002; Yanggen et al. 2003).

The transition from more traditional crops and agricultural methods (mixes of grains, legumes, and potatoes) to market-oriented intensive production (primarily horticulture, with some potatoes) has been associated with greater pesticide-related symptoms and lower financial benefits (Orozco et al. 2007). The low level of government investment in agriculture extension activities (less than 1% of the national budget, Planning Office, Instituto Nacional Autónomo de Investigaciones Agropecuarias, Quito) and deep rural impoverishment (61% in chronic poverty, Guzmán 2002), have led farmers to use highly and moderately hazardous pesticides (types Ib and II, WHO 2005). Farmers use these pesticides not only because they are considered more efficient than less hazardous substances, but because of their lower cost (Orozco et al. 2009).

Research carried out in Carchi Province between 1998 and 2002 (Ecosalud I) showed that community-based educational activities (e.g., farmer-field schools, women's groups, school activities, and radio spots) and the application of alternative

---

*Formerly, Project Coordinator and Principal Investigator, respectively, Ecosalud II.

F.A. Orozco (✉)
Instituto de Saúde Coletiva, Universidad Federal da Bahia, Rua Basílio da Gama, Salvador, Brazil
e-mail: fady5o@yahoo.es

D.C. Cole
Dalla Lana School of Public Health, University of Toronto, Toronto, ON, Canada

International Potato Center, Lima, Peru

crop-management technologies could improve understanding of ecosystem dynamics, reduce highly hazardous pesticide use and unsafe pesticide-related practices, and ultimately improve neurological functioning among farmers and their families (Cole et al. 2007). The success of taking an ecosystem approach to health in three small communities prompted the International Potato Center research team to scale up the experience from Carchi to include Chimborazo and Tungurahua, two provinces with larger indigenous populations and somewhat different production systems. Running from 2004 to mid-2008, the Ecosalud II project sought to work with multiple stakeholders, who were referred to as actors to emphasize their key role. The aim of Ecosalud II was to tackle the complex drivers of inappropriate use of highly hazardous pesticides (Orozco et al. 2009), with the long-term goal of greater agro-ecosystem sustainability, including better human health.

Implementing Ecosalud II

With an ecosystem approach to health (Cole et al. 2006) coupled with knowledge translation and exchange approaches (Parry et al. 2009), the project team worked to integrate different actors into an action-research process (Table 4.1). Everyone actively contributed to the research design, data gathering, results analysis and interpretation, dissemination of results, planning and implementation of interventions, and assessment of changes over time.

Cross-Sector and Multi-Disciplinary Approaches

Working across sectors with health and agriculture authorities was a distinguishing feature of Ecosalud I, which was achieved primarily through the engagement of the provincial health council of Carchi and a multi-stakeholder development forum. In Ecosalud II, "potato platforms" were used as social spaces that encouraged diverse

Table. 4.1 Actors integrated into Ecosalud II process by level of action

Level	Actors involved and institutions
Provincial	Potato producers' platforms: farmers, community leaders, local governments, NGOs, technical personnel from the National Agricultural Research Institute
	Provincial Office of Ministry of Health: managers from the Epidemiology Department
Municipal	Municipal Government: policymakers and decision makers
	Canton Office of the Ministry of Health: health-care providers and program managers
Parish and Community	Farmers and farm family members
	Community leaders
	Health care providers

actors with different knowledge, experience, and decision-making power to gather monthly to address issues related to potato farming. These actors included small-scale farmers, leaders of community organizations, technical staff of nongovernmental development organizations (NGOs), staff of various municipal governments, provincial government representatives, and provincial university faculty members. As the project evolved, so did the focus of the potato platforms: from potato productivity and marketing platforms to opportunities to champion human health and agroecosystem sustainability.

In parallel, district and provincial Ministry of Health staff were invited to cross-sector health and agriculture meetings. The project began raising awareness about the human health consequences of current agricultural practices. Preliminary findings from a baseline survey on health and nutrition status, household pesticide management practices, and agricultural production were disseminated to agriculture and health-sector managers and service providers, and suggestions for joint action were discussed, designed, and implemented. These activities included training of health-sector staff and community-based field days.

The information demands from the actors in both the potato platforms and cross-sector meetings, and the observations made by the research team during participant observation, generated questions like: What powers do municipalities have to regulate pesticide use? What distribution channels and markets are available for small farmers interested in more sustainable and diversified production? Answers to these questions required contributions from other disciplines. Graduate students from fields of agricultural sciences, nutrition, law, and social sciences joined Ecosalud II to address these gaps. The cross-fertilization of ideas and methods, and the engagement of diverse actors, led to enriched transdisciplinary thinking among the students (Orozco et al. 2008b). They developed a shared understanding of the problem of pesticides in agriculture, and jointly searched for solutions such as a workable program for healthy and sustainable agriculture in one municipality.

Data Collection

Two cross-sectional surveys sampled approximately 20 volunteer farm families in each of 24 communities in three provinces, before and after Ecosalud II interventions. Pesticide-related knowledge, practices, and health-outcome measures were drawn from earlier Ecosalud I instruments (Yanggen et al. 2003; Cole et al. 2007). Qualitative data regarding farmers' perceptions of use and household management of hazardous pesticides were collected through focus groups and participatory observations (e.g., Orozco et al. 2009), complementing quantitative data (see Table 4.3 discussed later). In-depth interviews were conducted with policy decision makers and community leaders as part of the efforts to influence local policies and programs (Orozco and Cole 2008). Project implementation (history, challenges, and responses) was documented in aide-memoires from all meetings. An assessment of the action-research process as a whole was carried out at the end of the project. It used self-evaluation questionnaires completed by participants in the potato platforms and cross-sector groups, and was complemented by in-depth interviews with these participants and additional actors connected to Ecosalud II.

Interventions

Work with the potato platforms was essential to mobilize human and modest financial resources to extend Ecosalud II's interventions to different settings. In keeping with community-based research models (Viswanathan et al. 2004), interventions were developed at multiple levels that corresponded to the key actors' mandate or scope (community, municipal, and provincial) (Orozco and Cole 2006).

At the community level, numerous field schools and field days were held for farmers to allow them to share their understanding of ecosystems and their experiences with traditional and new agricultural practices, including reducing highly hazardous pesticide use. Other interventions included community theatre, health-promotion groups, and puppet shows preformed at schools, during which questions were raised about current agricultural production, its impacts on health, and the reasons for pesticide use. Efforts were made to develop joint responses to changing practices, in keeping with "radical" approaches to health education (Oliveira 2005).

At the municipal level, ordinances were jointly drafted by the research team and local policymakers to promote local policies on training and agricultural extension related to alternative crop-management practices (Orozco and Cole 2008). As an off-shoot of the project, a farmers' organization teamed up with a municipality to obtain NGO support to set up a local store to sell less hazardous products and provide information on integrated pest management. The store has developed into a Centre for Agriculture and Livestock Services (Coagro-Q), which is led by farmers' organizations and has farmers as the main partners along with the municipality and nongovernmental organizations. Currently, more than 300 farmers are involved in some way with the Centre – seeking advice, purchasing products, receiving training in healthy and sustainable agriculture, or providing consultation to other farmers.

At the provincial level, Ecosalud II assisted Ministry of Health staff with peer-to-peer interchanges. For example, an emergency physician from Carchi came to Chimborazo's regional hospital and trained emergency and health centre staff in the diagnosis and treatment of pesticide poisonings using protocols jointly developed with Ecosalud II. To bolster surveillance for acute pesticide-related poisonings, reporting forms and a jointly developed health-information system were shared, and epidemiology staff in each province were trained, with the assistance of a Canadian epidemiology student.

Putting Knowledge into Action

Descriptive findings from the cross-sectional surveys were discussed in the potato platforms and in cross-sector meetings with the actors and participating communities. The research findings also enabled the team to better understand gender-related issues and their implications (Orozco et al., accepted). They discussed the greater exposure and health burden of pesticides on men, and the lack of training on health effects and integrated pest management among women (Orozco et al. 2008a). The findings helped broach issues of inequity and power imbalances in the development of public policies that result in unsustainable and unhealthy agro-ecosystems, poor agricultural practices, and families who must bear a heavy financial and health burden (Orozco et al. 2009). The effort bolstered actors' involvement in the design and planning of interventions to reduce pesticide use.

Findings and Achievements

In moving from the initial project site in Carchi to all three provinces in the later phase, the research team found a similar complex pattern of drivers that affected the livelihoods and health of farm families. These drivers included cheap and readily available highly hazardous pesticides; farmers' lack of knowledge about handling pesticides and reducing their exposure in the field and at home; poor general awareness of the extent of health impacts among both NGO and government actors; and weak policy responses to promote alternative crop-management technologies and practices that favour the sustainability of agro-ecosystem and farmers' health. The strategy of the research team was to tackle these issues based on knowledge production, capacity building, and multi-stakeholder collaboration.

Building Capacity in Transdisciplinary Research on Agro-Ecosystems

The long-term goal of fostering sustainable agriculture and healthy rural communities requires the building of dialogue among disciplines and the creation of opportunities for young researchers to gain experience. Ten students (six Ecuadorian undergraduates and four Canadian master's students) participated in Ecosalud II. Working with the coordinator and their own academic supervisors, they explored a range of topics (Orozco et al. 2007; Orozco and Cole 2008): temporal studies on ecological degradation through agricultural development; changes in agricultural technology and production; child malnutrition associated with intensive potato production; challenges in implementing farmers' field schools (Tracy 2007); forms of participation with different levels of social capital (Rubio 2007); and legal frameworks for pesticide sale and use (Orozco and Cole 2008). Being members of a multidisciplinary team helped them develop a richer conceptualization of the specific problem each was studying. Their studies provided important insights to the project.

Improving Diagnosis and Treatment of Pesticide Poisonings

Because primary and secondary health-care providers had limited knowledge of the adverse health effects of pesticides, the training provided by the project improved the ability of staff from the provincial Ministry of Health to more accurately record the burden of pesticide poisonings[1] (Chamorro et al. 2006).

The national Ministry of Health also adapted the information, training, and reporting tools developed by the project to set up a regional surveillance program for acute pesticide-related poisonings. The Ministry hired the project's health professional trainees during the implementation of the program. Work with the health authorities to improve diagnosis and surveillance of pesticide poisonings was instrumental in gaining the attention of the broader health sector and encouraging a renewed examination of the links between agriculture and health.

Engaging Actors in Cross-Sector Work

Before Ecosalud II, most actors engaged in the project had heard about pesticide-related problems and their impact on human health, but had no idea of the extent of the problem or ways to address it across sectors. Table 4.2 summarizes the feedback of key actors obtained during the project's multi-sector consultations (potato platforms and cross-sector meetings) on the nature and facilitation of cross-sector work. As one agricultural sector actor said:

> Sometimes we understand the concepts but we don't know how to apply them in practice. The topic of pesticides was important for us but we did not know how to address it. With Ecosalud, we learned how to translate what you call Healthy Crop Management[2] into practice.

---

[1] The period September 2005 to September 2006 showed a 1000% increase in pesticide-related poisoning cases in the province of Chimborazo – from 5 cases the year before the implementation of the surveillance system (2004) to 55 cases up to September 2006. The increase in Carchi was 400%, from 13 cases in 2003 to 60 cases up to September 2006 (Source: Provincial Health Directorates of Chimborazo and Carchi).

[2] Healthy crop management refers to production practices that seek to decrease the health risks associated with external input use and to promote sustainable ecosystems from a social, ecological, and economic viewpoint.

Table 4.2 Feedback from key actors who participated in the platform and cross-sector meetings

Factors	Characteristics that facilitated cross-sector work
Main facilitators of organizational links with Ecosalud II	Joint planning meetings
	Ecosalud II leadership in coordination
	Timely communications of advances and results
	Ecosalud II technical and logistical support
Characteristics of their organization that facilitated their work with Ecosalud II	Shared concerns and interests in healthy and sustainable agricultural production
	Support from their organization's leadership for joint work
Nature of participation with Ecosalud II	Participation was a collective co-learning process, and Ecosalud facilitated learning among different actors
Principal Ecosalud II contributions to their organization	Information on health impacts of pesticides
	Technical training on healthy crop management
	Learning to interact as a team with people from other disciplines
Utility of the information generated from Ecosalud II research	Most actors found the information extremely useful for their organization's planning of its own work

Sources: Self-evaluation questionnaires ($n=18$) and in-depth interviews ($n=22$) with key actors in the potato platforms and cross-sector meetings in Chimborazo and Tungurahua provinces

Public health professionals described how they previously focused more on providing health services than on addressing the determinants of the health of rural residents. After Ecosalud II, they understood better the links between agriculture and human health (e.g., talking about the implementation of alternative pest-management methods with people from agriculture). They also became more strongly committed to improving the health and well-being of farm families, including promoting ways to reduce pesticide use.

Promoting changes in structural factors

The project actively informed farmers about the Food and Agriculture Organization's International Code of Conduct on the Distribution and Use of Pesticides (FAO Food and Agriculture Organization 2003), which led to the development of a Charter of "Farmers' Rights" related to pesticide use in Ecuador (Orozco et al. 2009). The document is currently recognized by the National Council of Citizen Participation as an important tool for improving farmers' living conditions. The Charter has empowered farmers to monitor pesticide-use by governments, companies, and distributors, and to protect their health and that of their families, as part of another project supported by the IDRC Global Health and Leadership Program called "Governance and Social Capital". There is potential for widespread use of the Charter in Ecuador and the region given its generic components.

Social innovation, understood as the use of social values for the generation and implementation of new ideas in improving health and quality of life (Global Forum for Health Research 2009), depends on both existing governance processes and previous investments. For example, Ecosalud II support and local governments' and farmers organizations' financial inputs to start-up an alternative pest-management store, helped nurture and sustain social action and achieve social benefits. Municipalities that were willing to engage in joint activities with the project were later more likely to pass ordinances and to invest in community health and improved agricultural production (Orozco and Cole 2008). Picking up on community-based initiatives started by Ecosalud II, one municipality formed a Department for Agricultural Production and Development tasked with providing agricultural extension services to promote alternative crop-management practices. Another municipality was instrumental in the creation of the Centre for Agricultural Services (mentioned earlier) to provide access to alternative inputs, such as less hazardous pesticides and integrated crop-management products.

In contrast to the well-organized agrochemical industries, these locally based, social initiatives provide ecologically sound and health-friendly advice to farmers. They serve as examples of how local governments and farming communities can work together to support healthier rural livelihoods.

Promoting Changes in Farm-Households (Awareness, Knowledge, and Practices on Pesticides)

The results of the repeat cross-sectional surveys (Table 4.3) showed that participating farm households significantly improved their awareness of integrated pest management (one aspect of healthy crop management). Also improved was their overall knowledge of pesticides and their appropriate handling. Further, the reported use of highly hazardous "type 1b" pesticides per crop cycle declined (for further detailed comments on methods and results see Orozco et al., 2011).

Unlike the previous phase, Ecosalud II did not have the resources to obtain detailed quantitative information on crop yields at the farm level (Yanggen et al. 2003). However, most farmers anecdotally indicated that overall profitability was maintained with less pesticide use, as was found in medium-intensity production systems (Orozco et al. 2007). With improved knowledge and practices about pesticides, symptoms of acute poisoning declined in all three provinces. This decline was recorded despite the improved capacity to recognize and record pesticide-related symptoms. Changes in chronic exposure to pesticides were evaluated through a test of neurobehavioural performance, called Digit Span. This test evaluates cognitive functions such as memory, concentration, and attention. Mixed results across the province were found (Table 4.3). Carchi and Tungurahua showed positive changes, but not Chimborazo. Among factors explaining the results were the intensity of interventions in each province (Orozco et al., 2011), levels of education (Orozco et al. 2009), especially in Chimborazo, and socio-economic conditions,

Table 4.3 Principal quantitative indicators of changes in farm household members' information, knowledge, practices, and health status related to pesticides, pre (2005) and post (2007) interventions, by province (overall n = 465)[a]

Indicators	Carchi, Mean (SD)		Chimborazo, Mean (SD)		Tungurahua, Mean (SD)	
	2005	2007	2005	2007	2005	2007
Crop Management						
Information on IPM (integrated pest management)[b]	1.39 (0.49)	1.56* (0.50)	1.31 (0.46)	1.52* (0.50)	1.22 (0.42)	1.60* (0.49)
Knowledge of colours of pesticide labels[c]	5.78 (4.60)	6.51 (3.12)	2.40 (3.87)	5.14* (4.31)	2.78 (3.94)	5.50* (3.41)
Knowledge of symptoms of pesticide poisoning[c]	8.67 (2.21)	9.56* (0.90)	7.97 (2.46)	9.52* (1.21)	7.44 (1.87)	8.36* (2.46)
Knowledge of practices that can increase contamination during mixing and spraying[c]	8.56 (1.68)	9.02* (1.43)	8.32 (1.73)	9.51* (1.12)	7.36 (1.93)	8.79* (1.67)
Personal protective equipment used during spraying[d]	4.28 (1.84)	4.52 (1.89)	3.18 (1.96)	4.34* (1.88)	4.47 (1.77)	4.34 (1.70)
Hand washing after use of pesticides[d]	7.56 (2.48)	8.94* (2.66)	7.45 (1.99)	9.26* (3.26)	6.68 (1.98)	8.26* (3.47)
Use of highly hazardous type 1b pesticides over one crop cycle (mean kilograms/hectare)	3.47	1.22*	0.59	0.91	2.81	0.63*
Health impacts						
Reported pesticide poisoning symptoms[e]	1.60 (2.46)	0.90* (1.11)	2.49 (2.40)	1.22* (1.63)	1.98 (1.80)	1.10* (1.40)
Neurobehavioural performance[f]	4.42 (1.19)	4.86* (1.42)	4.33 (1.65)	3.89* (1.36)	4.40 (1.72)	4.78 (1.46)

*$p<0.05$ for difference across years
[a] Authors and Ecosalud II research team
[b] Values closer to 2 indicate better information
[c] Scale from 0 = not known to 10 = perfect knowledge
[d] Scale from 0 = poor practices to 10 = very good practices
[e] Scale from 0 = no symptoms to 10 = many symptoms
[f] Values close to 10 signify better neurobehavioural performance, as measured by Digit Span (Cole et al. 2007)

although unmeasured factors such as lifetime pesticide use were likely also relevant.

Overall, the project confirmed the feasibility of up-scaling changes in awareness, knowledge, and practices of farmers related to pesticide use and management from three communities to three provinces. The success of this up-scaling depended on cross-sector collaboration between actors at different levels (community, municipal, provincial, and national) and the use of community-based education approaches to improve crop-management knowledge and practices among small-scale farmers.

More generally, Ecosalud II's work with multiple actors and different societal levels over time both co-promoted and was complemented by the involvement of other actors (e.g., the Humanist movement, the Pesticide Action Network, and journalists). Jointly these actors maintained pressure for change on national political authorities. Such placement on the political agenda resulted in a legal decree that cancelled the registration of type Ia and Ib pesticides in June 2010 (Registro Official No. 224).

Challenges Encountered

Involving key actors to influence policies and structures (Public Health Agency of Canada 2007) is essential to the success of a project like Ecosalud II that is seeking policy change, but it poses some challenges. The project sought to identify potential producers and users of research results: potato growing communities; men and women farmers; community leaders; technical staff working in agriculture and social development; NGOs; local politicians; and managers of the health system. The project invested substantial time, financial resources, and intellectual effort to engage with, understand, and mediate between the diverse priorities and viewpoints of these actors. With time, collaboration between different actors and their contributions were better incorporated into the project, thanks in large part to the building of trust and the evidence generated by the research. For example, economic incentives that determine crop-production choices were used to introduce health and environmental considerations more strongly into the dialogue about intervention options. To do so successfully required an ability to lead and manage teams and appreciate the potential technical and operational contributions of each actor. For example, those most involved in the potato platform created an entire train-the-trainer program and developed marketing channels that eventually grew into outlets for potatoes produced in healthier and more sustainable ways. Applying innovative negotiation techniques was crucial to encourage the actors to remain enthusiastic and active in the project. Maintaining flexibility and openness to help find the right balance between research priorities and the organizational priorities of the actors was also essential.

Conclusions

Because human health is often highly valued, research that explores people's livelihood and health links can contribute to public health interventions (Hawe and Potvin 2009). This project helped rural communities concretely address their development needs by responding to their priorities. It also generated new knowledge about health and environmental risks associated with the use of highly toxic pesticides, and empowered rural households to adopt healthy farming practices. The strategic communication and use of research results with farmers, key government officials, NGOs, and other stakeholders was effective in fostering change in potato-production systems in the three provinces targeted by the project, and helped make potato production a safer income-generating activity with less damage to farmers' health and their environment. The concrete actions taken by the different actors enabled them to work on new projects that also sought social transformation, and provided an optimal social return on Ecosalud II's overall investment.

Acknowledgments We thank the members of the Ecosalud II field team (Cecilia Pérez, Jacqueline Arevalo, Leticia Guaman, and Byron Arevalo) for their enthusiasm and social commitment, and the community leaders and the women and men farmers who opened their homes and gave their valuable time to participate. Colleagues from collaborating organizations also deserve our thanks: Fortipapa; Plataforma de la Papa de Chimborazo; CONPAPA Quero – Guano; Ayuda en Acción Chimborazo; CESA Chimborazo; Diócesis de Riobamba; Proyecto UDOCACH; Proyecto Punín; Fundación Marco; las Direcciones de Salud Provincial de Carchi y Chimborazo; la Escuela Politécnica del Chimborazo; los Gobiernos Municipales de Quero y Guano; and al Programa de Papa del INIAP. Finally, thanks to our colleagues at the International Potato Center for hosting Ecosalud I and II and for having an intersectoral vision. IDRC support for Ecosalud I and II was provided through the projects 004321, 101816, and 101810.

References

Cole, D.C., Crissman, C., Orozco, A.F. (2006). Canada's International Development Research Centre's Eco-Health Projects with Latin Americans: Origins, Development and Challenges. Canadian Journal of Public Health, 97(6), 8–14.

Cole, D.C., Sherwood, S., Crissman, C., Barrera, V., and Espinosa, P. (2002). Pesticides and Health in Highland Ecuadorian Potato Production: Assessing Impacts and Developing Responses. International Journal of Occupational and Environmental Health, 8(3), 182–190.

Cole, D.C., Sherwood, S., Paredes, M., Sanin, L.H., Crissman, C., Espinosa, P., and Muñoz, F. (2007). Reducing Pesticide Exposure and Associated Neurotoxic Burden in an Ecuadorian Small Farm Population. International Journal of Occupational and Environmental Health, 13(3), 281–289.

Chamorro, P., Jácome, N., Baca, M., Castillo, G., Villareal, M., Castillo, J., and Narváez, N. (2006). Proyecto de Vigilancia y Control de Intoxicaciones por Plaguicidas en la Provincia del Carchi. Red Ecuatoriana de Epidemiología, 21–24.

FAO (Food and Agriculture Organization). (2003). International Code of Conduct on the Distribution and Use of Pesticides (Revised Version) adopted by the Hundred and Twenty-Third Session of the FAO Council in November 2002. FAO, Rome, Italy.

Guzmán, L.M. (2002). Cálculo de la pobreza en el Euador (Primera parte). Sistema Integrado de Indicadores Sociales del Ecuador (SIISE). Available at: http://www.siise.gov.ec/Publicaciones/calpob.pdf.

Global Forum for Health Research (2009). Innovando para la salud de todos. La Habana, Cuba, 16–20 November 2009. Available at: http://www.globalforumhealth.org/.

Hawe, P., and Potvin, L. (2009). What is Population Health Intervention Research? Canadian Journal of Public Health, 100(1), 8–14.

Oliveira, D.L. (2005). A "Nova" Saúde Pública e a Promoção da Saúde via Educação: Entre a Tradição e a Inovação. Revista Latino-Americana de Enfermagen, 13(3), 423–431.

Orozco, F., and Cole, D.C. (2006). Salud Humana y Cambios en la Produccion Tegnologica de la Papa. 11th World Public Health Congress, Rio de Janeiro, Brasil. Available at: http://idl-bnc.idrc.ca/dspace/handle/10625/45405 and http://idl-bnc.idrc.ca/dspace/handle/10625/45406.

Orozco, F., and Cole, D.C. (2008). Development of Transdisciplinarity Among Students Placed with a Sustainability for Health Research Project. EcoHealth, 5(4), 491–503.

Orozco, F., Cole, D.C., Muñoz, V., Altamirano, A., Wanigaratne, S., Espinosa, P., and Muñoz, F. (2007). Relationship Among Production Systems, Preschool Nutritional Status and Pesticide Related Toxicity in Seven Ecuadorian Communities: A Multiple Case Study Approach. Food and Nutrition Bulletin, 28(2), 247–257.

Orozco, F., Cole, D.C., and Munoz, F. (2008a). Farm Household Gender Roles, Differences in Crop Management and Health Implications. XVIII IEA World Congress of Epidemiology, 20–24 September, Porto Alegre, Brazil. Revista Saude Publica Brasiliera (Supplement).

Orozco, F., Cole, D.C., Muñoz, F., Ibrahim, S., Perez, C., Wanigaratne, S., Arevalo, J., and Guzman, L. (2008b). Multidisciplinary Action Research to Reduce Hazardous Pesticide Use. XVIII IEA World Congress of Epidemiology, 20–24 September, Porto Alegre, Brazil. Revista Saude Publica Brasiliera (Supplement).

Orozco, F., Cole, D.C., Forbes, G., Kroschel, J., Wanigaratne, S., and Arica, D. (2009). Monitoring Adherence to the International Code of Conduct: Highly Hazardous Pesticides in Central Andean Agriculture and Farmers' Rights to Health. International Journal of Occupational and Environmental Health, 15(3), 255–269.

Orozco, F., Cole, D.C., Ibrahim, S., Wanigartne, S. (2011). Health promotion outcomes associated with a community based project on pesticide use and handling among small farm households. Health Promotion International, Advance Access published February 2011.

Orozco, F., Cole, D.C., Muñoz, F. (*accepted*). Gender relations and pesticide-related knowledge, crop management practices, and health status among small farmers in highland Ecuador. International Journal of Occupational and Environmental Health.

Parry, D., Salsberg, J., and Macaulay, A.C. (2009). A Guide to Researcher and Knowledge-User Collaboration in Health Research. Available at: http://www.cihr-irsc.gc.ca/e/39128.html#1.

Public Health Agency of Canada (2007). Crossing Sectors: Experiences in Intersectorial Action, Public Policy and Health. Public Health Agency of Canada in collaboration with Health Systems Knowledge Network of the World Health Organisation's Commission on Social Determinants of Health and the Regional Network for Equity in Health in East and Southern Africa (EQUINET), Canada, vi–24.

Rubio, F. (2007). Health Education and Collective Action: A Case Study in the Central Ecuadorian Andes. Master of Arts, The Norman Paterson School of International Affairs, Carleton University Ottawa, Ontario, Canada.

Sherwood, S.G. (2009). Learning from Carchi: Agricultural Modernisation and the Production of Decline. Wageningen University, Wageningen, The Netherlands, 286 pp.

Tracy, T. (2007). Papas, Plaguicidas y Personas: The Farmer Field School Methodology and Human Health in Ecuador. Master of Arts in International Development Studies, Saint Mary's University, Halifax, Nova Scotia, Canada.

Viswanathan, M., Ammerman, A., Eng, E., Gartlehner, G., Lohr, K.N., Griffith, D., Rhodes, S., Hodge, S., Maty, S., Lux, L., Webb, L., Sutton, S., Swinson, T., Jackman, A., and Whitener, L. (2004). Community-Based Participatory Research: Assessing the Evidence (Evidence Report/Technology Assessment Number 99). Agency for Healthcare Research and Quality, U.S. Department of Health and Human Services, AHRQ Publication No. 04–E022-2, pp. 1–100.

WHO (World Health Organization). (2005). The WHO Recommended Classification of Pesticides by Hazard and Guidelines to Classification: 2004. WHO, Geneva, Switzerland, 57 pp.

Yanggen, D., Crissman, C., and Espinoza, P. (Editors). (2003). Los Plaguicidas: Impactos en Producción, Salud y Medio Ambiente en Carchi, Ecuador. Centro Internacional de la Papa (CIP), Instituto Nacional Autónomo de Investigaciones Agropecuarias (INIAP), and Ediciones Abya-Yala, Quito, Ecuador, pp. 199.

Chapter 5
Coping with Environmental and Health Impacts in a Floricultural Region of Ecuador

Jaime Breilh

In the early 1990s, like many other countries, Ecuador experienced fast economic growth that resulted in both rapid concentration of wealth and social exclusion that marginalized the poor. Rapid economic development during this time resulted in environmental degradation, which in turn affected human health and amplified poverty (Breilh and Tillería 2009).

Today in rural areas of Ecuador, the expansion of the agro-industry is evident, mainly in enclaves of high technology floriculture and horticulture enterprises. In many cases, these agri-businesses occupy ancestral agricultural lands where indigenous and *mestizo* communities, along with traditional *haciendas* and other, middle-sized farms, have operated for centuries. As these new, high technology farms continue to sprout up, there are fewer opportunities for traditional communities in these regions to produce food for local and national consumption, or to sustain their livelihoods (SIPAE 2004).

New economic, social, and cultural relationships brought about by industrialized agriculture for export markets have created a "new rurality" – an increased number of large agro-industrial units at the expense of smallholder agriculture. Although agro-industry provides employment opportunities and regional economic development, it also poses social, health and ecological challenges to communities, scientists and policymakers. The case of floriculture farms in Ecuador is a good example of this challenge (CEAS 2005; Breilh and Tillería 2009).

Modern floriculture brings with it technologically intensive activity that is imposed on the low-technology context of traditional agriculture in Ecuador. The contrast is not only technological. This agricultural transformation exacerbates unequal access to both land and water resources. Floriculture farms abundantly use pesticides and water, and have little incentive to apply alternative pest management

J. Breilh (✉)
Universidad Andina Simón Bolívar, Quito, Ecuador
e-mail: jbreilh@uasb.edu.ec

and water treatment or conservation methods. An overly permissive pesticide policy environment allows excessive and uncontrolled use of pesticides. Smallholder farmers in the highlands also overuse pesticides, particularly on potato crops. Medium- to high-toxicity pesticides are widely available and inexpensive (Breilh et al. 2005). Pesticide use is also detrimental to soils. The accumulation and persistence of pesticide residuals in soils increases as more pesticides are applied, which reduces soil microbial mass and diversity (Aguirre 2004).

The city of Cayambe is located in the Granobles watershed in the Andes region in northern Ecuador. There are 147 floriculture farms in this watershed, or about 38% of all such farms in the country. In 2001, the Health Research and Advisory Center (CEAS) was invited by different community organizations and community leaders based in the Cayambe region to discuss the emerging issue of industrialized floriculture. A multistakeholder workshop took place in May 2001 in Quito to discuss the different views, needs, and knowledge gaps around this issue. Stakeholder representatives from the CAMAREN consortium (http://www.cap-net.org/) for natural resources management, the technical school in Cayambe, the indigenous community of Cangahua, the ECUANARI[1] indigenous organization, the UNOPAC (http://unopac.org/) peasant organization of Cayambe, the President of the Environment Committee of the Municipality of Cayambe and other municipal authorities, the local health services, the workers' floriculture union, the School of Chemistry, and CEAS met over 3 days to discuss the main goals for a participatory research project that would assess the social, cultural, environmental, and health impacts of floriculture in the region.

Communities and their representatives had already been discussing the pros and cons of floriculture, and they shared their views during the initial workshop. Floriculture was seen to provide employment opportunities with salaries slightly higher than average, but was suspected of causing pesticide contamination that affected both humans and ecosystems. Some community elders also claimed that community bonds were being affected, and that negative "western" patterns of consumerism were resulting from these changes. A collaborative project was developed to assist the community in building the knowledge and evidence needed to guide and promote community-based awareness and action, and to strive for policy change.

Getting Started

At the outset, the community suspected that chemical contamination was widespread. There was anecdotal evidence of environmental changes such as strong smell of sulfurous vapours near greenhouses, and changes in the colour of surface waters. Some people described how local animal species and endemic insects were

---

[1] Confederación de Pueblos de la Nación Kichwa de Ecuador.

beginning to dwindle in number. Rising social unrest and negative behaviour patterns, such as increased drug use, were also causing alarm among some community members. Finally, health complaints included headaches and lack of concentration in school-children neighbouring the floriculture farms, and self-reported symptoms such as recurring headaches, stomach cramps, and drowsiness among workers in the flower industry.

Pesticide-intensive agriculture like floriculture and other activities in Ecuador relies on easily accessible, inexpensive, and poorly regulated chemicals, especially organophosphate and carbamate products, which are designated as classes I (extremely or highly hazardous) and II (moderately hazardous) by the World Health Organization (2005). Floriculture workers are exposed to pesticides by contact, inhalation, or ingestion while they work in the fields, greenhouses, or refrigerated rooms where the flowers are processed. The complex exposure patterns in Cayambe can be characterized as chronic low-dose exposure to multiple products from many sources (Breilh et al. 2009). Reports of acute pesticide toxicity are rare. This chronic pattern of exposure produces diverse physiological and clinical impacts in workers (Alavanja et al. 2004, 2999; Wesseling et al. 1997): reduced neurotransmitter enzymes (erythrocyte acetylcholinesterase (AChE) and plasma AChE (buChE)); elevated levels of liver enzymes (including alanine aminotransferase (ALT) and aspartate aminotransferase (AST)); bone-marrow suppression with decreased levels of hemoglobin and reduced white blood cell (WBC) counts; neurobehavioral deficits; and self-reported symptoms like ear, nose, and throat irritation, irritable character, headaches, dizziness, unexplained sweating, and weakness.

Potentially widespread pesticide contamination in Ecuadorian floricultural areas is suspected to arise from: wind-borne drift of chemicals from cut-flower plantations; careless disposal of used pesticide containers in the environment; household pesticide use; and reuse of the pesticide-laden plastic sheeting salvaged from cut-flower greenhouses. Contamination was confirmed in samples of irrigation water running off from flower farms. Analyses demonstrated the presence of many toxic chemical residuals, including organophosphates (malathion, diazinon, and cadusafos); carbamates (carbofuran, methomyl, and oxamyl); and chlorinates (chlorothalonil and endosulfan) (Breilh 2007).

Project Methodology and Results

The project goal was to study the relationship between floriculture and the health of floriculture workers, their communities, and the surrounding ecosystem. As a first step, key stakeholders were identified. Community organizations were vital to providing support for project activities, and the platform needed for stakeholders to debate conflicting views around floriculture. Stakeholders included local leaders, representatives of municipal government, experts from the Ecuadorian ministries of health and environment, peasant women's organizations, employees in the regional health system, entrepreneurs from the flower industry, and flower workers. As the project progressed, others became involved. These included local water network leaders, Ecuadorian universities under the leadership of Universidad Andina Simón Bolívar, the Flower Label Program (FLP)[2], and faculty from the University of British Columbia, Canada.

The principles of ecohealth – multidisciplinarity, stakeholder participation, and gender and social equity (Lebel 2003) – provided the approach needed to tackle the complex problems related to expanded floriculture production and human and ecosystem health. The team complemented this approach with a political-economy perspective, which included the analysis of inequitable power relations defined by class, ethnicity, and gender (Breilh 2004).

Pesticide Dynamics in the Watershed

The Granobles watershed was the focus for assessing pesticide dynamics in the ecosystem. The watershed was subdivided according to the agricultural water system used, and sampling sites were systematically identified to detect chemical residuals in the water and sediments. Sites were selected upstream and downstream from traditional agricultural areas and flower farms. Four different sampling series (2004–2007) were taken at 28 irrigation-water sites through different seasonal stages (related to climate and agricultural variations). Geographical identifiers were recorded using hand-held GPS. Sample collection underwent standardized quality assurance procedures to avoid external contamination and chemical degradation and to assure cold-chain conservation before chromatography and mass spectroscopy were used to identify nearly 20 of the most frequent agricultural contaminants.

Health Exposure and Impacts

In 2008, pesticide exposure in the worker population was assessed using a cluster sample of working-age adults in each of two communities of the Granobles River basin. After various discussion sessions with the leaders and community groups, the project obtained the participation of 69 families in Cananvalle and 35 families in San Isidro. The study collected clinical and socio-economic data. One economically active member from each family was recruited to participate in the survey of social and cultural practices related to life style and pesticide-exposure patterns. This was supplemented with information gathered during focus groups. San Isidro represented a low-pesticide exposure area, located upstream from the floriculture area,

---

[2]For more information on the Flower Label Program see: http://www.fairflowers.de whose certification adheres to the International Code of Conduct for the Production of Cut-Flowers.

above 3,000 m in the highlands. This area is characterized by agriculture crops (particularly potatoes), which can represent a different source of pesticide exposure. The community of Cananvalle (at 2,200–2,500 m) is in the valley, and was presumed to be more exposed to pesticides because it was closer to the floriculture growing areas.

Data were obtained from physical examination and clinical tests performed by medical personnel (CEAS 2005) with informed consent of the patients. Urine samples were collected and tested for pesticide residuals and residuals from phthalate components of plastics. Analysis was conducted using a gas chromatographer coupled to a mass spectrometer. Solid-phase extraction (SPE) was based on EPA 8270 and employed National Institute of Standards and Technology (NIST), Wiley, and National Bureau of Standards (NBS) libraries. Throughout project planning and fieldwork, community members participated actively in validating the representativeness of samples, and in organizing and conducting surveys.

The 2008 field work built on two earlier investigations of pesticide exposure. In 2003, floriculture workers in two farms were studied: those in a more modern, higher technology production system ($n=51$) and those who were part of an older less-developed system ($n=110$). Occupational pesticide exposure was assessed in these groups of floriculture workers. Their health records were obtained, and additional data were gathered using questionnaires that addressed social and occupational information, exposure patterns, use of protective gear, and vulnerability. In 2005, an international collaborative study of the neurobehavioural impact of pesticide exposure among children 3–61 months old was undertaken. Survey data were collected from mothers of children in this age range who had lived for at least 1 year in the study area. Three communities were included in the study based on a range of likely pesticide contamination levels and, because they were known to the researchers, the likelihood that the communities would participate and collaborate in the research. An adapted version of the Ages and Stages Questionnaire (ASQ) – a neurobehavioural screening instrument – was applied to assess communication, fine motor, gross motor, problem solving, and personal social-traits in the children (Handal et al. 2007).

Children 3–23 months old who resided in high-exposure communities scored lower in gross motor, fine motor, and social skills. Children 24–61 months old from these same communities scored lower in gross motor skills. The relation between high exposure to organophosphates and carbamates and poorer neurobehavioral development was robust, after controlling for other variables linked to social determinants of delayed development (Handal et al. 2007). This study showed that children are one of the most vulnerable populations, especially those who had higher levels of exposure and were living in the low valley. The project also responded to the need expressed by the Technical College of Cayambe, which was one of the first community organizations to request a scientific study to assess children's exposure to pesticides and the corresponding health impacts.

The project explored the sensitivity of a wide range of monitoring tests, and recorded clinical signs and symptoms to assess exposure to pesticides. To diagnose the impacts of chronic exposure to pesticides, the project developed a Basic Battery of tests, which combined neurobehavioural assessment with blood and urines tests. The Basic Battery test was combined with an acetylcholinesterase (AChE) test in adults to assess flower workers (2003) and the two community groups (2008). Flower workers were also assessed using computer-assisted neurobehavioural evaluation tests such as the Neurobehavioral Evaluation System (NES2) for reaction time, fine coordination and finger tapping, hand eye coordination, and symbol digit operations.

Environmental Results

Pesticide residuals were found in 67.9% of the 28 water and soil sampling sites in the Granobles River basin. The organophosphate malathion was the most common, and tended to accumulate in river sediments; but other organophosphates (diazinon and cadusafos); carbamates (carbofuran, methomyl, andoxamyl); and chlorinates (chlorothalonil and endosulphan) were also detected. High levels of organophosphates were found in most samples (in highland and valley sites), which shows the mixed pattern of pollution caused by agriculture crops (mainly caused by potatoes in the high valley) and floriculture production (in the low valley).

Project results also showed that tonnes of discarded greenhouse debris, including greenhouse plastic sheets and containers contaminated with pesticides, were being dumped in creeks or sold to poorer members of the community for reuse at home, in animal sheds, or for agricultural purposes. In fact, more than half of the families in Cananvalle used contaminated greenhouse plastic sheets and wooden splints in their houses and animal sheds.

Health Results

Overall results showed that flower farms workers and both highland and low basin communities were highly exposed to pesticides. Different health impact indicators yielded exposure gradients among the three populations. AChE suppression tended to be more pronounced in workers, followed by the low basin community and finally in the highland area (ANOVA, $p=0.000$). Other tests did not follow the expected gradient but evidence extended exposure throughout the region. As expected, the AChE test showed low levels of toxic exposure in comparison with other tests and compound indicators, due to its lack of sensitivity for assessing chronic exposure. A clear majority of flower workers presented symptoms related to chemical exposure (69.1% accumulated at least four of them); showing a consistent significant difference between high and low risk sections of the farm. Other chemical exposure screening tests showed considerably high positivity and also significant differences between high and low exposure sections. On average, a considerable number of the workers (58.5%) suffered AChE suppression, or had at least one positive blood test and at least seven symptoms associated with pesticides exposure. The impact rose to 82.6% in those who worked in the high exposure sections of the farm. The neurobehavioral assessment of flower workers (measured by NES2) also showed high impact levels and significant differences between farm sections (Breilh et al. 2009; Breilh et al. submitted).

Ten percent ($n = 69$) of urine samples from Cananvalle indicated pesticide residuals: diazinon (banned in USA for being an endocrine disruptor and causing bone-marrow toxicity); carbofuran (one of the most dangerous restricted-use carbamates); and malathion (relative low human toxicity but readily converts to malaoxon, a substantially more toxic metabolite). In all cases with pesticide-residual positive urine samples, blood AChE tests did not indicate any impairment confirming the poor sensitivity of this test for low-dose or chronic exposure to pesticides. Of the urine samples from San Isidro, 8.6% ($n = 35$) were positive for pesticide residuals (Breilh et al. 2009).

Plastics residuals (from greenhouses and chemical containers) were found in urine samples from people in Cananvalle, near the floriculture areas. The levels recorded were above those allowed by the Environmental Protection Agency (US-EPA). Of the 60 urine samples collected in Cananvalle, 51 (85%) showed evidence of exposure to phthalate components, such as Di-2-ethylhexyl phthalate (DEHP) and methyl glycol phthalate, from plastics. These are carcinogenic compounds that are also known to cause hormone disruption (US-EPA 2010). DEHP toxicity has been reported and its public-health significance needs to be studied further (Schulz 1989).

Social and Economic Results

Salaries slightly above the average rural wages attract indigenous and *mestizo* community members to jobs in floriculture. However, this comes at a cost. The research showed that floriculture workers are exposed to highly demanding, repetitive, routine, and stressful work. Workers are given insufficient breaks (especially during cycles of high-flower demand such as Valentine's Day and throughout the months of November to January), and suffer from chronic exposure to chemical, physical, and ergonomic hazards (Breilh et al. 2005).

Floriculture workers are predominantly young (18–30 years old), have partial high school education, and have little involvement with the indigenous community organizations (if indigenous) or unions. Gender roles and responsibilities, as can be expected, differ according to tasks. For example, women are preferentially hired for postharvest activities because they are assumed to have finer manual dexterity than men. Although the involvement of young women in floriculture eases the negative aspects of their former lives in traditional patriarchal communities, it also exposes them to new challenges from industrial power relations, particularly in their interactions with mainly male supervisors. Many female workers reported having been harassed and criticized for leaving their traditional roles.

Communities neighbouring floriculture farms are affected by the industry. Many young adults seek employment on flower farms. Consequently, pesticide residues reach their households by way of contaminated clothes, plastics, wood, and other material. Therefore, the unsafe and unhealthy use of pesticides has affected not only those who deal with these chemicals directly, but even those on the periphery of the industry. As noted above, children also demonstrate early symptoms of chronic low-dose exposure to pesticides.

The project helped identify a new economic incentive for floriculture farms to cut down on harmful pesticide use. The International Code of Conduct for Cut Flower Production requires farms to comply with a set of standards for: social protection; environmental, occupational, and health protection, including rigorous control of pesticide application, exposure, and worker protection; and gender and organizational rights. Flowers certified under this Code of Conduct command higher prices in export markets. Therefore, the project developed and implemented a farm-certification program for the floriculture industry in Ecuador. Part of this work involved developing a rigorous checklist for assessing on-farm compliance with the International Code of Conduct for Cut Flower Production, and a compliance monitoring system. As a result of this project, about 18% of floriculture farms in Ecuador now comply with the International Code of Conduct.

Conclusion

By leveraging knowledge, social networks, platforms for innovative research, and postgraduate training, the project opened policy space for better governance. For example, the project helped set up a Health Rights Network, which is coordinated by the Health Department at Universidad Andina Simón Bolívar of Ecuador. This network hosted a series of workshops and focus group discussions on sustainable rural development. It advocated for health and environmental rights, and was successful in having these principles incorporated into the new constitution of Ecuador (Republic of Ecuador 2008). The project was also pivotal in introducing ecohealth to academic programs at the universities of Cuenca (master's course on health with an ecosystem perspective) and at Universidad Andina Simón Bolívar of Ecuador (PhD program on health, environment, and society).

The project generated evidence of neurobehavioural impacts of pesticide exposure among community children and floriculture workers. It also led to the first certification program in Ecuador for the health of workers and the environment for agro-industrial floriculture farms. Several alternative clinical instruments for measuring pesticide exposure were also explored (Breilh et al. submitted). The project developed *Healthy Flowers*, a health care management software, that is now being tested in a small number of farms as a tool for epidemiological assessment of toxicity.

The project translated research into action that improved occupational and household health conditions for floriculture workers and their families and neighbours. As well, in the form of an international certification program, the project identified and institutionalized an economic incentive for the industry that may allow such changes to persist.

In the course of this work, the team has come to understand environmental sustainability as not only the capacity of society to satisfy its current and future basic needs. Sustainability is now seen to also include multidimensional links between health, society, and the environment (Breilh 2004). From this standpoint, there is a need to develop *sustainable capacity*, defined as society's capacity and aptitude to produce equitable, healthy, and dignified working and living conditions for all people. This is what the indigenous people of Ecuador call *sumac kawsay* or good living.

Acknowledgements We acknowledge contributions from CEAS team members (A. Campaña, F. Hidalgo, M. Larrea, O. Felicita, E. Valle, and L. Saranchi) and the community leaders of San Isidro and Cananvalle. This project helped establish a laboratory for solid-phase extraction (SPE), which was a joint initiative of CEAS, the University of British Columbia, Canada, and Andina University, Ecuador. IDRC support was provided through projects 100661 and 103697.

References

Aguirre, P. (2004). Effect of Pesticides on Soil Quality: The Case of Ecuadorian Floriculture. PhD Thesis. Universitat Göttingen, Göttingen, Germany.
Alavanja, M., Hoppin, J., and Kamel, F. (2004). Health Effects of Chronic Pesticide Exposure: Cancer and Neurotoxicity. Annual Review of Public Health, 25, 155–197.
Breilh, J. (2004). Epidemiología Crítica. (Second Edition). Lugar Editorial, Buenos Aires, Argentina.
Breilh, J. (2007). Nuevo Modelo de Acumulación y Agroindustria: Las Implicaciones Ecológicas y Epidemiológicas de la Floricultura en Ecuador. Ciencia e Saude Coletiva, 12(1), 91–104.
Breilh Paz y Miño, J.E., Campaña Karolys, M.A., Felicita Nato, O.M., Hidalgo Flor, F.X., de Lourdes Larrea Castelo, M., and Sánchez Navarrete, D.E. (2009). Informe técnico final : Consolidación del Estudio Sobre la Relación entre Impactos Ambientales de la Floricultura, Patrones de Exposición y Consecuencias en Comunidades de la Cuenca del Granobles (Sierra Norte, Ecuador). Centro de Estudios y Asesoría en Salud, Quito, Ecuador. (Project 103697). Final Technical Report to IDRC (in Spanish). Available at: http://idl-bnc.idrc.ca/dspace/handle/10625/45111
Breilh, J., and Tillería, Y. (2009). Aceleración Global y Despojo en Ecuador: El Retroceso del Derecho a la Salud en la Era Neoliberal. (Global Acceleration and Dispossession in Ecuador: Regression in the Health Rights During Neoliberal Decades). Universidad Andina y Abya Yala, Quito, Ecuador.
Breilh, J., Campaña, A., Hidalgo, F., Sanchez, D., Larrea, M.L., Felicita, O., Valle, E., Mac Aleese, J., Lopez, J., Handal, A., Zapatta, A., Maldonado, P., Ferrero, J., and Morel, S. (2005). Floriculture and the Health Divide: A Struggle for Fair and Ecological Flowers. In: CEAS (Editor). Latin American Health Watch. Alternative Latin American Health Report. Latin American Health Watch, Quito, Ecuador.
Breilh, J., Pagliccia, N., and Yassi, A. (*submitted*). Chronic Pesticide Poisoning from Persistent Low Dose Exposures in Ecuadorean Floriculture Workers: Towards Validating a Low-Cost Test Battery. International Journal of Occupational and Environmental Health.

CEAS (Centro de Estudios y Asesoria en Salud). (2005). Informe Tecnico Final del Projecto: Ruptura del Ecosistema Floricolae Impacto en la Salud Humana en Cayambe: Abordaje Participativo Hacia un Ecosistema Saludable. (Project 10066). Final Technical Report to IDRC. Available at: http://idl-bnc.idrc.ca/dspace/handle/10625/29296.

Handal, A., Lozzof, B., Breilh, J., and Harlow, S. (2007). Effect of Community Residence on Neurobehavioral Development in Infants and Young Children in a Flower-Growing Region of Ecuador. Environmental Health Perspectives, 115, 128–133.

Lebel, J. (2003). Health: An Ecosystem Approach. In Focus Series. International Development Research Centre (IDRC), Ottawa, Canada. Available at: http://www.idrc.ca/in_focus_health/.

Republic of Ecuador.(2008). Articles 14 (Right to Healthy Environment) and 32 (Right to Health). Constitution of 2008. Political Database of the Americas. Available at: http://pdba.georgetown.edu/Constitutions/Ecuador/ecuador08.html#mozTocId735388.

SIPAE (Sistema Integrado de la Problemática Agraria del Ecuador). (2004). El TLC y los Agroquímicos: La Urgencia de un Debate Sobre el Modelo Agrario. Quito, SIPAE, Quito, Ecuador. Available at: www.bvsde.paho.org/bvsacd/cd61/plaguicidas/prefa.pdf.

Schulz, C. (1989). Assessing Human Health Risks from Exposure to Di(2-Ethylhexyl.Phthalate (DEHP) and Related Phthalates: Scientific Issues. Drug Metabolism Reviews, 21, 111–120.

US-EPA (United States Environmental Protection Agency). (2010). Basic information about Di(2-ethylhexyl) phthalate in drinking water. Available at: http://www.epa.gov/safewater/pdfs/factsheets/soc/phthalat.pdf.

Wesseling, C., McConnell, R., Partanen, T., and Hogstedt, C.(1997). Agricultural Pesticide Use in Developing Countries: Health Effects and Research Needs, 27(2), 273–308.

WHO (World Health Organization). (2005). The WHO Recommended Classification of Pesticides by Hazard and Guidelines to Classification 2004. World Health Organization, Geneva, Switzerland.

Chapter 6
Dietary Diversity in Lebanon and Yemen: A Tale of Two Countries

Malek Batal*, Amin Al-Hakimi**, and Frédéric Pelat

One late winter afternoon in 2005 in the village of Arsaal in Northeast Lebanon, a meeting to discuss an upcoming project with the community members was wrapping up. An elderly lady from the community stood up to address the research team:

> I thank you for talking to us about wild edible plants. I'm now able to eat this food without shame … Young people yearn after pasta and canned tuna and dismiss our food; I'm happy the university is talking about our local food.

The researchers from the American University of Beirut (AUB) found this statement intriguing.

Her statement summarizes a problem facing many people in the world. Much of the food they consume today is from foreign sources. More than 70% of the food consumed in Lebanon (a relatively developed country) is imported (Customs and Ministry of Trade of Lebanon 2009; Nasreddine et al. 2006), as is 92% of cereals, chiefly in the form of refined wheat flour for bread making (FAOSTAT 2004).

---

*Malek Batal was team leader for the project: Wild Edible Plants: Promoting Dietary Diversity in Poor Communities of Lebanon, which was funded by IDRC (2004–2007).

**Amin al-Hakimi was team leader for the project: Traditional Yemeni Rural Diets and Local Food Systems: Enhancing Contributions to Health and the Environment, which was funded by IDRC (2005–2008).

M. Batal (✉)
Nutrition Program, University of Ottawa, Ottawa, ON, Canada
e-mail: Malek.batal@uottawa.ca

A. Al-Hakimi
Yemeni Genetic Resources Center, Sana'a University, Sana'a, Yemen

F. Pelat
Initiatives de Développement Durable et Equitable sur la base d'Actions Locales et d'Echanges de Savoirs (IDDEALES), Yemen Branch, Sana'a, Yemen

Likewise in Yemen (one of the poorest countries in the region), although only 13% of cereals consumed were imported in 1971, the figure rose to 61.5% in 1991 and 75.3% in 2002 and was coupled with a decrease in domestic cereal production (802 vs. 640 thousand tonnes in 1971 and 1991, respectively). These figures reflect an ever-increasing dependence on imports (FAOSTAT 2004).

Lebanon and Yemen are two countries in the predominantly arid Middle East that have markedly different human-development indicators. For example, the under five mortality rate (per 1,000 live births) was 30 in Lebanon and 102 in Yemen in 2005, and the two countries scored 0.772 and 0.508 on the human-development index (HDI), respectively (UNDP 2009). However, both are undergoing significant rural transformations that are affecting nutrition and health (Batal and Hunter 2007; Jumaan et al. 1989). Both countries have relied on agriculture for millennia and yet both are encountering changes to the availability and access to local traditional food (Al-Makhlafi 1999; Hamadeh et al. 2006; Hashim 1999).

Food production systems are different in the two countries. In the traditional agro-pastoral communities in the arid and marginalized region of Arsaal in Lebanon, conflict over land has been ongoing for many years. This region is also experiencing climate change (less rainfall) and lost soil fertility. To survive, many Arsaalis have switched from food crops to growing cherry trees. Some are engaged in more profitable stone quarrying rather than agriculture (Hamadeh et al. 2006). Others have abandoned the area altogether to seek jobs in the city. Why this is happening has not been thoroughly studied, but economic challenges associated with agricultural livelihoods may play a role.

In Yemen, rapid rural transformations began in the 1970s when water pumps were introduced to the lowlands and the lower to medium altitude plains (Varisco 1991). Like many other countries, Yemeni agriculture was transformed by irrigation, leading to expansion to previously uncultivated lands. This new water-supply practice encouraged the production of cash crops that consume significant quantities of water (World Bank 2007) and require chemical fertilizers and pesticides, which the poorer farmers cannot afford. Forty years of such practices have led to the depletion or salinization of groundwater resources. Rural transformations have also included expanded Qat production (a shrub whose leaves are widely chewed like tobacco to produce a mild euphoric effect), and fruit production for export (Aw-Hassan et al. 2000). As a result, the traditional rain-fed highland areas that used to produce subsistence food crops, such as sorghum and wheat, have dramatically changed: local foods traditionally grown in this area are disappearing.

Most food is being grown for the market, and poverty and food insecurity at the local level are on the rise. As food prices increase worldwide, poor farmers are challenged from two directions: they are unable to buy expensive food from the market and they are unable to grow enough food to feed their families. The very poor cope by eating less food or lower quality food (Al-Makhlafi 1999; Government of Yemen 2002). Yemen has the highest rates of child malnutrition and food insecurity in the

Table 6.1 Consumption pattern of different food groups in Lebanon

	1961–1963 (Hwalla (Baba) 1998)	1971–1973 (Hwalla (Baba) 1998)	1981–1983 (Hwalla (Baba) 1998)	1990–1992 (Hwalla (Baba) 1998)	2000 (Hwalla (Baba) 2000)
Kcal	2,396	2,319	2,844	3,144	3,196
Protein (g)	62.3	58.2	80.2	81.2	88.5
Animal protein (%)	29.8	32.6	37.5	31.6	39.1
Energy source (%)					
Cereals	49.3	45.7	39.9	36.4	37.2
Meats, fish, dairy, and eggs	10.9	11.1	14.4	10.5	21.4
Oils and fats	11.3	12.7	14.4	15.6	6.8

Middle East (Lofgren and Richards 2003) and problems of stunting and wasting are common (Raja'a et al. 2001). However, there are also signs that overweight and obesity are creeping in, particularly among certain socio-economic groups in the urban areas (Raja'a and Bin Mohanna 2005).

In Lebanon, research has shown how dietary habits over the years have changed: traditional healthy diets characterized by inherent diversity (Batal 2008; Batal and Hunter 2007; Issa et al. 2009; Jeambey et al. 2009) are fading away in favor of a more limited repertoire of food types, little of which is produced locally. A review of food consumption patterns from 1960 to 2002 (Table 6.1) reveals a decrease in cereal consumption and an increase in meat consumption, a typical indicator of economic development. Energy and protein availability have improved so much during that period than the Lebanese are now consuming a hyper-caloric diet, with an increased risk of cardiovascular diseases, obesity, and other noncommunicable diseases (Hwalla et al. 2005; Obeid et al. 2008; Sibai et al. 2003).

In the twentieth century, nutrition problems were linked to nutritional deficiencies (Hwalla (Baba) 1998), but today, 53 and 17% of Lebanon's adults and 19.3 and 5.3% of its children are identified as overweight or obese, respectively (Sibai et al. 2003). Figure 6.1 (Batal et al. 2007) describes the rates of overweight and obesity expressed through the body mass index for adults 40–60 years of age in the rural communities participating in the Ecohealth project in Lebanon. These communities were deemed to be of low socio-economic status compared with the rest of the Lebanese population. Published data show that children from lower socio-economic strata exhibit both mild and moderate stunting, a good indicator of undernutrition (Baba et al. 1991, 1996). Research has also confirmed micronutrient deficiencies among these populations – iron deficiency in 33% of women and 25.2% of children (Hwalla et al. 2004). Analyzed together, the data show the extent of social inequality. The rich are few and eat too much, and the poor are many and either eat too little or eat poor quality food (Melzer 2002).

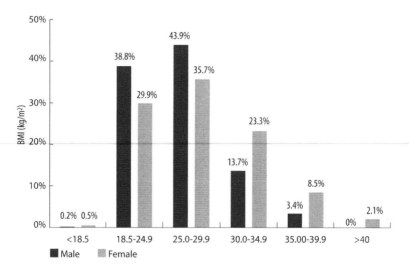

Fig. 6.1 Overweight and obesity of males (key) and females (key) expressed as body mass index (BMI) (kg/m^2) in Arsaal, Kuakh, and Chouf cluster (Batloun, Warhaniyeh, and Kfarnabrakh) in Lebanon (2005). Underweight <18.5; normal weight 18.5–24.9; overweight 25.0–29.9; obese class I 30.0–34.9; obese class II 35.0–39.9; and obese class III >40.0

When the Local Environment is no Longer the Source of Food

Many of the health and nutrition issues in Lebanon and Yemen can be linked to the lack of dietary diversity. Diversity is a trait of many traditional food systems. The current trend in food consumption is to eat food that is purchased from stores – food that is often less diverse than the traditional diet. Purchased foods are often imported and processed, are usually high in fat and sugar, and are low in fiber. The current diet in Lebanon is now limited to only a few staple foods. According to the World Health Organization (WHO), the average daily consumption of bread in Lebanon per capita per day is 350 g (WHO 1998, p. 45), or in terms of total cereals 141 kg per capita per year (FAOSTAT 2004). This is indicative of an imbalance in diet quality, but not necessarily of quantity, and thus is not an adequate indicator for food security.

The situation in Lebanon contrasts significantly with that in Yemen, where prices of cereals and cereal products increased by 20% in 2007 and 23% in 2008. Due to these price increases, and because cereal consumption is estimated at 166 kg/year and imports constituted 61.5% of cereal consumption in 1991 (FAOSTAT 2004), an estimated additional 6% of all Yemenis have dropped below the poverty line, joining the 40% that were already considered poor.

The population of Yemen is presently estimated at 21 million and expected to double in the next 20 years. According to the 2003 Food Insecurity and Vulnerability Information Mapping System Survey (Government of Yemen 2003), half a million households in Yemen are food insecure, which represents 21.8% of all households nationally. Rural households are the most vulnerable, especially those with high child-to-adult ratios. Household size and the number of children were also found to be risk factors associated with food insecurity. Two-thirds of all agricultural holdings are smaller than 1 ha (Ministry of Agriculture and Irrigation 2007), which worsens food security.

In the Yemeni highlands, the gradual deterioration of traditional food systems is intimately connected to the degradation of the natural resources on which communities have relied, and used sustainably, for centuries. Unsustainable water-supply policies, coupled with the degradation of the mountain terraces, increasing climate variability, and population pressure have pushed many poor families to adopt difficult coping strategies. For example, many men from the highlands have emigrated to neighboring Arab countries, where they are subjected to difficult work and living conditions, to be able to support their families in Yemen (Adra 1983).

In addition, to be able to cope with rising food prices and the threat of food insecurity, seed stocks are depleted to feed the family. Increased dependence on expensive market foods and consumption of unsuitable seeds are two of the main factors that increase household vulnerability in the rain-fed agricultural areas (WFP 2008). According to a World Bank study to understand women's adaptation strategies to climate change in rain-fed highlands (Al-Hakimi and Ya'ni 2008), the social and economic changes linked to the decline of agricultural household income (until 2007) have forced men to leave in search of jobs in cities. The women have been left behind to farm, but they face severe culturally based restrictions on movement and access to information. In some districts, women are permitted to secure livestock feed, but are discouraged from becoming engaged in cereal production (Al-Hakimi and Ya'ni 2008).

The serious health and nutrition challenges facing the highlanders of Yemen and the drylanders of Lebanon require innovative solutions to address these complex and interlinked problems. Greater reliance on local nutritious foods can be part of such solutions. For innovations to be successful, the entire community – both young and old men and women and youth, religious, political, and other traditional leaders – must be involved in developing long-term strategies to protect their health, ensure themselves sufficient quantities of food, safeguard their environment, and provide themselves with dignified livelihood options.

Linking Research and Action on Local Food Systems

Two ecohealth projects in Lebanon and Yemen during 2004–2008 investigated how social, political, economic, and ecological transformations had affected dietary diversity and how these transformations affected health. The purpose of the studies was to improve dietary diversification through increased reliance on local food systems, such as wild edible plants and traditional foods, to combat health problems associated with poor nutrition.

Using complex system analysis, both teams hypothesized that the agro-ecosystems were deteriorating, negatively affecting nutrition and health, especially among the poor (Al Hakimi et al. 2008; Batal 2008; Batal and Hunter 2007; Issa et al. 2009, 2011; Jeambey et al. 2009). Multiple and interlinked factors were at play at various levels. As well as drawing on concepts of sustainable use of biodiversity, the researchers employed an ecosystem approach to health to address these links and try to resolve these problems.

Thus, separately, two multidisciplinary teams, from the AUB[1] in Lebanon and from Sana'a University[2] in Yemen, began working in close collaboration with their target communities and other local actors. Each team characterized many facets of these ecosystems, seeking associations between dietary diversity, food security, and ecosystem management and between dietary diversity and various risk factors for chronic disease. The hypothesis was that a reliance on local foods would improve nutritional intake and sustain biodiversity in the ecosystem, and contribute to both human and ecosystem health. Encouraging the consumption of wild edible plants and food grown locally would also draw the community back to their traditional diet and regain pride and interest in the ecosystems that produce this food, resulting in better management of key (and in some cases threatened) natural resources.

Focus group meetings and in-depth interviews were carried out with community members in Lebanon to better understand indigenous knowledge around the collection, consumption, preservation, and health benefits of wild edible plants. All key informants identified by community members as knowledgeable were above the age of 55, and most of them were women. They reported that the younger generations were uninterested in maintaining and using this knowledge (Jeambey et al. 2009).

Community members reported that the problem of accessing local food was a major hindrance to its consumption. Traditional food suffered from an "image problem" – it does not have the prestige associated with foods advertised on television. The project helped to set up a Healthy Kitchen network in the villages of Arsaal, Kuakh, and Batloun. The working hypothesis was that improved image and perceived value of local food would lead to greater consumption of such foods both locally and in the city. Local food and traditional recipes passed down over generations were collected and tested by the Healthy Kitchen network. More than 25 women involved in the kitchens also collected wild edible plants. They shared and promoted recipes during income-generating catering events and trade shows, particularly in the city. They also became strong advocates for the natural

---

[1] In November 2004, the Department of Nutrition and Food Science in partnership with the Initiative for Biodiversity in Arid Regions (IBSAR) and the Environment and Sustainable Development Unit (ESDU) at the American University of Beirut (AUB) started the project Wild Edible Plants: Promoting Dietary Diversity in Poor Communities of Lebanon (WEP-DD).

[2] Through partnership between the University of Sana'a, the Yemeni Genetic Resource Center, and IDDEALES, the project Health and Dietary Diversity in Yemen – Traditional Yemeni Rural Diets and Local Food Systems: Enhancing Contributions to Health and the Environment was started in winter 2005.

environment from which these wild plants were being collected, and they became involved in conservation activities.

Women involved in the project received training in good manufacturing practices as well as in organizing cooperatives, marketing, and accounting. The project produced a food-safety manual in Arabic and a website[3] that contains extensive plant and recipe databases, a bilingual book that contains close to 40 local recipes, and records both indigenous and scientific information on the nutritional and health qualities of 15 wild edible plants (Batal 2008).

The network has played a key role in promoting wild plants and ecosystem protection. Village cooking festivals and catering events garnered national media coverage and contributed to the increased visibility of traditional foods. Traditional food appeared to be becoming more appealing to the urban elite. For example, the kitchens were asked to cater for more than ten events held in Beirut in 2007 alone. This phenomenon heightened the interest of the rural communities in their traditional knowledge and natural heritage. Many women were employed by this initiative, which enhanced their income but also, perhaps more importantly, empowered them in the community. In the words of one of the ladies involved with the Healthy Kitchen in Batloun:

> Women, who have been confined to their own homes within the borders of their villages are now traveling throughout Lebanon, meeting new people from different backgrounds, entering new markets, and taking responsibilities for business transactions.

In Yemen, the research used a similar participatory methodology because the intention was to foster scientific knowledge that integrated valuable indigenous experience. The systemic approach that addressed food production, rural community health, and environment connections in their socio-economic conditions was very new in the country and provided more holistic and dynamic knowledge on rain-fed agro-ecosystems. Scientific data were recorded and indigenous knowledge on local landraces, farming techniques, natural resources management, and cooking practices documented. Two neighboring communities, which lived under full rain-fed conditions 50 years ago but since then followed very distinct agricultural and food evolutions, were compared by using PRA (Participatory Rural Appraisal) tools, focus group meetings, formal surveys, medical consultations, and laboratory analysis.

The project produced maps of the agro-ecosystems (e.g., linking water and soil resources, topography, farming practices, rotation and intercropping systems, and cropping patterns).The results were shared widely to better understand the diversity of existing conditions and the way farmers have adapted decision making around crop choice and other practices to reflect this diversity in the fields and in their diets. Several booklets about traditional practices were produced and disseminated. Although farmers' perceptions on local varieties were not directly "quantified," communities reported that low productivity contributed to their negative

---

[3] See: http://www.wildedibleplants.org.

image when compared with high-productive seeds, which supposedly required uniform and simplified work according to what they had heard. Yet, some farmers were eager to revive landraces in cropping patterns provided that productivity was increased. Through the Yemeni Genetic Resource Center, the project acted as a seed bank. It collected relevant indigenous seeds and selected seeds from other Yemeni highlands that were available in the Yemeni Genetic Resource Center and had potentially good adaptability to local conditions. Seeds were distributed to farmers along with information about their growing and nutritional qualities, cultivation, and seed-selection methods. Comparative experiments were conducted with communities. The project's intention was to enhance the farmers' pragmatic "agro-biodiversity reflex," which has been used as an ancestral strategy to mitigate erratic changes in the environment and weather and to find the best solutions to current needs.

In parallel, ethnographic work was conducted with women in the communities to record more than 100 traditional recipes, along with local ingredients, and the utensils necessary for their production. These recipes were collated in a book published by the University of Sana'a Press (Ya'ni et al. 2008). The book is both an ethnographic and nutritional reference as it is the first cookbook on local recipes ever published in the country. Conserving traditional recipes requires the promotion of their diversity, quality, and particular tastes. Beside the first challenge of supporting local varieties in the field, a second one was found in the cooking pots. Elders' preference and attachment to traditional meals based on diversified local cereals clashed with younger generations' food habits who were partial to standardized preparations with white flour even in rural areas. Because young mothers were more and more attracted to easier and shorter food preparations, the cookbook was distributed primarily among them to raise awareness about the health virtues of traditional dishes and to disseminate the documented knowledge before it was lost.

As a result of these two projects, there is a renewed interest and pride in local knowledge about food in the community and beyond.

Achievements

The researchers and the community jointly developed a more complete picture of nutrition and health in rural villages in Lebanon and Yemen, and showed how these were linked to social, economic, political, and environmental factors at different levels and scales.

In Lebanon, project findings confirmed the high prevalence of overweight and obesity, dyslipidemia, and other chronic disease risk factors in the rural Lebanese population (Batal et al. 2007). The project identified, categorized, and documented wild plants and showed how they were linked to traditional rural diets and traditional cultural practices (http://www.wildedibleplants.org; Batal 2008). The project also identified some of the pressures exerted on the ecosystem, and noted the degraded state of the once-rich biodiversity. Most of the collection sites for wild edible plants in the surveyed communities were located in semi-natural habitats or abandoned agricultural lands. The most important threat to these wild edible plant species were two practices, over-harvesting and over-grazing, commonly encountered on such lands. In all sites, and for all plant species of interest, density was low (Batal et al. 2007). This is possibly due to the marginal status of the semi-arid highlands in Lebanon. The findings point to the urgent need to manage grazing and harvesting activities in these fragile ecosystems. Most importantly, the study showed, through chemical and nutrient analysis, that traditional Lebanese food, if consumed regularly has the potential to improve diets (Batal and Hunter 2007; Issa et al. 2009). Through close collaboration with communities, the project increased the value of local food resources and developed a heightened respect for the ecosystem that provides both nutritional food and valuable income. In the words of a member of the Healthy Kitchen in Kuakh:

> The kitchens have also allowed local suppliers from the agriculture and animal-farming sectors to sell their own products and increase their profits.
>
> One of the most important outcomes of the Healthy Kitchen was bringing attention to our otherwise forgotten community through the media. Our village is now a focal point for large-scale development projects initiated by international NGOs. The establishment of the kitchen opened new windows of opportunities for the people, especially the women to work in different areas, develop new relationships, and widen experiences.

In Yemen, increasing childhood overweight and obesity problems were uncovered, sometimes coexisting with parasitic infections and malnutrition in the same communities (Al Hakimi et al. 2008). Whatever the context, both in intensified monoculture systems that produce potatoes and rain-fed farms that produce very low yields, insufficient cereal production has increased all households' dependence on markets and shops. People buy food to meet rising needs for wheat and not to diversify food baskets with vegetables or fruits (with 93–57% of the inhabitants admitting buying wheat grains every year in both farming situations, respectively). Findings also confirmed that local wheat and barley varieties often had higher mineral, fiber, or protein contents than improved grains introduced in local systems or imported refined flours. However, in these very poor socio-economic conditions, degraded environments (garbage, unhealthy farming practices, polluted water), and weak domestic hygiene (including in kitchens) are common. Thus parasitosis (affecting more than 80% of the people in both communities) and new diseases due to chemicals are now local priorities that may diminish the potential positive effects of local grains and foods.

The project identified several advantages of local food sources. Traditional farming and cropping systems usually include several local wheat, barley, sorghum, or millet varieties and are based on crop rotation and intercropping that protect local genetic diversity and support local food habits. They also allow for the production of secondary crops such as legumes (lentils, fenugreek, and beans), which in turn contribute to more dietary diversity into a cereal-based diet. The research revealed that farmers reserved local landraces for use on rain-fed fields and in extensive organic farming because they would get optimal yields even when the season was drier than usual. Farmers also noticed that irrigated or intensified practices negatively affected the taste and nutritional quality of local seeds. Although less used than in the past, local bread recipes still use different wheat, as well as barley and sorghum varieties, and remain popular and valued for reasons of health, energy, and taste. The research discovered that there was still a wide range of traditional recipes from rural areas dedicated to diverse ways of consuming local food products. These recipes showed varied practices between regions and enhanced the demand for, and cultivation of, a diversity of local landraces. By using a systemic approach, the project highlighted interrelations between the main three elements of rural food systems: traditional agricultural practices; local varieties; and traditional dishes. They are intimately connected so that the loss of any of these elements leads to the deterioration of the others, and results in environmental degradation and diet simplification. Therefore, the project collected local seeds, recorded ancestral agricultural practices, and documented traditional Yemeni recipes.

The research in both Lebanon and Yemen characterized complex problems in a systematic fashion. Using an ecohealth approach, the complexity of the links at different scales and levels was better understood. Both projects showed how ecosystem health and human heath are interconnected. The projects demonstrated the link between nutrition and livelihoods, created stronger markets for local food, and influenced changes among the practices of producers and consumers. These results have the potential to encourage more sustainable agro-ecological practices and local agro-biodiversity to avoid jeopardizing the sustaining ecosystem, improve nutrition and health, and increase national food security. However, there is still scope for both projects to make the links more explicit and to produce the evidence that policymakers and other actors require to make long-lasting positive changes for the benefit of the poor. Though strong efforts were made to seek policy attention in workshops and through policy briefs, no significant policy influence was achieved by either project. In both countries, weak political institutions are typical. Civil strife, war, and poverty constrain governance structures in these countries from acting on evidence from research and initiating change. In both contexts, these issues challenge not only development and poverty-alleviation initiatives, but also efforts that support the sustainable and equitable use of scarce resources.

For improved food security, there is a need to evaluate agricultural and trade policies and assess current subsidies on bread, including the need to examine the impact of cereal imports on health and social equity. The rich traditions, local biodiversity, and indigenous knowledge in both Lebanon and Yemen are resources for local people, and may contribute to developing strategies for food security, nutrition, and sustainable development.

One issue that remains to be researched is the change in the lifestyle of rural communities. Research on lifestyles may have potentially important contributions to address health problems encountered in both countries. An ecohealth approach would help take into account the role of physical activity (work and leisure) and their relationship with both the environment and improved health for all.

Acknowledgments We thank the communities of Batloun, Kfarnabrakh, Warhaniyeh, Arsaal, and Kuakh in Lebanon and Al-Arafah, Ribat al Qalaa, Masyab, and Saber in Yemen for welcoming us in their midst. We also acknowledge the support of all of the researchers and other project stakeholders. The following people contributed to the material used in this paper: Anhar Yaani, Sadeq Sharaf, Adnan Al-Qubati, Mokhtar Dael, Ahmed Al-Samawi, Darine Barakat, Salma Talhouk, Shadi Hamadeh, Beth Hunter, Cynthia Farhat, Zeinab Jeambey, and Nader Kabbani. IDRC support for this research was provided through the projects 102692 and 103153.

References

Adra, N. (1983). The Impact of Male Migration on Women's Roles in Agriculture in the Yemen Arab Republic (Arabic, English). Inter-Country Expert Meeting on Women in Food Production in the Near East Region. Amman (Jordan), 22 October 1983. Food and Agriculture Organization of the United Nations, Rome.

Al Hakimi, A., Al Qubati, A., Al Hagami, A., Saed, S.S., Othman, M.D., Al Samawi, A., Yaani Abdulkarim, A. and Pelat, F. (2008). Health and Dietary Diversity in Yemen – Traditional Yemeni Rural Diets and Local Food Systems: Enhancing Contributions to Health and the Environment (Project 103153). Final Technical Report to IDRC. Available at: http://idl-bnc.idrc.ca/dspace/handle/10625/44794

Al-Hakimi, A., and Ya'ni, A.A. (2008). Women and Coping Strategies for Adaptation to Climate Change Using Agrobiodiversity Resources in the Rainfed Highlands of Yemen. Report Prepared for the World Bank, Washington.

Al-Makhlafi, H.K. (1999). Food security in Yemen; An analysis study of the current nutrition situation. Yemeni Journal of Science, 1(1), 1–16.

Aw-Hassan, A., Alsanabani, M., and Bamatraf, A. (2000). Impact of land tenure and other socioeconomic factors on mountain terrace maintenance in Yemen. CAPRI Working Paper. International Food Policy Research Institute, Washington.

Baba, N.H., Hamadeh, S., and Adra, N. (1991). Nutritional status of Lebanese school children from different socioeconomic backgrounds. Ecology of Food and Nutrition, 25, 183–192.

Baba, N., Shaar, K., El-Sheikh Ismail, L., and Adra, N. (1996). Comparison of nutritional status of pre-school children at day care centres and at home from different socioeconomic backgrounds in Beirut. Journal of Human Nutrition and Dietetics, 9(2), 89–103

Batal, M. (2008). The Healthy Kitchen: Recipes from Rural Lebanon. American University of Beirut Press, Beirut.

Batal, M. and Hunter, B. (2007). Traditional Lebanese recipes based on wild plants: An answer to diet simplification? Food and Nutrition Bulletin, 28(2), S303–S311.

Batal, M., Hamadeh, S., Hwalla, N., Kabbani, N., and Talhouk, S. (2007). Wild edible plants: Promoting dietary diversity in poor communities of Lebanon (Project 102692). Final Technical Report to IDRC. Available at: http://idl-bnc.idrc.ca/dspace/handle/10625/38050.

Customs and Ministry of Trade Lebanon. (2009). Trade Statistics. Available at: http://www.customs.gov.lb/customs/trade_statistics/yearly/search.asp.

FAOSTAT. (2004). FAO Statistical Databases. Food and Agriculture Organization of the United Nations, Rome, Italy. Available at: http://faostat.fao.org/faostat/default.jsp.

Government of Yemen. (2002). Poverty Reduction Strategy Paper (PRSP) 2003–2005. Government of Yemen, Sana'a.

Government of Yemen. (2003). Yemen Food Insecurity and Vulnerability Information Mapping System (FIVIMS) Survey Report. Government of Yemen, Sana'a.

Hamadeh, S., Haidar, M., and Zurayk, R. (2006). Research for Development in the Dry Arab Region: The Cactus Flower. Southbound, Penang, Malaysia, and IDRC, Ottawa.

Hashim, A.M.A. (1999). Food security and the nutritional gap in the Republic of Yemen. In: Le Ye'men Contemporain. Editions Karthala, Paris.

Hwalla (Baba), N. (1998). Food and dietary fiber consumption pattern in Lebanon. International Journal of Food Sciences and Nutrition, 49, 41–45.

Hwalla (Baba), N. (2000). Dietary intake and nutrition related disorders in Lebanon. Nutrition and Health, 14, 33–40.

Hwalla, N., Adran, N., and Jackson, R. (2004). Iron deficiency is an important contributor to anemia among reproductive age women in Lebanon. Ecology of Food and Nutrition, 43, 77–92.

Hwalla, N., Sibai, A., and Adra, N. (2005). Adolescent obesity and physical activity. In: Simopoulos, A.P. (Ed.), Nutrition and Fitness: Obesity, the Metabolic Syndrome, Cardiovascular Disease, and Cancer. World Review of Nutrition and Dietetics, 94, 42–50.

Issa, C., Darmon, N., Batal, M., and Lairon, D. (2009). The nutrient profile of traditional Lebanese composite dishes: comparison with composite dishes consumed in France. International Journal of Food Sciences and Nutrition, 60 (S4), 285–295.

Issa, C., Darmon, N., Salameh, P., Maillot, M., Batal, M., Lairon, D. (2011). A Mediterranean diet pattern with low consumption of liquid sweets and refined cereals is negatively associated with adiposity in adults from rural Lebanon. International Journal of Obesity, 35(2), 251–258.

Jeambey, Z., Johns, T., Talhouk, S., and Batal, M. (2009). Perceived health and medicinal properties of six species of wild edible plants in the northeast of Lebanon. Journal of Public Health Nutrition, 12(10), 1902–1911.

Jumaan, A.O., Serdula, M.K., Williamson, D.F., Dibley, M.J., Binkin, N.J., and Boring, J.J. (1989). Feeding practices and growth in Yemeni children. Journal of Tropical Pediatrics, 35, 82–86.

Lofgren, H. and Richards, A. (2003). Food security, poverty, and economic policy in the Middle East and North Africa. Research in Middle East Economics, 5, 1–31.

Ministry of Agriculture and Irrigation. (2007). Agricultural Statistics Yearbook 2007, Ministry of Agriculture and Irrigation, Republic of Yemen.

Melzer, K. (2002). Nutritional status of Lebanese school children aged 6–9 years from different socioeconomic backgrounds. MS Thesis, American University of Beirut, Beirut.

Nasreddine, L., Hwalla, N., Sibai, A., Hamze, M., and Parent-Massin, D. (2006). Food consumption patterns in an adult population in Beirut. Public Health Nutrition, 9(2), 194–203.

Obeid, O.A., Al-Khatib, L., Batal, M., Adra, N., Hwalla, N. (2008). Established and suspected biomarkers of cardiovascular disease (CVD) risk in pre-menopausal Lebanese women. Ecology of Food and Nutrition, 47(3), 298–311.

Raja'a, A., Sulaiman, S.M., Elkarib, S.A., and Mubarak, J.S. (2001). Nutritional status of Yemeni schoolchildren in Al-Mahweet Governorate. Eastern Mediterranean Health Journal, 7(1/2), 204–210.

Raja'a, A., and Bin Mohanna, M.A. (2005). Overweight and obesity among schoolchildren in Sana'a City, Yemen. Annals of Nutrition and Metabolism, 49, 342–345.

Sibai, A., Hwalla, N., Adra, N., and Rahal, B. (2003). Prevalence and covariates of obesity in Lebanon: Findings from the first epidemiological study. Obesity Research, 11(11), 1353–1361.

UNDP (United Nations Development Programme). (2009). Arab Human Development Report. UNDP, New York.

Varisco, D.M. (1991). The future of terrace farming in Yemen: a development dilemma. Agriculture and Human Values, 8, 166–172.

World Bank. (2007). Making the most of scarcity: Accountability for better water management results in the Middle East and North Africa. World Bank, Washington.

WFP (World FoodProgram, Yemen). (2008). Impact of food security situation on vulnerable children in Yemen. Survey Report, UNICEF Symposium on Child Poverty. Available at: http://sites.google.com/site/globalstudy2/Impactoffoodsecurityonvulnerablechil.ppt?attredirects=1.

WHO (World Health Organization). (1998). Fortification of flour with iron in countries of the Eastern Mediterranean, Middle East and North Africa, Annex 1. Eastern Mediterranean Regional Office, Cairo. Available at: http://whqlibdoc.who.int/hq/1998/WHO_EM_NUT_202_E_G.pdf.

Ya'ni, A., Al-Hakimi, A., Al-Qubati, A., Saed, S.S., Othman, M.D., Al-Samawi, A., and Pelat. F. (2008). Traditional Rural Yemeni Dishes (In Arabic). Sana'a University Press, Sana'a.

Part II
Natural Resources, Ecosystems, Pollution, and Health

Chapter 7
Introduction

Ana Boischio and Zsófia Orosz

Natural resources are a key element of any country's economic growth and development. Extraction and transformation of natural resources have impacts on ecosystems and human health, as well as both positive and negative social and economic effects. Formal and informal economic activities based on natural resources generate access to goods and services through barter or cash income. A wide diversity of social benefits and public goods like employment, political voice, and decision-making power are associated with natural resource development. However, these may come at the cost of negative ecosystem, occupational, and community-level health impacts. These costs are not always considered in economic development schemes. History has shown that neither the benefits nor the costs are shared equally.

The case studies in this section share a common entry-point on environmental pollution associated with mining. Environmental pollution from natural-resource extraction, whether from mining or ecosystem-transforming land-use, poses significant health threats to people. These threats are transmitted directly through occupational exposures (highlighted in the stone dust in India and metals in Ecuador case studies) and environmental exposures (manganese in Mexico and metals in Ecuador), and indirectly through transformation in the ecosystem (mercury in the Amazon and in Ecuador).

Other sections of this book also address pollution's impacts on ecosystems and human health. In Part I, pesticides are assessed, and in Part IV, different pollution sources and exposures are considered. Environmental pollution from natural-resource extraction, and in particular from mining, illustrates the challenges of managing the trade-offs between economic development of natural resources and environmental damage. Problems of human health present tough challenges in such contexts – challenges that can be addressed by considering the interdependent behaviour of the social-ecological system.

A. Boischio (✉) • Z. Orosz
International Development Research Centre, Ottawa, ON, Canada
e-mail: ecohealth@idrc.ca

Pollution refers to excessive amounts of chemical, physical, or biological hazards that a given system cannot absorb or incorporate without causing adverse effects in its structure and function (Klassen 1996; Odum 1975). Human exposure to chemicals and metals has been linked with biological alterations, subtle adverse health effects, and eventually, illness, major disabilities, and sometimes death (Mergler et al. 1999). The toxicological health effects of chemical and metal pollutants are dose-dependent. The effect depends on the amount of substance, duration of exposure, and time in life when the exposure occurs.

Certain pollutants can damage the neurological system with irreversible effects. Therefore, early detection of subtle damage is important to prevent long-term disability. Neurobehavioural assessments were used in several of the case studies to evaluate subtle adverse health effects in association with relatively low levels of exposures to metals. Subtle neurobehavioural impairments like reduced dexterity or slower reaction time are rarely noticed by affected people, who are often struggling with other competing health problems such as nutrition and infectious diseases.

Early detection of the subtle health effects associated with low levels of exposure to pollutants poses other challenges. Such effects may also be associated with, or exacerbated by, other factors such as education, social context, genetics, nutrition, and underlying infectious diseases. The type of test (and its appropriateness to the cultural context) may also influence the result.

Perception of pollution varies widely according to knowledge, awareness, trade-off opportunities, health concerns, and willingness to change. These different perceptions need to be considered when assessing hazards or designing interventions. People often do not recognize the threats to health posed by chemicals or physical hazards they encounter every day. These perceptions are influenced by the often subtle nature of early biological changes associated with relatively low-dose long-term exposures. Populations often face other, more obvious and discrete health concerns. Perceptions regarding impacts on ecosystems can also vary. Certainly in many cases, pollution of the ecosystem can be very clearly observed and yet, as these case studies illustrate, some pollutants that are invisible or present in very small quantities (e.g., methyl-mercury in fish consumed for food) can still be harmful to human health. Variations in suspended matter in the river below gold-mining sites in Ecuador, and dust from manganese mining in Molango, Mexico, were indicators of ecosystem disturbance. This disturbance was clearly recognized by the communities and connected to health concerns to varying degrees. However, fish polluted by mercury was not perceived by local people as a concern in the Brazilian Amazon. Fish look and taste the same, no matter the level of methyl-mercury contamination.

An ecosystem approach to health broadens the scope of research beyond exposure assessment and epidemiological studies. Transdisciplinarity and participatory methods allow research teams to understand social drivers of exposure as well as how people relate to their environment. Research can unearth ways to make positive changes in the local situation, as well as uncover new avenues for research.

The case studies in this section illustrate the importance of systems thinking in ecohealth research. Case studies follow various potentially toxic substances through the ecosystem, assess various social and economic pathways of human exposure, and integrate these with action plans for reducing pollution and preventing exposure. In Ecuador, metals were traced to pollution from the mining process but also were found to be coming from natural sources in the soil and other sources like cookware. In the Amazon case study, considerable biogeochemical investigation was required to show that mining was not the main source of mercury in the Tapajos River. In the stone quarrying and crushing industry case study from India, investigation of poor respiratory health pointed to both occupational dust exposure and other domestic sources of pollution.

The case studies also highlight common governance challenges and illustrate how, even in politically strained situations, an ecohealth research project can engage decision makers to achieve positive change. In manganese mining in Mexico and stone-crushing in India, large private-sector companies were convinced to address pollution problems. Due to their substantial resources, changes could be implemented to protect health. Both of these case studies emphasize the links between multistakeholder participation and knowledge-to-action. In contrast, controlling pollution from informal small-scale gold mining in Ecuador posed different challenges of governance. Change was challenging to achieve in this context. It required buy-in from large numbers of small- and medium-size enterprise owners and many actors from local authorities. Small- and medium-size enterprises are key to economic development but are not easily regulated, and as illustrated in Ecuador's gold mining region, can have substantial negative environmental, occupational, and other social impacts. Opportunities to foster dialogue among business owners, workers, policymakers, and communities can help better balance these trade-offs and mitigate adverse health effects. The stone quarrying project in a very poor region in India illustrates the achievements from an action-oriented dialogue among the parties, built on existing relationships among stakeholders. As the case study describes, owners of stone quarrying and crushing operations started to cooperate by putting in place occupational and environmental safety measures.

The economic benefits derived from natural resource extraction and associated activities come at a sometimes high cost to human health and ecosystem conditions. Ecohealth research that systematically includes dialogue with a diversity of stakeholders can lead to locally appropriate changes that have a better chance of making a lasting difference to the health of local communities.

References

Klassen, C.D. (1996). Casarett and Doull's Toxicology: The Basic Science of Poisons. Fifth Edition. McGraw Hill, New York.

Mergler, D., Baldwin, M., Bélanger, S., Larribe, F., Beuter, A., Bowler, R., Panisset, M., Edwards, R., de Geoffroy, A., Sassine, M.P., and Hudnell, K. (1999). Manganese Neurotoxicity, a Continuum of Dysfunction: Results from a Community Based Study. Neurotoxicology, 20, 327–342.

Odum, E. (1975). Fundamentals of Ecology. Saunders, Philadelphia.

Chapter 8
An Ecosystem Study of Manganese Mining in Molango, Mexico

Horacio Riojas-Rodríguez and Sandra Rodríguez-Dozal

The region of Molango in the state of Hidalgo, Mexico, has one of the largest manganese (Mn) ore deposits in the world. The region covers approximately 1,250 km^2 and has proven reserves of 32 million tonnes of Mn ore, plus another 250 million tonnes categorized as probable. Manganese is one of the five most abundant minerals on the planet and is valued for its use in manufacturing steel alloys. Other uses include manufacturing batteries and ceramics. In some countries, Mn can be used as an anti-knock additive in gasoline.

The mining company Autlán began Mn extraction and processing work in Molango in the 1960s. Since then, Mn nodules have been produced and transported for commercialization and subsequent use in steel manufacturing. Extraction methods include both subterranean shafts and open pits or trenches that are located near communities. To process the ore, Autlán operates a nodulizing plant and a plant that produces battery-grade Mn.

Mining has transformed the landscape of this formerly agricultural and forested area with open-face ore extraction and additional roads. Mining activities have also brought services, and they initially created expectations among local communities that were hoping to gain economic benefits. However, in the mid-1980s, communities began to complain to the mining company and the government about damage to their homes, crops, livestock, and eventually their own health (Paz 2008a).

Because of public complaints in the late 1980s, the state government commissioned the nongovernmental organization (NGO) Health, Environment, and Work Institute (Instituto de Salud, Ambiente y Trabajo) to conduct a preliminary study on health risks in this mining area. Manganese concentrations in outdoor and indoor areas were observed to be up to three times higher than values from non-polluted

---

H. Riojas-Rodríguez (✉) • S. Rodríguez-Dozal
Dirección de Salud Ambiental, Instituto Nacional de Salud Pública,
Cuernavaca, Morelos, México
e-mail: hriojas@insp.mx

areas. The average Mn concentration in blood (Mn-B) from 73 people (14–93 years old) was 17.7 μg/L (range: 7.5–88.0 μg/L). Normal background levels of Mn-B in adults are 4 to 15 μg/L (ASTDR 2008). Neurobehavioural tests showed deficits associated with increased Mn-B, and increased air concentrations of Mn (Santos-Burgoa et al. 2001). These results were shared with Autlán, the state government, and the environmental authorities. Due to the preliminary nature of the project, the results were questioned, especially by Autlán, and a more comprehensive study using ecosystem approaches to health started in 2001.

Research Using Ecohealth Concepts

The focus of the ecohealth study was to update knowledge about Mn exposures, pathways, and health effects, especially among children in the Molango area. Using an ecohealth framework, the study was built on a multidisciplinary research team, multistakeholder approaches, and systems analysis to determine exposure pathways and health effects associated with mining operations. Consultations, meetings, and workshops with representatives of state and federal government institutions, Autlán, and, less intensively, community representatives were organized throughout the study. A multidisciplinary team with expertise in toxicology, epidemiology, neuropsychology, geology, chemistry, and social sciences (sociology and anthropology) was assembled to better integrate research questions, objectives, and workplans.

Multistakeholder consultations conducted during project planning confirmed the need for more knowledge on the fate of Mn in this particular ecosystem. New national regulations would be needed to limit Mn emissions.

Initially, this study focused on the environmental fate of Mn in the Molango area. Manganese concentrations in water, soil, sediments, and atmospheric particulate matter (PM) were analyzed. Water and soil samples were analyzed with an atomic absorption spectrometer, sediment concentrations were determined by X-Ray fluorescence, and atmospheric PM by a gravimetric method. The manganese in PM was measured using PIXE method (Protons induced X-Ray Emission). Emission sources and pathways were characterized through air measurements and dispersion analysis. A Geographic Information System (GIS) was used to map relevant data.

In parallel with the environmental studies, the project mapped and analyzed interactions among stakeholders to evaluate the feasibility of new agreements to reduce manganese pollution. A stakeholder analysis (Ramírez 1999) was used to investigate relationships among these different stakeholders. Many documents were reviewed – minutes of community meetings, internal documents from Autlán, and government archives and media material. Semi-structured interviews with key stakeholders from the government, the community, and Autlán enriched the existing information.

Table 8.1 Risk-management plan for Mn exposure – roles and responsibilities

- Ecological land-use management: SEMARNAT–COEDE[a]
- Comprehensive deposit management: SEMARNAT–COEDE
- Epidemiological surveillance and integral management of populations at risk: SSH
- Regulatory framework: PROFEPA–INSP–SEMARNAT–SSH (Federal and State)
- Monitoring and surveillance systems PROFEPA–INSP–COEDE–AUTLAN
- Communication and environmental education INSP–SEMARNAT

[a]SEMARNAT – Environment and Natural Resources Department (Secretaría del Medio Ambiente y Recursos Naturales); COEDE – State Ecology Council (Consejo Estatal de Ecología); SSH – Hidalgo Health Services (Servicios de Salud de Hidalgo); PROFEPA – Federal Environmental Protection Agency (Procuraduría Federal de Protección al Ambiente); and INSP – National Public Health Institute (Instituto Nacional de Salud Pública)

Facilitated by the open dialogue of the consultation process, communication strategies tailored to the different target audiences were developed and implemented. Preliminary research results were disseminated during workshops and field work, with the participation of representatives of the state, the municipalities, and the communities. In parallel, a risk-management plan to reduce Mn exposures was developed using the results and recommendations (Table 8.1).

Project Results

Manganese in People

This study deepened our understanding of the impacts of Mn mining in Molango. The mean outdoor airborne Mn level was 0.10 $\mu g/m^3$ (range 0.003–5.86 $\mu g/m^3$ from 25 samples), double the US-EPA recommended level (ASTDR 2000). About 37% of the study population ($n=288$) had Mn-B concentrations between 10 and 15 $\mu g/L$, and 12% had levels above normal background (4–15 $\mu g/L$) (ASTDR 2008). Moreover, a significant association between high Mn concentrations in the air (above 0.099 $\mu g/m^3$) and the results of neurobehavioural tests, specifically motor tests, among 288 adults, were consistent with a previous study (Santos-Burgoa et al. 2001) in the same mining area (Rodriguez-Agudelo et al. 2006; Solis-Vivanco et al. 2009). Inhalation was confirmed as the main pathway for human exposure.

Figures 8.1 and 8.2 present results from neurobehavioural evaluations. From a cross-sectional design, the sample included 95 children aged 7–11 years old from Chiconcoac-Tolago who were considered exposed; and 100 children from the same age range from the control area – communities from the Agua Blanca district located 80 km southeast of the manganesiferous basin. These communities were selected because they showed similar socio-economic conditions as the

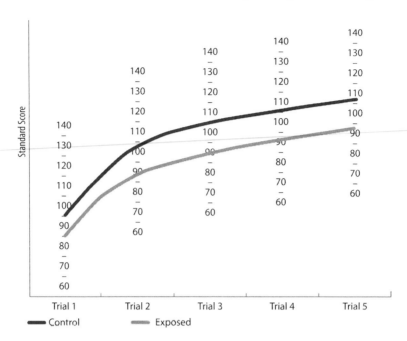

Fig. 8.1 California verbal learning test-two – children's version (Dean et al. 1994) standard scores expressed as means among children from exposed ($n=95$) and control groups ($n=100$)

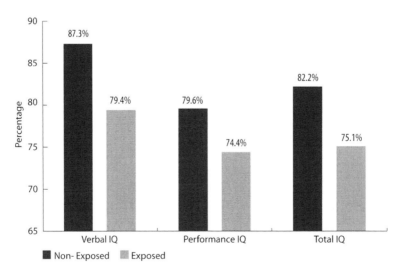

Fig. 8.2 Comparison of intelligence quotients between children from exposed ($n=95$) and control ($n=100$)

Table 8.2 Levels of manganese in blood (Mn-B) and hair (Mn-H) in exposed and control children

Manganese in biomarkers	Control group ($n=93$)	Exposed group ($n=79$)
Mn-H (µg/g), median (range)	0.6 (0.06–3.6)*	12.6 (4.2–48)*
Mn-B (µg/L), median (range)	8.0 (5.0–14.0)*	9.5 (5.5–18.0)*
Mn-B ≥10[a] (%)	24.7*	49.4*

*$p<0.001$
[a]MnB ≥10: percentage of children with Mn-B levels equal to, or above, 10 µg/L

exposed group, according to the marginalization index assigned by the Mexican National Council for Population (CONAPO 2005). These children were evaluated with a battery of neuropsychological tests. Motor, sensory, and cognitive functions were explored. Significant differences among children from the study and control areas were observed ($p <0.05$).

These findings are consistent with results from a previous study in the area and from other parts of the world (Hafeman et al. 2007; Menezes-Filho et al. 2009). Developmental delays in children proved to be a very important finding for stakeholders, and resulted in a veritable wake-up call for governmental authorities and mine owners who had previously been reluctant to deal with the communities' health complaints. Documenting the health effects of child Mn levels in the air was a significant achievement of this study. Recommendations have been made to use the assessment of children's developmental milestones in other similar conditions.

Table 8.2 shows blood (Mn-B) and hair (Mn-H) concentrations comparing the exposed and control groups. The median Mn-H concentration was 20-fold higher in the exposed children compared with the control group. The association between Mn-H and Intelligence Quotient (IQ) was negative.

Age and gender significantly modify the effect of Mn-H. A stratified analysis by gender showed that the negative effect was stronger in girls (Riojas-Rodríguez et al. 2010). It is possible that this effect was due to the fact that girls in these communities are apparently more marginalized than boys, according to a risk-perception study (Catalán-Vasquez et al. 2010). Other research on adults from this region found the median of Mn-B in women to be significantly higher than in men (9.20 vs. 8.00 µg/L). This difference could be related to iron levels in women, which are frequently lower than in men, and to the hypothesis of Mn–iron antagonism (Montes et al. 2008).

Manganese in the Ecosystem

To better characterize emission sources and pathways, the air concentrations of PM and manganese were measured in the mine's smokestack, near roads, and in other locations, including the indoor environment. The main findings of the emission

Table 8.3 Manganese concentrations in the air ($\mu g/m^3$)

Community	Mn in PM$_{2.5}$ (indoors)				Mn in PM$_{2.5}$ (outdoors)			
	n	Mean	Range	Median	n	Mean	Range	Median
Exposed	55	0.09	0.01–0.03	0.07	78	0.12	0.01–1.5	0.07
Control	8	0.02	0.004–0.03	0.02	25	0.03	0.001–0.08	0.02

sources and pathways study are shown in Table 8.3. Resuspended dust from vehicular traffic and unpaved road erosion were significant sources of atmospheric Mn (Jazcilevich, personal communication 2008). The exposed community had higher average indoor and outdoor PM$_{2.5}$ levels than the control community (51.6 vs. 34.2 $\mu g/m^3$ indoors; 43.6 vs. 17.9 $\mu g/m^3$ outdoors, respectively). Also, the exposed community had higher indoor and outdoor mean Mn levels in PM$_{2.5}$ compared with the control community (0.09 vs. 0.02 $\mu g/m^3$ indoors; 0.12 vs. 0.03 $\mu g/m^3$ outdoors, respectively). These results indicate that there is a significant difference in both PM$_{2.5}$ and Mn concentrations between exposed and control communities (Cortez-Lugo, personal communication 2008). In addition, the fact that 88% of Mn_PM$_{10}$ were Mn_PM$_{2.5}$ indicates that almost all Mn particles were coming from combustion sources, very likely from the chimney of the processing plant.

Based on the highest air Mn levels found in the two sampling periods of the project (5.86 $\mu g/m^3$ in 2005 and 1.5 $\mu g/m^3$ in 2009), safety guidelines for new allowable limits of Mn levels in air were proposed by the research team. If adopted, over 5 years, these safety guidelines would facilitate meeting the target maximum permissible level of 0.05 $\mu g/m^3$ as recommended by US-EPA (ASTDR 2000). Both the project findings and data from other studies support the recommendation that this concentration would be sufficient to protect the vulnerable population. At writing, the Ministry of Health of the State of Hidalgo was reviewing the content of these safety guidelines.

Adverse health effects observed among children and adults from the initial study (Santos-Burgoa et al. 2001, Rodríguez Agudelo et al. 2006, Solis-Vivanco et al. 2009, Riojas-Rodríguez et al. 2010) clearly indicate the need to lower Mn exposure from all sources and routes. The project recommended the following refinements to the risk-management plan presented in Table 8.1:

- Control and reduce the emission of sulfur dioxide and particles at the nodule plant.
- Implement health-monitoring activities among exposed communities in the mining area.
- Pave roads, especially those that have been covered with Mn-related materials, and enhance green areas.
- Legislate and enforce safe Mn transportation and waste disposal.
- Set up health campaigns on Mn toxicity and other mining-related issues.
- Implement the risk-management plan as a collective duty.
- Strengthen institutional actions (at local, state, and federal levels) to confront the negative effects of the global economy.

- Strengthen the roles of the Environmental Management Roundtable of the Mining District of Molango (MIGA).
- Legislate new safety limits for Mn exposure.

Multistakeholder Governance

For the purpose of this project, governance was defined as the interaction between government and society to construct public policies in democratic conditions (Aguilar 2005; Kooiman et al. 2008; Natera 2005). Understanding the socio-political environment of Molango was critical to achieve meaningful improvement in the lives of the affected communities.

The relationship between Autlán and the communities was paternalistic (i.e., authority based on relations among the parties, or client-based). This situation was likely fostered by both parties, even when the communities had ongoing complaints about Autlán. Clearly, Autlán encouraged the paternalistic approach to facilitate its work and avoid contested claims and public demonstrations (Paz 2008b). For example, the community of Naopa blocked an open trench mine site for years, pressuring Autlán to pave the road servicing the site. The relationship was also characterized by a high degree of mutual distrust – this is typical of a mutual dependence relationship and can be highly detrimental to the community (Paz 2008b).

The ties between municipalities and Autlán manifested themselves in different ways. No documentation of formal negotiations between Autlán and municipal councils could be found. In some municipalities, the relation was limited to the company paying property taxes, undoubtedly very important for municipal coffers. In other cases, there were closer links. At its most blatant, an officer from Autlán held the position of municipal clerk in Molango. However, communities felt they could appeal to their municipal authority for assistance in their disputes with the mining company, although mainly as individuals operating in a paternalistic system. Municipal audiences had been used to request individual assistance, especially financial. Ultimately, mayors tended to protect Autlán, and emphasized its economic importance especially as an employment generator in the area. The project recommended that civil society be represented in the municipal government and in the MIGA (Environmental Management Roundtable).

The relationship between Autlán and the state government was found to be unclear, especially in the area of enforcing government regulations. This lack of transparency, and the perception that the government was protecting Autlán's interests, likely created distrust and undermined the government's protective efforts.

MIGA was set up by the state government in 1995, with the participation of state institutions, some federal institutions, and more recently, municipal governments and some communities. It provided a platform for discussion of common environmental problems in the Molango region. Decisions, follow-up actions, monitoring, and the communication of research results have been part of MIGA's mandate. Initially a coordination agency, by 2005 MIGA had become a multistakeholder agency whose decisions were binding for MIGA participants. The project provided ongoing results to MIGA, while the project was helped by ongoing MIGA meetings where proposals for exposure management were discussed.

Although a useful multistakeholder space, MIGA also faced problems related to uneven power distribution. For example, in practice, most decisions were still made by state government agents who were strongly influenced by political agendas. The findings of this study were shared with MIGA, including the recommendation that MIGA should be formalized as an institution with clear definitions of the rights and responsibilities of its members. Together with transparent internal accountability, this would free MIGA from the vagaries of politics and would make it a truly useful venue for arbitration and action for environmental protection.

Both barriers and opportunities were found during the processes of building collective agreements (Paz 2008b). Difficulties involved in obtaining long-term agreements among the actors for the benefit of the population could be partially explained by national history. The manganese issue could not be separated from the centralized and hierarchical structures built in the last 50 years by social, economic, and political processes in Mexico. These aspects should be considered in the future when new interventions are planned or coalitions are built to improve the lives of people in communities in the region.

Advantages of Using an Ecohealth Research Framework

The ecohealth framework suited the complexity of the situation. The multidisciplinary team included people with expertise in environmental, health, and social sciences. For example, toxicological knowledge was generated mostly on Mn exposure pathways, levels of exposure, associated adverse health effects, and the fate of Mn in the environment. A social lens was used to characterize the relationships among involved parties – communities, the mining company Autlán, and state and municipal governments. Historic conflicts among stakeholders were analyzed to better pinpoint the difficulties in decision making.

Continued interactions among researchers provided significant contributions for consistency and rigor in methods, data analysis, and results dissemination. Knowledge gaps and priorities were refined after multistakeholder consultations in a multidisciplinary team environment, and led to a common sampling framework that was useful for the whole team. Intermediate and final research results were discussed jointly by the team, as were priorities and recommendations.

The ecosystem approach was also useful to ensure that mechanisms for consultation and results dissemination were in place throughout the project. The relatively long period of project implementation (since 2000) allowed the research team to disseminate research results not only in different academic settings, such as peer-reviewed journals and conferences, but especially in the meetings with communities, governmental institutions, and Autlán. The feedback received was used to further refine project implementation.

Challenges

Looking at complex links between an ecosystem and human health, in a changing economic setting, can be challenging. In the case of Molango, adverse health effects were hard to document, especially because their severity can be influenced by a range of risk factors (exposure pathways and the severity and frequency of often subtle health effects) and because extensive collection of field data was required.

Working for 10 years in Molango, both the team and the relationships have evolved. A deepening understanding of the situation helped to strengthen the voice of these relatively underserved rural communities in their negotiation with different levels of government and with a well-established, globally significant mining company. A stronger involvement of social sciences for both knowledge refinement and participatory methods would provide even more integrated results. As well, better documentation of the effects of Mn on local ecosystems, for example, on aquatic biodiversity, would have added value to the strongly anthropocentric perspective with which the ecohealth framework was interpreted in this project.

Even when recommendations were provided based on the best knowledge available, and were painstakingly collected over a long time, willingness to act was dependent on the political conditions, which were driven by electoral interests. Competing interests between private and public priorities have prevented legislative decisions from being made and health protection and promotion programs from being developed. Similarly, no significant measures have been taken to move forward with environmental health monitoring coordinated between the health and environment ministries, although this has been recommended consistently since 2005 by MIGA.

Conclusion

This project emerged as a result of public pressure. Therefore, an emphasis was placed on how the different viewpoints of stakeholders could be considered when the primary objective was to improve the lives of communities without further destroying the ecosystem they depended on for their livelihoods (farming and mining). A large group of researchers, who brought together different disciplinary expertise, worked together with the communities, different levels of government, and the company for almost a decade. Significant new knowledge on the health impacts of Mn pollution was generated and interpreted in a way that could be meaningful for policy makers. This knowledge guided the various risk-management plans that were presented by the project. A range of policy and intervention recommendations were put forward – some of these have been implemented, such as the paving of some roads. However, other aspects were not accomplished, including emission reductions, waste management, and epidemiological surveillance.

In addition to advancing general knowledge on the health impact of manganese exposure, this project has allowed the research team to experiment with an ecosystem approach. This approach was useful because it targeted integrated knowledge, participatory methods, and equity issues. At the same time, it was critical for teams to conceptualize the research problem in a way that balanced ecosystem health, human health, and development.

This applied research project generated considerable new knowledge, but also demonstrated that having evidence does not necessarily mean solving a problem. However, the project's long-standing commitment to the communities has allowed the team to find signs of gradual change in the attitude of the mining company. Examples include the increased involvement of managerial staff (when in the past only local personnel had been involved in MIGA) and their participation in the new phase of the risk-management plan now under development (2010–2013). This plan includes further paving of the nodulation plant's access roads and other roads in the communities, and new strategies for dust and emission control. The team hopes that this gradual deepening of the relationship and increased interest bode well for eventual implementation of the project's recommendations.

Acknowledgments We thank the members of our multidisciplinary research group for their important contribution to this study: Yaneth Rodríguez A., Fernanda Paz, Marlene Cortez Lugo, Christine Siebe, Minerva Catalán, Irma Rosas, Rodolfo Solis V., Sergio Montes, Camilo Ríos, David Hernández Bonilla, Alejandra Mondragón, Astrid Shilmann H, Eva Sabido Pedraza, Aron Jazcilevich, Jaqueline Martínez and Rafael Santibañez. We thank the members of the community for their participation. IDRC support was provided through the projects 100552, 100662, 102379, and 103052.

References

Aguilar, L.F. (2005). América Latina: Sociedad Civil, Democracia y Gobernanza. El Futuro de las Organizaciones de la Sociedad Civil: Incidencia e Interés Público. Memorias del Coloquio Internacional, México, DECA, Equipo Pueblo, A.C., pp. 64–74.

ASTDR (Agency for Toxic Substances and Disease Registry). (2008). Draft for Public Comment (Update). Toxicological Profile for Manganese. Agency for Toxic Substances and Disease Registry, Atlanta.

ASTDR (Agency for Toxic Substances and Disease Registry). (2000). Toxicological Profile for Manganese. US Department of Health and Human Services, Agency for Toxic Substances and Disease Registry, Atlanta.

Catalán-Vasquez, M., Shilmann A., Riojas-Rodriguez, H. (2010). Perceived Health Risk of Manganese in the Molango Mining District, Mexico. Risk Analysis, 30(4), 619–634.

CONAPO (Consejo Nacional de Población). (2005). Índices de marginación. Consejo Nacional de Población, México. Retrieved from: http://www.conapo.gob.mx/index.php?option=com_content&view=article&id=126&Itemid=392

Dean, C.D., Kramer, J.H., Kaplan, E., and Ober, B.A. (1994). California Verbal Learning Test®–Children's Version (CVLT®–C). Retrieved from: http://www.pearsonassessments.com/HAIWEB/Cultures/en-us/Productdetail.htm?Pid=015-8033-957.

Hafeman, D., Factor-Litvak, P., Cheng, Z., van Geen, A., and Ahsan, H. (2007). Association between manganese through drinking water and infant mortality in Bangladesh. Environmental Health Perspectives, 115(7), 1107–1112.

Kooiman, J., Banvinck, M., Chuenpagdee, R., Mahon, R., and Pullin, R. (2008). Interactive Governance and Governability: An Introduction. Journal of Transdisciplinary Environmental Studies, 7(1), 1–12.

Menezes-Filho, J.A., Bouchard, M., Sarcinelli, P.N., and Moreira, J.C. (2009). Manganese Exposure and Neuro-Psychological Effect on Children and Adolescents: A Review. Pan American Journal of Public Health, 26(6), 541–548.

Montes, S., Riojas-Rodríguez, H., Sabido Pedraza, E., and Rios, C. (2008). Biomarkers of Manganese Exposure of a Population Living Close to a Mine and Mineral Processing Plant in Mexico. Environmental Research, 106, 89–95.

Natera, A. (2005). Nuevas estructuras y redes de gobernanza. Revista Mexicana de Sociología, 4, 755–791.

Paz, M. F. (2008a). Del Caciquismo a la Gobernanza. Desafíos en la Construcción de Acuerdos en un Distrito Minero en México. In: Bustamante, T. and Weiss, J. (Eds.). Ajedrez Ambiental. Manejo de Recursos Naturales, Comunidades, Conflictos y Cooperación. FLACSO (Facultad Latinoamericana de Ciencias Sociales), Ecuador.

Paz, M.F. (2008b) Tensiones de la gobernanza en el México rural. Política y Cultura, 30, 193–208.

Ramírez, R. (1999). Stakeholder Analysis and Conflict Management. In: Buckles, D. (Ed.). Conflict and Collaboration in Natural Resource Management. IDRC, Ottawa, and the World Bank, Washington, pp. 101–126.

Riojas-Rodríguez, H., Solís-Vivanco, R., Schilmann, A., Montes, S., Rodríguez, S., Ríos, C., Rodríguez-Agudelo, Y. (2010). Intellectual Function in Mexican Children Living in a Mining Area and Environmentally Exposed to Managanese. Environmental Health Perspectives, 118(10), 1465–1470.

Rodriguez-Agudelo, Y., Riojas-Rodriguez, H., Rios, C., Rosas, I., Sabido Pedraza, E., Miranda, J., Siebe, C., Texcalac, J.L., and Santos-Burgoa, C. (2006). Motor Alterations Associated with Exposure to Manganese in the Environment in Mexico. Science of the Total Environment, 368(2–3), 542–556.

Santos-Burgoa, C., Rios, C., Mercado, L.A., Arechiga-Serrano, R., Cano-Valle, F., Alatorre Eder-Wynter, R., Texcalac-Sangrador, J.L., Villa-Barragan, J.P., Rodriguez-Agudelo, Y., Montes, S. (2001). Exposure to manganese: health effects on the general population, a pilot study in Central Mexico. Environmental Research, 85(2), 90–104.

Solis-Vivanco, R., Rodriguez-Agudelo, Y., Riojas-Rodriguez, H., Rios, C., Rosas, I., and Montes, S. (2009). Cognitive Impairment in an Adult Mexican Population Non-Occupationally Exposed to Manganese. Environmental Toxicology and Pharmacology, 28(2), 172–178.

Chapter 9
Ecohealth Research for Mitigating Health Risks of Stone Crushing and Quarrying, India

Raghwesh Ranjan, K. Vijaya Lakshmi, and Kalpana Balakrishnan

Bundelkhand comprises six districts of Madhya Pradesh and seven districts of Uttar Pradesh. It is home to the famous Khajuraho temple, the fort of the Queen of Jhansi, and many other glorious monuments. It is also one of the poorest regions of India. Over the years, rising population has led to the fragmentation of family land holdings. In addition, the growth of private land ownership and past mismanagement of natural resources has caused a rapid decline in forest cover and reduced traditional sources of fuel, fodder, and food. These factors, combined with limited rainfall and water resources, have resulted in low agricultural productivity.

Increasing water scarcity is a particular concern: government records show there was one drought every 16 years during the eighteenth and nineteenth centuries. From 1968 to 1992, however, the region saw a drought every 5 years, while 7 of the first 10 years of the twenty-first century were besieged by drought (Inter-Ministerial Central Team 2008; Khurana and Mohapatra 2008). Many families are no longer able to meet their subsistence needs. In such abject poverty, women and children are particularly vulnerable. Limited scope for agriculture, coupled with stagnant industrial growth, has led to temporary or long-term migration of family members, predominantly males from rural areas, in search of alternative sources of livelihood.

In addition to this "distress migration," Bundelkhand communities have resorted to sometimes hazardous occupations. Bundelkhand is rich in granite reserves: the stone quarrying and crushing (SQC) sector is the second largest source of employment after agriculture in the region. SQC operations are booming due to the surge in infrastructure development in India. This sector employs largely rural, migrant, and unskilled workers, both men and women, and offers both primary employment and opportunities for supplementary income for farmers.

R. Ranjan (✉) • K.V. Lakshmi
Development Alternatives, New Delhi, India
e-mail: rranjan@devalt.org

K. Balakrishnan
Department of Environmental Health Engineering, Sri Ramachandra University, Chennai, India

Still, poverty coupled with narrow livelihood options in Bundelkhand has eroded the resilience of communities involved in SQC operations. This research grew out of previous work in vulnerable communities of the region by Development Alternatives (DA), which emphasized the integration of human and ecosystem health considerations into development processes in the region.

SQC units are located in many states of India. DA identified a cluster located within the Tikamgarh district of Bundelkhand for the piloting of interventions to mitigate health risks, the core of this project. In Tikamgarh, more than 80% of the 12 million people live in rural areas, under poor conditions, as illustrated by the 93 infant deaths per 1,000 births (Government of Madhya Pradesh 2006). Reliable statistics on SQC units, capital, production, and number of employees are lacking due to the sector's informal nature. Field investigation revealed that most of the stone-crushing units in Tikamgarh were labor-intensive and had been set up with capital investments of INR50–100 Lakh (USD120,000–240,000), which is low by international industry standards. Unofficially, the district has more than 250 SQC units, which are small- and medium-scale and provide direct employment to about 7,500 people and indirect employment to more than 25,000 people.

Through numerous interactions with diverse stakeholders, a range of problems were identified:

- SQC involves a variety of processes that potentially expose the workforce to a wide range of physical, chemical, and ergonomic hazards. Tikamgarh lacks a formal recording system for occupational and environmental health and safety for this sector.
- The SQC sector attracts a large number of people seeking work, because of a lack of alternate livelihood opportunities and unstable farm incomes.
- The SQC sector is not organized; workers seem not to have social security benefits enjoyed by workers in other sectors in India.
- The majority of SQC unit owners are locally politically influential and powerful. This appears to inhibit workers from actively seeking improvements from quarry management.
- The dusty environments of the SQC units pose health risks for workers and local communities.

Ecohealth Multidisciplinary Research

The research issues surrounding worker and community health in the SQC industry of the Tikamgarh region were diverse and called for a multidisciplinary team and collective action. DA and Sri Ramachandra University (SRU) joined efforts to build on their combined track records in occupational health, safety and risk assessments, and intervention design, and to capitalize on their well-established technical support in the field.

An ecosystem approach to health was used to design and implement participatory processes that would lead to action by diverse stakeholders to improve the health and social conditions of the communities dependent on SQC units. The study site included 13 stone-crushing units in the Tikamgarh district (all within approximately 1 km^2) and two villages. The study design was a qualitative one, supported by air-quality measures and baseline health assessments. The approach was based on an initial baseline assessment and subsequent design and implementation of interventions. The communities that depend on these units for their livelihoods were part of the community-level assessments and participated actively throughout the project.

The research sought to identify, through participatory engagement of study communities and other stakeholders, the major environmental, social, economic, cultural, and political risks to health that were posed by the SQC industry. Focus group discussions were used for community-related information. Open-ended interviews were tailored to each stakeholder or group. The environmental quality of air, water, and soil were assessed against standards recommended by the American Public Health Association (APHA) and the Bureau of Indian Standards. These assessments were used to establish an environmental baseline, population exposures, and their consequent health impacts. Baseline health status was assessed by clinicians in a cross-sectional sample of workers and community residents. The analyses aimed to establish potential contributions to health status from living conditions and from the social, gender, and occupational environment. The study also sought to test the implementation of participatory intervention strategies to reduce environmental and health risks, and support long-term participatory interaction among stakeholders.

The project faced multiple constraints in the beginning. DA had, over the years, gained the trust and confidence of communities over more than two decades of work in the region. However, the SQC unit owners were initially reluctant to participate in the research because of usually tense relations with regulatory authorities. They were concerned about facing possible reprisal if they were found to be somehow transgressing pollution-control regulations. The research team established and maintained close links with the local Pollution Control Board (PCB) officials to facilitate dialogue with the SQC owners, for a constructive process of improving understanding of regulations, improvement of conditions for workers, and ultimately, adherence to regulations. Confidentiality of the information gathered during the research was maintained. A small advisory group, including representatives from DA, the unit owners' association, and a PCB official, was constituted to take stock of progress, information collected, and reports generated. This mechanism was effective because, over time, SQC gradually increased their participation and eventually made investments in piloting interventions to improve air quality. The unit owners also facilitated access to their units and workers for a series of environmental and health assessments that were an important part of the research.

The project used a combination of conventional environmental monitoring and epidemiological methods that were integrated with well-established social science research methods (such as mapping, metrics, and focus group discussions) to identify opportunities for interventions. Project objectives were further specified with active engagement of stakeholders. Throughout the project, participatory research processes were undertaken to build strong rapport with the communities. Assessments were conducted on several topics – social maps, historical changes, health-seeking behavior, perceptions of well-being, and access to, and control over, infrastructure and facilities.

Results of Participatory Assessment

The health assessment included 269 men and 203 women from both villages. Because the populations of the villages were relatively small (about 914 people), sampling was based on individual volunteer responses with informed consent to open invitations for free health examinations. Thus, the findings reflect the conditions of this sample, and are indicative of problems in the villages but cannot be extrapolated to the level of the community. Both villages had fragmented societies that were strongly divided by socio-economic status, which resulted in apparently low levels of group collaborative action to mitigate common concerns. This was more pronounced in the case of women, who were not as well represented as men in our sample. Both villages lacked social institutions or organizations.

Both village communities perceived a significant decline in forest cover and rains in the area over the last three decades (about 90% referred to degradation in forest cover and 80% referred to deterioration in annual rainfall over the last 30 years). Water scarcity is a major problem in this region, which depends on rain-fed agriculture. Decreased rainfall would be associated with declines in agriculture and livestock productivity. There were mixed opinions about the costs and benefits of SQC units among both village communities. Some felt that the SQC units had destroyed the ecological strength of the area, but others acknowledged that they provided reliable incomes and infrastructure improvements in their villages.

Health-related perception studies indicated that people depended on home remedies as the first level of health-service access for nearly all categories of general ailments, with the formal health-care system being accessed only for critical conditions. Trust in government health services was poor among all sectors of society represented in the sample. Children were accorded the highest priority for health-related expenditures, followed by men. Women's health merited the least amount of attention. Respondents reported that institutions such as panchayat (local government units at the town level) were not adequately delivering expected services. Respondents appeared often to be poorly informed about how to access public services. Lack of sanitation, poor mosquito control, smoking, and no access to vaccinations were frequently reported by respondents, or observed by the study team.

Measurement of Environmental Health Risk

Noise measurement followed protocols established by the American National Standards Institute (ANSI 1991, 1996). Noise intensities were >90 dB (maximum tolerable level as per Indian Factories Act is 90 dB) in the vicinity of the primary

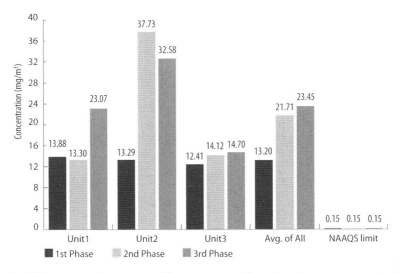

Fig. 9.1 PM_{10} concentrations in three different stone-crushing units without dust control system

crushing unit, secondary crushing units, and vibrators. Occasionally, high noise levels were recorded in other peripheral areas due to the noise coming from adjacent crushing units and the frequent running of trucks within the units.

Measurements of respirable dust in the workplace were made for 10 and 4 μm particulate matter (PM_{10} and PM_4) to allow comparison with relevant health standards. Respirable dust is composed of particles that penetrate to the unciliated airways of the lung (the alveolar region), and is generally represented by a cumulative log-normal curve having a median aerodynamic diameter of 4 μm, standard deviation 2 μm in humans (Nordberg et al. 2004). The 24 h concentrations of PM_{10} were measured using the protocol of the National Institute of Occupational and Safety Health (NIOSH 1998; Sivacoumar et al. 2006). Results exceeded national guidelines by nearly 100-fold (Fig. 9.1). The area concentrations of PM_4 also exceeded the guidelines in many locations across units (Fig. 9.2).

Because emissions from stone crushers are important not only for workers within the units but for community residents in adjoining areas, PM_{10} levels were monitored through high-volume air sampling in both villages. The levels of respirable dust averaged up to 0.8 mg/m³ in village A and up to 0.5 mg/m³ in village B, as compared to National Ambient Air Quality Guidelines maximum level of 0.15 mg/m³ (NAAQS 2009) and WHO guidelines of 0.02 mg/m³ (WHO 2005). Silica levels were measured in a subset of samples. The subset selection criteria included observed proportions of silica in quarry stones from different depths and sites during different seasons, the observed proportion of silica in indoor air in the communities, and the estimated silica exposure of workers. However, levels of silica did not prove to be a major concern within the community.

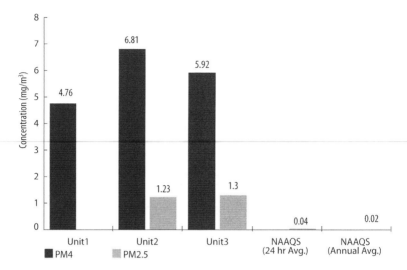

Fig. 9.2 Average day shift PM_4 and $PM_{2.5}$ particulate concentrations in three unit sites during first three-phase study

Indicators of Health Status

Health indicators are presented in Table 9.1. Almost half of the people sampled had a low Body Mass Index. Distribution of hypertension among non-smoking men was significantly higher among SQC workers than among workers in other sectors.

Assessments of pulmonary function revealed that the distribution of respiratory impairment in women was significantly higher than in men despite the fact that a much higher proportion of men were smokers and were employed in SQCs. This suggests that occupational exposure to dust is not the primary source of lung impairment in this group. Other factors, such as household air pollution from use of biomass fuels, may be important contributing factors. Most households (close to 90%) rely on biomass fuels.

Taken together, the findings of these assessments describe poor rural communities in Tikamgarh facing challenges with their livelihoods and health conditions. The findings of the perception survey indicated that due to environmental degradation, farming is not sufficient as a sole source of income or livelihood, placing workers in a dependent condition with their SQC employers. The communities face a number of health challenges, including some occupational health problems. Unacceptably high levels of noise and dust in the workplace were measured. The study sample group was generally under-weight, pointing to broader nutritional and food security issues. Hypertension and impaired respiratory function were observed. Most severe respiratory impairment was noted in women, who had little or no exposure to SQC workplace or to smoking.

Table 9.1 Percentage distribution of health indicators assessed through clinical examination and survey (n = number of individuals)

Body Mass Index = weight/height² $(n=468)$ Reference values: <18 Underweight; 18–25 Normal; >25 Over weight; and >30 Obese							
Normal				57	Underweight		43
Anemia Reference values: 10–12 g%[a] Normal; <10 g% Anemia							
Male ($n=267$) Mean Hb% 13.2 g% Anemia 15				Female ($n=191$) Mean Hb% 11.5 g% Anemia 12			
Hypertension Reference values >130/80 mmHg							
Male						Female	
Smokers ($n=98$)			Non-smokers ($n=171$)			Smokers ($n=6$) and Non-smokers ($n=197$) 4	
SQC ($n=73$) 5	Farming ($n=110$) 17	Others[b] ($n=24$) 6	SQC ($n=76$) 13	SQC + Farming ($n=41$) 4	Others[b] ($n=95$) 2	Homemaker ($n=76$) 17	Student ($n=3$) 3
Abnormal Pulmonary Function Tests (PFT) across occupations[c]							
Male						Female	
Smoker			Non-smokers				
SQC ($n=112$) 15		Others[b] ($n=14$) 14	SQC ($n=70$) 10		Others[b] ($n=85$) 3		
Abnormal PFT in Men across occupations							
SQC ($n=67$) 13		Others[b] ($n=14$) 14					
Abnormal PFT by Sex across occupations							
SQC			Others[b]				
Male ($n=136$) 11		Female ($n=16$) 31	Male ($n=98$) 5		Female ($n=135$) 22		
Abnormal PFT in Women across fuel and kitchen type							
Separate Kitchen ($n=20$) 25		Biomass ($n=131$) 25	No Separate Kitchen ($n=111$) 31		LPG ($n=20$) 5		

[a] g% refers to the percentage of hemoglobin in grams
[b] Others include homemakers (where biomass exposure is likely) and occupations such as farming where dust exposures are likely as well
[c] Pulmonary function was assessed by spirometry with forced vital capacity (FVC), forced expiratory volume in 1 s (FEV1) and FEV1/FVC. Reference values were taken from normograms available for Indian populations. More than 80% of the predicted value was considered normal (NIOSH 1997) and <80% predicted value as abnormal

Developing Interventions

Based on the results of health and environmental assessments, several interventions were developed and piloted with participation from different stakeholders. The project built capacity among different actors in the health-care delivery system to improve recognition and treatment of various environmental and occupational health problems. During initial project activities in 2005, a gap between community health needs and health services was noted, especially regarding respiratory diseases. The project trained physicians in primary health centers and in the local medical college responsible for local communities, on how to address local environmental and occupational hazards, especially those related to respiratory health. Community-level health-care extension workers and paramedical staff were also trained and sensitized to respond to the needs of the communities.

To facilitate efforts for hazard recognition and subsequent facility-level assessment by SQC unit management, a safety manual was prepared in English (Wagner et al. 2009), with plans to translate into local languages. Accompanied by pictures of unsafe working conditions, the manual provides a guide on how to recognize and manage hazards.

The air-quality assessments found that exposure to dust was a potential risk for workers. The project, in collaboration with SQC management, designed, developed, and installed cost-effective and environmentally friendly dust abatement systems at selected SQC units. Unit owners and PCB representatives presented four requirements to the project team for developing a suitable dust-abatement system:

- It should not interfere with current operating processes.
- It should be energy efficient and operate using available surplus power.
- It should not consume water (because the region has limited water resources).
- It should preferably not exceed an investment of INR400,000 (USD8,000).

The project team developed a system that was installed and tested at two participating SQC units, which each invested about INR175,000 (USD3,500). A 30–40% reduction in respirable dust was achieved, but assessment of long-term efficacy has not yet been conducted.

Several priority needs identified by community members through the participatory assessment were addressed by the project. A youth club, an adolescent girls club, a farmers' group, and kitchen-garden groups were formed in the village. These groups also provided community-based oversight for some other interventions. For example, long-term water scarcity in the region led the community to call for a geo-hydrological study to be included in the intervention plan. This study identified a reliable perennial water source. A handpump was installed in one of the villages, and the community invested about 30% of the cost of installation. Several women were trained in kitchen gardening and 12 such gardens were planted during the project. Several educational activities in agriculture, food and nutrition, organic agriculture, and suitable cropping systems for semi-aid zones were conducted.

Project Learning

The project drew on ecosystem approaches to health (Lebel 2003) to take this study beyond a typical occupational-health assessment, and to consider social relations among the community, the employer, and the government. It fostered the development and piloting of interventions that will help reduce the magnitude of environmental and occupational-health risks that face the SQC sector and its supporting communities. It also built awareness and ownership of several of these improvements and activities, in a bid for their longer-term sustainability and impact, after the project team has gone.

Several changes were noted during the research. The project's interaction platform for discussing the environment and its links to health was a first for these communities. This provided a unique setting wherein the assessments and recommendations were weighted by community needs and not by the investigators' intent. Women and youth, in particular, were mobilized for environmental management to promote health and well-being.

Important changes were noted among stone-quarry owners. Previously, this group had been resistant to any form of communication or consultative engagement about reforms in the sector. The project managed to bring them to the discussion table. Although dust abatement remained their main concern, they were sensitized to other workplace hazards, notably to safety issues.

A participatory multistakeholder research process allows researchers to contribute with their particular expertise, but demands "out-of-the box" thinking to integrate several disciplines and stakeholders. This provokes innovation that might not otherwise be reached through single-discipline research. In this study, this innovation led to the successful engagement of SQC unit owners and the workers to work together to reduce a substantial health problem in the industry. The project was also successful in its ability to generate multistakeholder action, accomplish what was intended, and garner ownership and participation by an array of actors, while building in flexibility to adapt over time.

Sustaining the Interventions and Collaborative Action

The project represents a first effort by these investigators to develop a comprehensive understanding of a complex social setting in which management of environmental risks involves the interplay of social, economic, and political forces. A single short-term project faces many challenges in this context, trying to assist the concerned communities. And yet, several small but effective interventions were initiated as a part of this research. With knowledge and the learning attained, the team anticipates that further action in the development of these communities will include participatory and multistakeholder processes. DA and SRU are already applying lessons from this project to other development challenges elsewhere in India.

Acknowledgments We acknowledge the support extended by the communities of the two villages and the three crushing units that participated in the study, the officials of local Pollution Control Board, the physicians and staff of the Primary Health Centre and colleague researchers. We would also like to sincerely thank Norbert L. Wagner, University of South Florida for his continued guidance and encouragement during the implementation of the research. IDRC support was provided through project 103055.

References

ANSI (American National Standards Institute). (1991). Maximum Permissible Ambient Noise Levels for Audiometric Test Rooms. S3.1-1991. American National Standards Institute, New York, NY, USA.

ANSI (American National Standards Institute). (1996). Specifications for Audiometers. S3.6-1996. American National Standards Institute, New York, NY, USA.

Government of Madhya Pradesh. (2006). National Family Health Survey (2005–2006). Key Indicators for Madhya Pradesh. Available at: http://www.nfhsindia.org/factsheet.shtml, and Government of Madhya Pradesh Website. Available at: http://www.mp.gov.in/.

Inter-Ministerial Central Team. (2008). Report on Drought Mitigation Strategy for Bundelkhand Region of Uttar Pradesh and Madhya Pradesh, Government of India.

Khurana, I., and Mohapatra, R. (2008). Water and Sanitation Perspective 01. Half Full, Half Empty. WaterAid India, New Delhi, India. Available at: http://www.wateraid.org/documents/perspectivebundelkhand.pdf

Lebel, J. (2003). Health: An Ecosystem Approach. In Focus Series. International Development Research Centre (IDRC), Ottawa, Canada. Available at: http://www.idrc.ca/in_focus_health/.

NAAQS (National Ambient Air Quality Standards). (2009). India Environment Portal, Centre for Science and Environment. Available at: http://www.indiaenvironmentportal.org.in/reports-documents/revised-national-ambient-air-quality-standards-naaqs-2009.

NIOSH (National Institute of Occupational Safety and Health). (1998). Manual of Analytical Methods (NMAM), Fourth Edition. Particulates Not Otherwise Regulated: Method 0600, Issue 3, 15 January 1998, page 2 of 6. National Institute of Occupational Safety and Health, Atlanta, GA, USA.

NIOSH (National Institute of Occupational Safety and Health). (1997). Spirometry Guide for Occupational Settings, ATS, January 1997. National Institute of Occupational Safety and Health, Atlanta, GA, USA.

Nordberg, M., Duffus, J.H., and Templeton, D.M. (2004). Glossary of Terms used in Toxicokinetics. Pure and Applied Chemistry, 76, 1033–1082.

Sivacoumar, R., Jayabalou, R., Swarnalatha, S., and Balakrishnan, K. (2006). Particulate Matter from Stone Crushing Industry: Size Distribution and Health Effects. Journal of Environmental Engineering, 132(3), 405–414.

Wagner, N., Nithiyananthan, M., and Farina, L. (Editors). (2009). Safety and Health in the Stone Crushing Industry: A Practical Manual for Preventing Accidents, Preserving Health and Keeping a Company Profitable. Occupational Knowledge International, San Francisco, CA, USA. Available at: http://www.okinternational.org/docs/Safety%20and%20Health%20in%20Stone%20Crushing%202008.pdf.

WHO (World Health Organization). (2005). Air Quality Guidelines. Available at: http://www.who.int/phe/health_topics/outdoorair_aqg/en/.

Chapter 10
A Virtuous Cycle in the Amazon: Reducing Mercury Exposure from Fish Consumption Requires Sustainable Agriculture

Jean Remy Davée Guimarães and Donna Mergler

The Caruso Project[1] started in 1994 as a new proposal, but built on a pre-existing collaboration agreement between Université du Québec à Montréal (UQAM) and the Federal University of Pará (UFPA) in Belem, Brazil. The objectives of the project were to characterize sources, transmission routes, and effects of mercury on communities in the lower Tapajós River region. The upper Tapajós had been heavily impacted by artisanal gold mining (Akagi et al. 1995; Malm and Guimarães 1999). The initial hypothesis of the Caruso Project was that there would be a downstream gradient of mercury exposure and effects from the gold-mining region, situated south of São Luis do Tapajos (Fig. 10.1).

The Brazilian gold rush period, which peaked during the 1980s, was fertile ground for research projects on mercury in the environment and its impacts on humans, especially in the Madeira and Tapajós River basins. These projects either focused on environment, or health, or other disciplines. Both high mercury emissions, of up to 100 t of metallic mercury per year in the more productive periods (Pfeiffer et al. 1993), and occupational exposure to mercury vapour in gold fields and gold shops (Malm et al. 1997) had been described. The first data on total mercury in sediments and soils showed high levels close to the sources of the emissions (Bastos et al. 1999), and mercury in fish and humans was high throughout the Tapajós region (Akagi et al. 1995; Malm et al. 1997). When the Caruso Project began, the causal link between mercury emissions from gold mining and the

---

[1] http://www.unites.uqam.ca/gmf/caruso/caruso_home.htm. Original proposal presented by Marc Lucotte, Marucia Amorim, and Donna Mergler.

J.R.D. Guimarães (✉)
Biophysics Institute, Federal University of Rio de Janeiro, Rio de Janeiro, Brazil
e-mail: jeanrdg@biof.ufrj.br

D. Mergler
Centre de recherche interdisciplinaire sur la biologie, la santé, la société et l'environnement (CINBIOSE), Université du Québec à Montréal, Montréal, QC, Canada

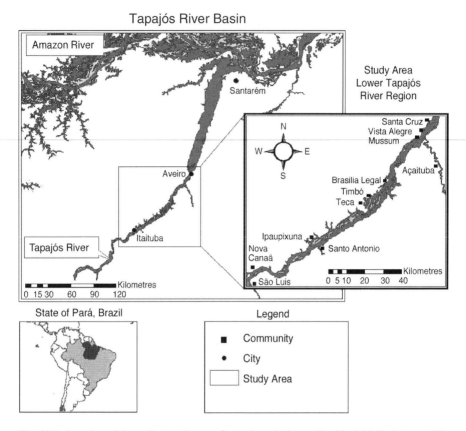

Fig. 10.1 Location of the project study area, focused on the lower Tapajós, initially between São Luis and Santarém then focused in the highlighted area. Gold-mining fields, upstream the study site, are concentrated on the Tapajós and its tributaries, south of São Luis

presence of mercury in the environment and in humans was generally assumed. However, there were many inconsistencies in the scientific findings, and numerous questions remained unanswered.

The first studies of the Caruso Project were carried out in and around the village of Brasília Legal, located at one of the few stops of the large river transport boats between the cities of Santarém and Itaituba. Researchers from UFPA and Fernando Branches, a medical doctor from Santarém, had been active in this community and were worried about the possible health effects of human exposure to mercury in this region. Around Brasília Legal, river-transport conditions were adequate at all seasons, and there was a gradient in land use, from pristine forest areas to farming plots of different ages. It was also a convenient location from which to study the 300 km river stretch from São Luis do Tapajós to Santarém to evaluate the impact, if any, of the main Tapajós gold-mining fields – all situated on tributaries upstream of São Luis do Tapajós. Furthermore, just across the river from Brasília Legal, there were several floodplain lakes, surrounded by small farming settlements, which provided an ideal location to study the relationships between land use and livelihoods, and their possible role in the health effects of mercury.

The Fruits of Researcher–Community Collaboration

At the beginning of the 1990s, approximately 500 people lived in Brasília Legal, and a large proportion of the community was involved in either commercial or subsistence fishing. Most houses had gardens and manioc was grown for the home production of "farinha" (manioc flour). There was a health post with a nursing aide and midwife, and practitioners of traditional medicine. A well-organized and fully staffed primary school served this and neighbouring communities. Unlike many other villages, Brasília Legal boasted commercial establishments, including a few small general stores, two bakeries, several bars, a gasoline pump for boats, and a solar-powered telephone. Electricity was available for a few hours in the evening from a community-run diesel generator. The "docks" were very lively with small boats from nearby villages bringing children to school, families coming to intercommunity soccer matches, and people seeking health care and shopping. Farming areas were located behind the riverfront village, including a "colonia" implemented by the national program for agrarian colonization, where newly arrived migrants from Brazil's northeast and villagers exchanged fish for rice and vegetables.

The Caruso team was built over the years progressively to include researchers from the social, health, and natural sciences from UQAM, UFPA, and the Universidade Federal do Rio de Janeiro (UFRJ). Scientists from the Universidade de São Paulo – Ribeirão Preto (USP–RP), the Universidade de Brasília (UnB), and the Université de Montréal (U de M) later joined the group. The several annual campaigns, initiated in 1994, included many Brazilian and Canadian graduate students, as well as the project leads and disciplinary experts. Working together helped to progressively build transdisciplinarity, which is a hallmark of implementing an ecosystem approach. Numerous undergraduate students from the outreach UFPA campus in Santarém participated actively in the fieldwork, and several went on to do graduate work in mercury-related topics.

Villagers in Brasília Legal were initially hesitant about participating in yet another study project. Other research groups from Brazil and elsewhere around the world had come by, taken hair and blood samples, and never explained their presence or returned with the results of their analyses. The Caruso team endeavoured to work differently, by offering to involve the villagers throughout the study. The project adopted research-action methodologies and community participation was intense. Planning campaigns and workshops were organized with women and men, fishers, farmers, teachers, health agents, and at times, local authorities to discuss the project's objectives and methods. Where possible, community members were involved in the actual fieldwork. These characteristics were essential to the success of the Caruso Project: when project staff returned to share the first results with the community, the barriers went down and, from that moment forward, the long-standing collaboration with the community was solidified and progressively grew into friendships. As the study advanced, the input from village meetings and discussions oriented many of the new research questions. The village midwife and some of the teachers and fishers played key roles throughout the research. The evolution of community participation in the project is described by Mertens et al. 2005.

The First Scientific Discoveries

During the first phase of the study, environmental mercury measurements were performed on a 300 km section of the river, while community-based research on health and social representation were concentrated in the villages of Brasília Legal and Cametá. The study found no detectable gradient in mercury concentration in the river when moving away from gold-mining river areas, neither in suspended particles, bottom sediments, soil, nor vegetation. Surprisingly, some small tributaries draining pristine forest areas had higher mercury levels than the Tapajós itself (Roulet et al. 1998a, b; 2001). These findings were inconsistent with a hypothesis of mining-derived mercury contamination.

Detailed soil studies showed that the soils of the region had high levels of mercury of natural origin. Emissions from gold mining, even when considering the worst-case pollution scenarios, could not explain more than 3% of the soil-mercury burden. The sedimentary records in floodplain lakes clearly showed mercury enrichment in the more recent layers, but sediment dating indicated that this enrichment was related to the onset of human settlement and exploitation of the basin in the 1950s, rather than to the gold-mining rush of the 1980s. Comparisons of mercury levels in forest soils and in crop soils indicated that deforestation for agricultural purposes was associated with depleted soil mercury levels, especially on slopes (Roulet et al. 1999; 2000a, b).

Mercury exposure of humans reflected the mercury concentrations in the environment. Here too, no gradient was observed between villages located up and down river, despite them being at various distances from gold-mining areas. Biomarkers of mercury exposure in humans were tightly linked to mercury in the environment and to fish consumption. Mercury levels in human hair were considerably higher than those reported elsewhere in the world (Passos and Mergler 2008). Sequential mercury analysis along human hair strands clearly showed a strong seasonal variation in mercury exposure. The hypothesis was that mercury exposure varied because of the importance of fish consumption year round and the seasonal variations in the availability of various fish species, particularly piscivorous versus non-piscivorous fish (Dolbec et al. 2001; Lebel et al. 1997). Indeed, the entire subsistence diet mirrored the changes in the dynamics of the local ecosystem associated with the 5 m rise and fall of the water level over a yearly cycle (Passos et al. 2001). Depending on the time of year, different fish species predominate.

Mercury in fish is found mainly as methylmercury, and investigations of mercury methylation sites and factors revealed high net methylmercury production and accumulation in floating meadows (areas of grassy vegetation growing in floating mats in the river) and seasonally flooded forest soil and litter. These environments are essential in the life cycle of most local fish species as spawning grounds, nurseries for young fish, and hunting grounds for mature fish. The input of fresh labile organic matter brought by the seasonal flood creates favourable conditions for fish development and for methylmercury production in these environments (Guimarães et al. 2000a, b; Roulet et al. 2000b; Miranda et al. 2004).

Due to the overwhelming diversity of fish species, mercury bioaccumulation patterns in fish were challenging to analyse, and the direct relation between fish size or age and its mercury content, commonly found in temperate fish species, was far from obvious. In fact, repeated fish sampling, at different phases of the flood cycle, revealed various bioaccumulation patterns: linear or ladder-like; increasing or decreasing; or remaining unchanged with fish size. Analysis of nitrogen-isotopic ratios showed that fish from the same species and size, caught in different but neighbouring lakes, could be one or more trophic levels apart (Sampaio da Silva et al. 2005, 2009). These findings were important in shaping the messages with the local villagers: emphasis was placed on the fact that fish that eat other fish have more mercury, a generalization that was mostly true.

Health studies among the communities sought to determine whether there was an association between mercury exposure and neurophysiological functions, particularly of the motor and visual systems. The disasters in Minamata (Japan) and Iraq, where large populations had been exposed to high levels of methylmercury over a very short period, had shown that the nervous system is particularly vulnerable to methylmercury poisoning (WHO 1990). In the studies performed with the villagers of the Tapajós, sensitive tests were administered to evaluate manual dexterity, motor speed and accuracy, colour discrimination, and near visual contrast sensitivity. Over the weekends, Dr Branches joined the team in the field. He performed neurological examinations, including a test he had developed to assess coordination. The results showed clear dose-related deficits in motor and visual functions, with increasing hair mercury concentrations (Lebel et al. 1998; Dolbec et al. 2000). The studies also revealed damage to cell reproduction (Amorim et al. 2000).

It was not feasible for the villagers to avoid fish entirely, because it is a key source of protein in their diet. At the village meeting where the results about the source and effects of mercury were presented and discussed, villagers and researchers developed a campaign based on the slogan: *Eat more fish that don't eat other fish* – a positive campaign that encouraged eating fish for their nutritious value, while reducing mercury intake. Posters and pamphlets were developed and distributed (see Caruso Project website). Community perception and communication were followed by social scientists who studied villagers' social representations of health, fishing, and mercury from a gender perspective. Earlier studies had indicated that women's role in decision making, albeit invisible, was key. This was later confirmed and extensively analysed in Caruso studies of social-communication networks (Mertens et al. 2005; 2008).

The campaign advising people to eat more non-piscivorous fish was quite successful. Preliminary results of a follow-up study on exposure and health carried out 5 years after the initial evaluation showed that mercury exposure decreased. This study also noted an improvement in motor function but not in visual performance, which continued to decline in proportion to the amount of mercury before the reduction.

Dietary issues caught women's attention and, during a workshop, some women asked whether there was anything else in the local diet that would help reduce mercury toxicity. Not knowing the answer, the team proposed a study with the women to find out. In 2000, a food-frequency survey was designed and 26 women volunteered to keep a daily food diary. The village midwife coordinated the work. Because of seasonal differences in food availability, the women pursued this activity for an entire year. At the end of the year, they provided the researchers with hair strands that could be cut and analysed for mercury concentration month by month. The results showed that for the same amount of fish consumption, those who ate more fruit had less mercury in their blood and hair, giving rise to the hypothesis that fruit consumption may influence mercury absorption (Passos et al. 2003).

Scaling-Up

Widespread mercury contamination from deforestation activities, notably "slash and burn" agricultural practices, and its effects on human health required a more regional approach to the problem. In June 2003, the Caruso team invited farmers and key persons from 12 villages on the lower Tapajós to a meeting in São Luis do Tapajós. The participants from Brasília Legal played an important role. They explained to others how the researchers collaborated with communities, and described the results of the initial phases of the study. Following this meeting, a series of projects were initiated with these villages. These projects focused on short-term solutions through the identification of dietary factors that could influence mercury absorption, metabolism, and toxicity and sought more long-term solutions based on a better understanding of the relation between land use and mercury contamination.

Regional epidemiologic studies confirmed the modifying effect of fruit consumption on the relation between fish consumption and indicators of mercury exposure (Passos et al. 2007). This population is also characterized by high selenium levels, which were linked to the consumption of Brazil nuts and certain fish (Lemire et al. 2006; 2009). Recent studies indicate that selenium may also independently offset some of the toxic effects of mercury, notably with respect to motor performance and cataract formation (Lemire et al. 2010). However, mercury continues to affect visual and motor functions in these populations, and a positive association between mercury exposure and blood pressure was found (Fillion et al. 2006). Further studies on the interrelationships between mercury toxicity and possibly protective effects of dietary fish consumption are ongoing. The Caruso project also uncovered another source of toxic exposure. People in this area have high levels of lead in their blood, and the probable source lies in the low-quality metal plates that are widely used and heated in the production of manioc flour, or farinha (Barbosa et al. 2009).

Continued soil and sediment studies confirmed the hypothesis from the first phase of the study: mercury in the soil is of natural origin, and deforestation and soil erosion caused by erosive agricultural practices, such as slash and burn, lead to its transfer to rivers and lakes. Recent sediment layers were enriched in mercury, carbon, and nitrogen, as well as lignin markers, which indicate increased sedimentation of material of terrestrial origin, namely trees and forest soils (Farella et al. 2001). This is caused especially by slash and burn practices. Ash introduces large amounts of reactive cations in soils, dislodging mercury, phosphorus, and nitrogen from the few available soil-binding sites. These unbound elements are then washed away by heavy rainfall on unprotected soils. In this process of mercury remobilization, the initial deforestation is the main contributor. Subsequent land uses and changes appear to have a lesser impact on mercury (Farella et al. 2006; Béliveau et al. 2009).

Working at the interface of ecosystems and human health presented some unexpected conundrums. Carbon and nitrogen losses by soils reduce their fertility, and the consequent carbon and nitrogen enrichment in rivers and lakes increases methylmercury production: a veritable vicious cycle. But these findings also suggest an opposite, virtuous cycle: reducing mercury in fish requires sustainable agriculture practices at the watershed level (Farella et al. 2007). This is a great challenge in a region increasingly dominated by logging, ranching, and slash and burn agriculture.

The challenge was taken up by Poor Land Use, Poor Health (PLUPH), an agroforestry project (www.pluph.uqam.ca), led by the Université du Québec à Montréal and the University of Brasília, with funding from the Teasdale-Corti Foundation and IDRC. The project is testing the efficiency of crop varieties and alternative farming techniques to reduce soil erosion, their acceptability by local smallholders, and the effect of these different land-use patterns on soil fertility and mercury dynamics in soil and downstream lakes and rivers. The PLUPH project addresses the connections between land use and health issues related not only to mercury exposure but to the prevalence of Chagas disease (given the ecology of its vectors, pathogens, and wildlife hosts in forest-based human settlements).

Conclusion

The Caruso Project constitutes a long and still ongoing journey of investigation and findings. By asking new questions and using appropriate methods, the presumed mercury source and hence the project interventions shifted from gold mining to agriculture practices. The mobilization of mercury by slash and burn agriculture, and its exposure pathways and health effects in humans, are mediated by complex socio-cultural and political processes. These realities must be taken into account when attempting to change practices and policies (Fillion et al. 2009; Mertens et al. 2008).

The choice of the lower Tapajós as the project area had many implications for project development. The scarcity of administrative and service infrastructure and the high turnover of administrators did not facilitate changes in local policymaking. As well, regional decision makers did not place high priority on mercury issues. Direct and sustained contact with local communities turned out to be an important element that helped reveal the complex links between mercury in the environment and human environmental exploitation (Lucotte et al. 2004), as well as being key to effecting change in local communities. This close relationship was also critical to the design of effective interventions to mitigate slash and burn practices – interventions that respected local knowledge and practices. The active participation of the Brazilian Agriculture and Livestock Research Agency (Empresa Brasileira de Pesquisa Agropecuaria – EMBRAPA), in the PLUPH project brings good prospects for the further uptake and impact of the considerable findings amassed over the past 15 years.

Acknowledgements We thank the communities of the lower Tapajós who participated in the Caruso project. We acknowledge the many project investigators of the past 15 years, in particular Marc Lucotte. This paper is dedicated to the memory of Fernando Branches, a cardiologist from Santarem who brought the issue of mercury and health in the Amazonian region to the attention of the world, and Marc Roulet, the biogeochemist who identified deforestation as the source of mercury contamination in this region. IDRC support to Caruso was provided through projects 001300, 003323 and 101416.

References

Akagi, H., Malm, O., Branches, F.J.P., Kinjo, Y., Kashima, Y., Guimarães, J.R.D., Oliveira, R.B., Haraguchi, K., Pfeiffer, W.C., Takizawa, Y., and Kato, H. (1995). Human Exposure to Mercury Due to Goldmining in the Tapajós River Basin, Amazon, Brazil: Speciation of Mercury in Human Hair, Blood and Urine. Water, Air and Soil Pollution, 80, 85–94.
Amorim, M.I., Mergler, D., Bahia, M.O., Dubeau, H., Miranda, D., Lebel, J., Burbano, R.R., Lucotte, M. (2000). Cytogenetic Damage Related to Low Levels of Methyl Mercury Contamination in the Brazilian Amazon. Anais da Academia Brasileira de Ciências, 72, 497–507.
Barbosa Jr., F., Fillion, M., Lemire, M., Passos, C.J.S., Rodrigues, J.L., Philibert, A., Guimarães, J.R., and Mergler, D. (2009) Elevated Blood Lead Levels in a Riverside Population in the Brazilian Amazon. Environmental Research, 109, 594–599.
Bastos, W.R., Silva, A.P., Guimarães, J.R.D., Malm, O., and Pfeiffer, W.C. (1999). Mercury Concentration in Suspended Particulate Matter and Bottom Sediment Samples of the Rato River, Itaituba, Pará State, Brazil. Proceedings of the 5th International Conference on Mercury as a Global Pollutant, 23–27 May, Rio de Janeiro, Brazil, p. 514
Béliveau, A., Lucotte, M., Davidson, R., do Canto Lopes, L.O., and Paquet, S. (2009). Early Hg mobility in cultivated tropical soils one year after slash-and-burn of the primary forest, in the Brazilian Amazon. Science of the Total Environment, 407, 4480–4489.
Dolbec, J., Mergler, D., Larribe, B., Roulet, M., Lebel, J., and Lucotte, M. (2001). Sequential Analysis of Hair Mercury Levels in Relation to Fish Diet of an Amazonian Population, Brazil. Science of the Total Environment, 271, 87–97.

Dolbec, J., Mergler, D., Sousa Passos, C.J., Sousa de Morais, S., and Lebel, J. (2000). Methylmercury Exposure Affects Motor Performance of a Riverine Population of the Tapajós River, Brazilian Amazon. International Archives of Occupational and Environmental Health, 73, 195–203.

Farella, N., Davidson, R., Lucotte, M., and Daigle, S. (2007). Nutrient and Mercury Variations in Soils from Family Farms of the Tapajós Region (Brazilian Amazon): Recommendations for Better Farming. Agriculture, Ecosystems and Environment, 120, 449–462.

Farella, N., Lucotte, M., Davidson, R., and Daigle, S. (2006) Mercury Release from Deforested Soils Triggered by Base Cation Enrichment. Science of the Total Environment, 368, 19–29.

Farella, N., Lucotte, M., Louchouarn, P., and Roulet, M. (2001). Deforestation Modifying Terrestrial Organic Transport in the Rio Tapajós, Brazilian Amazon. Organic Geochemistry, 32, 1443–1458.

Fillion, M., Mergler, D., Passos, C.J.S., Larribe, F., Lemire, M., and Guimarães, J.R.D. (2006). A Preliminary Study of Mercury Exposure and Blood Pressure in the Brazilian Amazon. Environmental Health, 5, 29.

Fillion, M., Passos, C.J., Lemire, M., Fournier, B., Mertens, F., Guimarães, J.R., and Mergler, D. (2009). Quality of Life and Health Perceptions Among Fish-Eating Communities of the Brazilian Amazon: An Ecosystem Approach to Well-Being. EcoHealth, 6(1), 121–134

Guimarães, J.R.D., Roulet, M., Lucotte, M., and Mergler, D. (2000a). Mercury Methylation Potentials Along a Lake-Forest Transect in the Tapajós River Floodplain, Brazilian Amazon: Seasonal and Vertical Variations. Science of the Total Environment, 261, 91–98.

Guimarães, J.R.D., Meili, M., Hylander, L.D., Castro e Silva, E., Roulet, M., Mauro, J.B.N., and Lemos, R.A. (2000b). Mercury Net Methylation in Five Tropical Flood Plain Regions of Brazil: High in the Root Zone of Floating Macrophyte Mats but Low in Surface Sediments and Flooded Soils. Science of the Total Environment, 261, 99–107.

Lebel, J., Roulet, M., Mergler, D., Lucotte, M., and Larribe, F. (1997). Fish Diet and Mercury Exposure in a Riparian Amazonian Population. Water, Air and Soil Pollution, 97, 31–44.

Lebel, J., Mergler, D., Branches, F., Lucotte, M., Amorim, M., Larribe, F., and Dolbec, J. (1998). Neurotoxic Effects of Low-Level Methylmercury Contamination in the Amazonian Basin. Environmental Research, 79, 20–32.

Lemire, M., Mergler, D., Fillion, M., Sousa Passos, C.J., Guimarães, J-R., Davidson, R., and Lucotte, M. (2006). Elevated Blood Selenium Levels in the Brazilian Amazon. Science of the Total Environment, 366, 101–111.

Lemire, M., Mergler, D., Huel, G., Passos, C.J.S., Fillion, M., Philibert, A., Guimarães, J.R.D., Rheault, I., and Norman, G. (2009). Biomarkers of Selenium Status in the Amazonian Context: Blood, Urine and Sequential Hair Segments. Journal of Exposure Science and Environmental Epidemiology, 19, 213–222.

Lemire, M., Fillion, M., Frenette, B., Mayer, A., Philibert, A., Passos, C.J.S., Guimarães, J.R.D., Barbosa, Jr., F., and Mergler, D. (2010) Selenium and Mercury in the Brazilian Amazon: Opposing Influences on Age-Related Cataracts. Environmental Health Perspectives, 118, 1584–1589.

Lucotte, M., Davidson, R., Mergler, D., Saint-Charles, J., and Guimarães, J.R.D. (2004). Human Exposure to Mercury as a Consequence of Landscape Management and Socio-Economical Behaviors. Part I: The Brazilian Amazon Case Study. RMZ — Materials and Geoenvironment (Materiali in geookolje), 51, 668–672.

Malm, O., Guimarães, J.R.D., Castro, M.B., Bastos, W.R., Viana, J.P., and Pfeiffer, W.C. (1997). Follow-Up of Mercury Levels in Fish, Human Hair and Urine in the Madeira and Tapajós Basins, Amazon, Brazil. Water, Air and Soil Pollution, 97, 45–51.

Malm, O., and Guimarães, J.R.D. (1999). Biomonitoring Environmental Contamination with Metallic and Methylmercury in Amazon Gold Mining Areas, Brazil. In: Azcue, J.M. (Ed.), Environmental Impacts of Mining Activities, Springer-Verlag, Berlin, Heidelberg, Chapter 4, 41–54.

Mertens, F., Saint-Charles, J., Lucotte, M., and Mergler, D. (2008). Emergence and Robustness of a Community Discussion Network on Mercury Contamination and Health in the Brazilian Amazon. Health Education and Behavior, 35, 509–521.

Mertens, F., Saint-Charles, J., Mergler, D., Passos, C.J., and Lucotte, M. (2005). Network Approach for Analyzing and Promoting Equity in Participatory Ecohealth Research. EcoHealth, 2(2), 113–126.
Miranda, M.R., Guimarães, J.R.D., Roulet, M., Acha, D., Coelho-Souza S.A., Mauro, J.B.N., and Iniguez, V. (2004). Mercury Methylation and Bacterial Activity in Macrophyte Associated Periphyton in Floodplain Lakes of the Amazon Basin. RMZ — Materials and Geoenvironment (Materiali in geookolje), 51(2), 1218–1220.
Passos, C.J., Mergler, D., Gaspar, E., Morais, S., Lucotte, M., Larribe, F., and Grosbois, S. (2001) Caracterização Geral do Consumo Alimentar de uma População Ribeirinha na Amazônia Brasileira. Revista Saúde e Ambiente, 4, 72–84.
Passos, C.J., and Mergler, D. (2008). Human Mercury Exposure and Adverse Health Effects in the Amazon: A Review. Cadernos de Saúde Pública, 24 (Suppl 4), s503–s520.
Passos, C.J., Mergler, D., Fillion, M., Lemire, M., Mertens, M., Guimarães, J.R.D., and Philibert, A. (2007). Epidemiologic Confirmation that Fruit Consumption Influences Mercury Exposure in Riparian Communities in the Brazilian Amazon. Environmental Research, 105, 183–193.
Passos, C.J., Mergler, D., Gaspar, E., Morais, S., Lucotte, M., Larribe, F., Davidson, D., and de Grosbois, S. (2003). Eating Tropical Fruit Reduces Mercury Exposure from Fish Consumption in the Brazilian Amazon. Environmental Research, 93, 123–130.
Pfeiffer, W.C., Lacerda, L.D., Salomons, W., Malm, O. (1993). Environmental Fate of Mercury from Gold Mining in the Brazilian Amazon. Environmental Review, 1, 26–37.
Roulet, M., Lucotte, M., Canuel, R., Rheault, I., Tran, S., Gog, Y.G.D., Valer, S.D., Passos, C.J., De Jesus Da Silva, E.D., Mergler, D., and Amorim, M. (1998a). Distribution and Partition of Total Mercury in Waters of the Tapajós River Basin, Brazilian Amazon. Science of the Total Environment, 213, 203–211.
Roulet, M., Lucotte, M., Saint-Aubin, A., Tran, S., Rhéault, I., Farella, N., De Jesus Da Silva E.D., Dezencourt, J., Sousa Passos, C.J., Santos Soares, G., Guimarães, J.R.D., Mergler, D., and Amorim, M. (1998b). The Geochemistry of Hg in Central Amazonian Soils Developed on the Alter-Do-Chão Formation of the Lower Tapajós River Valley, Pará State, Brazil. Science of the Total Environment, 223, 1–24.
Roulet, M., Lucotte, M., Farella, N., Serique, G., Coelho, H., Passos, C.J., De Jesus Da Silva, E.D., de Andrade, P.S., Mergler, D., Guimarães, J.R.D., and Amorim, M. (1999). Effects of Recent Human Colonization on the Presence of Mercury in Amazonian Ecosystems. Water, Air and Soil Pollution, 112, 297–313.
Roulet, M., Lucotte, M., Canuel, R., Farella, N., Courcelles, M., Guimarães, J.R.D., Mergler, D., and Amorim, M. (2000a). Increase in Mercury Contamination Recorded in Lacustrine Sediments Following Deforestation in Central Amazonia. Chemical Geology, 165, 243–266.
Roulet, M., Lucotte, M., Guimarães, J.R.D., and Rhéault, I. (2000b). Methylmercury in the Water, Seston and Epiphyton of an Amazonian River and its Floodplain, Tapajós River, Brazil. Science of the Total Environment, 261, 43–59.
Roulet, M., Lucotte, M., Canuel, R., Farella, N., De Freitos Goch Y.G., Pacheco Peleja, R.J., Guimarães, J.R.D., Mergler, D., and Amorim, M. (2001). Spatio-Temporal Geochemistry of Hg in Waters of the Tapajós and Amazon Rivers, Brazil. Limnology and Oceanography, 46 (5), 1141–1157.
Sampaio da Silva, D., Lucotte, M., Paquet, S., and Davidson, R. (2009). Influence of Ecological Factors and of Land Use on Mercury Levels in Fish in the Tapajós River Basin, Amazon. Environmental Research, 109, 432–446.
Sampaio da Silva, D., Lucotte, M., Roulet, M., Poirier, H., Mergler, D., de Oliveira Santos, E., and Crossa, M. (2005). Trophic Structure and Bioaccumulation of Mercury in Fish of Three Natural Lakes of the Brazilian Amazon. Water, Air and Soil Pollution, 165, 77–94.
WHO (World Health Organization). (1990). Methylmercury. Environmental Health Criteria 101. World Health Organization, Geneva, Switzerland.

Chapter 11
Impacts on Environmental Health of Small-Scale Gold Mining in Ecuador

Óscar Betancourt, Ramiro Barriga, Jean Remy Davée Guimarães, Edwin Cueva, and Sebastián Betancourt

With the discovery of mining and metal-working techniques in ancient times, metals, metals pollution, and human health were linked (Nriagu 1996). Adverse health effects associated with exposure to metals have been at the root of occupational health since Ramazzini's work in the seventeenth and eighteenth centuries (Franco 1999). Everywhere, but particularly in the developing world in the context of small-scale mining, whole communities are deeply affected by mining not just because their livelihoods depend on mining, but because of the proximity of their dwellings to the mine operations (PRODEMINCA 1998).

The mining regions of Portovelo and Zaruma are located in the Puyango River basin in the southwestern part of Ecuador, bordering Peru. The Puyango crosses into Peru, where it becomes the Tumbes River, and drains into the Pacific Ocean at Tumbes, Peru. The extraction of gold and silver has taken place here for more than 500 years (since the time of the Incas), and this is reflected in the name of the province, El Oro. Throughout most of the twentieth century, large mining companies (foreign and domestic) dominated this region. In the economic crisis that ravaged the country in the 1980s, the remaining large mining companies closed, creating unemployment and worsening poverty. This same crisis pushed many people to invade abandoned mines, and led to the emergence of widespread and informal small-scale mining activities. Approximately 60,000 Ecuadorians (mostly men) were employed in small-scale gold mining in 2000 (Sandoval 2002).

Ó. Betancourt (✉) • E. Cueva • S. Betancourt
Health, Environment, and Development Foundation (Fundación Salud Ambiente y Desarrollo, FUNSAD), Quito, Ecuador
e-mail: oscarbet@gmail.com

R. Barriga
National Polytechnical School (Escuela Politécnica Nacional, EPN), Quito, Ecuador

J.R.D. Guimarães
Biophysics Institute, Federal University of Rio de Janeiro, Rio de Janeiro, Brazil

Ecuadorian small-scale mining operations are run with minimal organization, have limited technological development, and existing legislation targeting the sector is rarely enforced (Sandoval 2002). As a consequence, mineral exploitation is inefficient, the environment is often negatively impacted, and miners work in hazardous conditions and are poorly paid (OIT 1999). Because settlements are often close to or downstream from the mining sites, whole communities are impacted by mine-related environmental pollution.

The main health concerns from small-scale mining relate to pollution from metals and chemicals used to extract precious metals from the raw ore. But mercury, lead, arsenic, manganese, and other substances known to be toxic are also naturally found in the rocks and soils. They may be released into rivers with mining wastes and tailings, or because of soil disturbance from agriculture or erosion (Appleton et al. 2001; Betancourt et al. 2005). In this region, mercury and cyanide are commonly used in mining. Mercury's properties make it a solvent for most metals, which allows the miners to separate small amounts of precious metals from sediments, finely ground rock, and other sources. Sodium cyanide (NaCN) solution is also used to dissolve precious metals from their source materials. Lead nitrate is sometimes added to increase efficiency of the extraction. The precious metals are then precipitated from solution by adding zinc or other amalgamating agents.

Small- and medium-size gold-processing plants that use mercury amalgamation or cyanidation are situated along the tributaries of the Puyango River, mainly in the upper part of the basin. Ecuadorian informal gold extraction is estimated to produce between 5 and 6 tonnes of gold per year (Sandoval 2002; Velasquez-Lopez et al. 2010). Mercury released to the river can be converted by bacteria to methylmercury, which is much more toxic than inorganic mercury and is strongly biomagnified in food chains. Cyanide is toxic to humans and wildlife. It may increase the biological availability of mercury by dissolving metallic mercury, but it may also reduce methylmercury formation because it has a toxic effect on methylating bacteria.

The study area was in the Portovelo and Zaruma mining region situated in the upper part of the Puyango basin (Fig. 11.1). Typical of mining in Ecuador, there are many small-scale extraction operations for gold and silver in this region, but the total production has been declining over the years. The study focused on a stretch of river beginning 25 km upstream of the mining area to 115 km downstream on the semi-arid coastal plain. Portovelo and Zaruma are the main towns in the mining area, and the small communities of Gramadal and Las Vegas (approximately 30 households each) are situated 115 km downstream. The larger town of Puyango Viejo is in-between, about 80 km downstream. Human population density is unevenly distributed. Along the upper basin, Zaruma and Portovelo counties have a population of about 42,000 inhabitants; whereas, along the middle and lower basin, human settlements are small and scarce. In the lower basin communities are more dependent on the river resources. Agriculture, livestock, and mining are the major economic activities along the study area (Betancourt et al. 2005).

In the past 20 years, people living along the river have expressed concerns about possible health problems associated with upstream mining activities. Their main concerns related to pollution of the river, which is the main source of drinking water

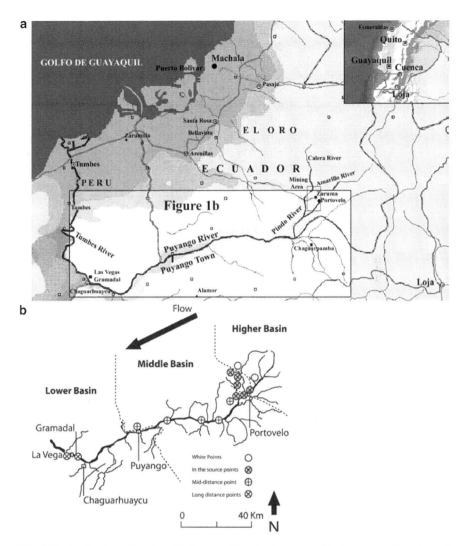

Fig. 11.1 (a) Southern Ecuador. (b) Puyango River basin: geographical zones and locations of villages included in the study and monitoring points. White points refer to control sites upstream from mining (reprinted from Betancourt et al. 2005 with kind permission from Springer Science+Business Media B.V.)

for the downstream communities of Gramadal and Las Vegas. Despite these concerns, water pollution has not been seriously addressed, partly due to deficiencies in the institutional control of mining pollution and weak environmental and health-management plans.

Half a ton of mercury is released annually with tailings, and NaCN (1.5–5 kg per tonne of tailings) is used for cyanidation, which produces cyanide-laden wastes with 200–300 mg/L of residual free cyanide that are released directly to the river (Velasquez-Lopez et al. 2010).

The research project was initiated in 1999. Initial consultations with the community helped identify the main objective: assess mining-related environmental pollution and associated health impacts and increase awareness among key actors of these effects. It quickly became clear that the communities were not very aware of the potential environmental and health hazards they were facing. Bolstered by exciting work coming out of Brazil (Malm 1998) on similar small-scale gold mining, the team was eager to experiment with ecosystem approaches to health (Lebel 2003).

The project's initial phase sought to explore the impact of mining on the ecosystem and on indicators of human health. The trandisciplinarity, multistakeholder participation, social equity, and gender-analysis aspects of ecohealth were emphasized. The project unfolded in two phases. This case study focuses on the results of the second phase, but key aspects and results of the first phase are presented.

The Social Dimensions of the Problem

From its inception, the project built on multistakeholder consultation and participation. The objective of the participatory process was to encourage members of the community to not only take part in the investigative process, but also to use the information to analyze the situation and identify, execute, and evaluate potential solutions. The project was implemented as a collective learning experience that involved academics, stakeholders, and community representatives (Torres 2001; Lebel 2003).

The roles and responsibilities of stakeholders were determined through the collection of qualitative data (2004) and workshops that discussed project issues and results (2006). Focus groups, workshops, and key-informant interviews were organized with miners, their families (men and women), government representatives, nongovernmental organizations, and community members to discuss project activities and findings, and to map potential strategies to reduce exposure to toxic substances. Figure 11.2 describes the roles of stakeholders involved in the project.

Leaders of the communities in Gramadal and Las Vegas (the downstream communities) expressed their concerns about the health implications of pollution from upstream mining. This community pressure influenced the later implementation of pollution-reduction initiatives by community organizations and local authorities in 2005 and 2006. The project invested time and effort to achieve a common understanding about a joint approach to community engagement and empowerment for action. This was a change for communities whose relationship with service providers was traditionally based on paternalism.

Key informants where interviewed to determine their knowledge about mining and its impacts on the environment and health. Initially, few people in Zaruma and Portovelo (the mining area) seemed concerned with environmental management

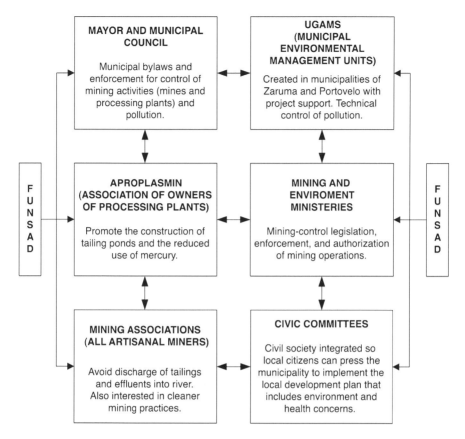

Fig. 11.2 Roles and responsibilities of project stakeholders (FUNSAD: Health, Environment, and Development Foundation)

and health (although level of concern increased with educational level). In the lower basin (Gramadal and Las Vegas) concern was higher from the outset. Awareness and concern increased over the life of the project. For example, by 2006, municipal authorities in Zaruma and Portovelo had started to appreciate the need for pollution-control mechanisms, and this facilitated the creation and enforcement of bylaws (see below).

Gender issues were identified in Portovelo and Zaruma. Men work directly in mining and carry a larger burden of exposure to metals. However, women tend to have traditional domestic roles and have little income. Interviews confirmed that men's work was considered more important because it earned income, which gave them greater decision-making power at the household level. Women encountered domestic violence and other hardships, and felt they had no power to change the situation.

Assessing the Environmental and Health Impacts of Gold Mining (1999–2002)

The project focused first on assessing environmental distribution of metals – mercury, manganese, and lead – to assess the extent of the pollution problem. Metals were measured in samples of river water, sediments, and suspended particulate matter (SPM), in both dry and rainy seasons, in the Puyango River basin. Particulate matter was used as a surrogate for mining-related environmental disturbance. Mining wastes are released directly into the river and increase SPM. Human population exposure pathways were also studied using household surveys and measurement of biomarkers (hair, blood, and urine) for mercury and lead (Betancourt et al. 2005).

Mercury and lead, at toxic levels, inhibit the functioning of the nervous system. Neurobehavioural tests are used to assess motor, sensory, and cognitive functions. Poor performance on these tests has been linked to metal toxicity (van Wendel de Joode et al. 2000; Cattell and Cattell 2001; Raven 2003). In 229 adult men working in, or living near, the mining areas, elevated blood mercury and lead seemed to be associated with poor performance on neurobehavioural tests. There were differences between the communities in fish consumption and drinking water source. Of the population in the gold-mining area, 10% consumed local fish compared with 98% of the population in the lower basin. In mining areas, people did not consume the river water, but in downstream communities, the river water was the sole source of water for drinking and cooking.

Lead concentrations in blood were found to be relatively high (mean 22 μg/dL, S.D. 22) among 40% of 225 adults investigated in the study area (all five communities: Portovelo, Zaruma, Puyango Viejo, Gramadal, and Las Vegas). People from Puyango Viejo (mid-way downstream from the mining areas) ($n=70$) had the highest concentrations of blood lead (mean of 34 μg/dL, S.D. 24). This level exceeded by a large margin the safe reference value of 20 μg/dL (WHO 1995).

Total mercury concentration in blood was observed to vary by occupation and geographical location, which indicated different exposure pathways. Among 32 miners (who did not generally consume fish), blood-mercury concentration was elevated (mean of 11 μg/L, S.D. 6.7 μg/L); whereas, among 128 people living far downstream in Gramadal and Las Vegas, who regularly consumed fish from the river, mercury concentration in blood was elevated, but lower than in the miners (mean 3.9 μg/L, S.D. 3.6 μg/L).

The detailed results of this first phase are described by Betancourt et al. (2005). After this first phase of work, the research team was left with questions that required further research. Although there was evidence of metal contamination in the watershed and evidence of human exposure, the pathways of exposure and their public health significance were not clear. Some findings, like the high exposure to lead, could not be fully explained by environmental exposure.

Defining the Sources of Exposure to Metals (2003–2009)

The second phase of research set out to better define the source of human exposure to potentially toxic metals – whether from mining or a combination of mining with other effects – that was found in the initial studies of the Puyango River populations. It also sought to explain the absence of mercury from the same downstream populations. Using participatory methods, the researchers developed interventions to better manage the pollution and reduce human exposure in downstream communities. The findings summarized here were also presented in FUNSAD (2007).

It was important to assess in depth the pollution problem in the second phase. The SPM measured at the baseline points was very low (3 mg/L during the March rainy season and 1.6 mg/L in the May dry season). Near the processing plants, elevated SPM were detected (132 mg/L in the rainy season and 328 mg/L in the dry season). In river-bottom sediments, mercury concentrations never exceeded 0.061 µg/g upstream of mining activities, but reached a maximum of 0.730 µg/g near the mining areas.

Fish-density indicators were higher upstream from the mining sites, including the headwaters of the Puyango River and a few of its tributaries, and also downstream (115 km from the mining sites) (Barriga 1991). Concentrations of mercury, manganese, and lead were measured in all fish species captured, using the methods of the Environmental Quality Laboratory (Chincheros 2007). Some fish were contaminated and exceeded WHO-recommended safe exposure levels. Although most fish (70% of the 195 samples) were not heavily contaminated by mercury (less than 0.50 µg/g, the WHO 1990 safe-exposure level), high levels of mercury contamination (2.25 µg/g) were found in dorado fish (*Brycon americus peruanus*), an omnivorous species caught in the middle and lower basin. However, this species is known to be eaten infrequently by people in the Puyango Basin. Manganese concentrations in fish were also generally safe, with a mean concentration of 0.40 µg/g in 112 samples. The recommended maximum concentration of manganese in fish is 2.5 µg/g (ATSDR 2000). The highest concentrations (3.18 µg/g) were measured in shad fish or sábalo (*B. atrocaudatus*), a fish that is eaten by local people. Lead concentrations in fish were on average 0.8 µg/g, substantially higher than the WHO-recommended maximum of 0.1 µg/g (WHO 1995).

In this phase, the project focused on the exposure of children to metals, in both mining and nonmining communities. Measurements were made of metal concentrations in hair, and of neurobehavioural performance. Household surveys showed that 12% of 72 children from Portovelo and Zaruma were working in mining activities at the time of the research (August 2006), mainly helping their parents in the processing plants.

Only 6% of 94 children had mercury concentrations in their hair that exceeded 2 µg/g, within permissible levels (WHO 1990). Arsenic concentrations in hair were also negligible in these children (not exceeding 0.1 µg/g).

In 2006, 83 children (8–12 years old) were assessed using the same neurobehavioural tests used to assess adults in the earlier part of the research. The study included 72 children from the mining areas of Zaruma and Portovelo, and 11 from the downstream areas of Gramadal and Las Vegas. Increased concentrations of manganese in the hair of girls (2.9–7.4 µg/g) were associated with decreased scores on the cognitive Raven test ($p=0.009$) and the digits test ($p=0.03$). In children, increased concentrations of mercury in hair (0.1–4.3 µg/g) was associated with decreased performance on Santa Ana dexterity ($p=0.005$), digits ($p=0.01$), and finger tapping ($p=0.04$) tests. These low levels of exposure have not previously been reported to be associated with neurobehavioural impairment. Children were found to have been exposed to lead, but the levels detected in their hair remained within permissible WHO levels. Although indicative, there is a need to further validate these tests in children. Further investigation should address other potentially relevant causes of poor performance on neurobehavioural tests, including education, nutrition, household environment, and culture.

The previously detected high level of lead in adults was puzzling because lead levels in fish were moderate, and thus dietary exposure from fish consumption was assumed to be low. Further investigation of potential household sources of lead exposure found that 84% of 40 families in the lower and middle basins used metal kitchenware thought to contain high concentrations of lead. This was confirmed by measuring the lead concentration in two pots (both from local suppliers, and similar to those used throughout the region) from different households. Pots were found to contain lead concentration of 230 and 1,135 µg/g, hypothetically enough to contaminate food cooked in these pots.

Why Was There No Methylmercury in the Puyango?

Although the environmental distribution of metals was associated with mining activities (Betancourt et al. 2005), the concentration of methylmercury (MeHg) in river-bottom sediments from the main channel of the Puyango River was negligible (0–0.1 ng/g) in seven samples. Mercury contamination was occurring, but it was not leading to formation of toxic methylmercury or its accumulation in the food chain or in people downstream. The project sought to explain why this was occurring. Methylmercury enrichment is usually found downstream from mining areas that release mercury and leads to health risks, notably through fish consumption (Boischio and Henshel 2000; Guimarães et al. 2000; Roulet et al. 2000; Gray et al. 2004). The natural methylation process depends on the presence of bacteria that transform the mercury into methylmercury.

The study set out to measure bacterial activity along the river. The hypothesis was that a toxic substance, perhaps cyanide from the gold processing, was impairing bacterial activity. Sampling points for cyanide and bacterial activity in water and potential mercury methylation in sediments were identified. Data were collected upstream and downstream of mining activities according to methods similar to Guimarães et al. 1995 and Miranda et al. 2007.

Cyanide levels were undetectable (<1 µg/L) upstream of mining activities, peaked at 280 µg/L immediately downstream from mining activities, and eventually returned to <1 µg/L 115 km downstream. Bacterial activity (in sediments), and thus potential for mercury methylation to occur, was high upstream of mining activities, but nearly absent near mining sites, and again high further downstream at Gramadal and Las Vegas. It is hypothesized that high levels of cyanide in the water immediately downstream from the mining areas could be so toxic to bacteria that they prevent mercury methylation in the river. Paradoxically, cyanide has also been shown to help in the mercury methylation process, by increasing the dissolution of mercury. For this hypothesized toxicity to be important, the cyanide toxicity to bacteria would need to override the mercury dissolving effect of cyanide. This result is in contrast with what has been suggested by other studies in the Puyango River basin (Velasquez-Lopez et al. 2010).

Community Empowerment to Protect Human Health (The Interventions)

To reduce exposure to environmental contaminants, project interventions were developed and implemented among the communities of Gramadal, Las Vegas, and Puyango Viejo (downstream from the mining areas). These interventions included the installation of water filters for homes and schools, the provision of an alternative source of potable water, electrification, and road improvement. Eventually, public drinking-water systems were established at Gramadal, Las Vegas, and Puyango Viejo to eliminate the direct intake of river water. Efforts were also made to eliminate other sources of heavy-metal contamination, such as the use of lead-containing kitchenware.

In Zaruma and Portovelo (the mining district), the project acted to protect people from exposure and to reduce contamination of the river. Project discussions with stakeholders led to the creation, implementation, and enforcement of new municipal bylaws to control the installation and operation of processing plants. These bylaws included requirements that processing plants not be built near rivers and that tailings and effluents not be dumped in rivers. The project promoted and supported the creation of Municipal Environmental Management Units led by environmental engineering or mining professionals. Workshops and other meetings also influenced mining organizations to implement additional pollution-control measures. Changes in environmental and health management at different levels were observed in a diverse number of stakeholders (Fig. 11.2). In Zaruma and Portovelo, the project assisted with the organization and operation of the canton civic committee, which successfully lobbied to include environmental and health-management plans in the cantonal development plan. The community lobbying efforts enhanced by the project also exerted some influence on national environmental authorities, notably in the

Ministry of the Environment and Mining Reductions in the discharge of mining and urban wastes directly into the river resulted from a range of pollution-control measures, implemented mainly by the Ministry of Environment and Mining. Dumping of municipal solid waste directly into the Puyango River was formally prohibited in 2009.

New techniques have been adopted by miners to reduce contamination (e.g., settling ponds for tailings, and the use of enclosed apparatus for amalgam burning). Ecological clubs were created in schools and colleges in Zaruma and Portovelo to stimulate knowledge about health care and build awareness of environmental pollution among youth. There is strong ongoing interest in health and environment among communities in the lower river basin as well, a remarkable commitment considering their particularly challenging living and livelihood conditions. Local organizations were strengthened for environmental and health management that led to the use of filters to sanitize river water. A public drinking water system was established in some downstream communities.

Conclusion

The project found that mining wastes released into the river cause metal and cyanide pollution. This pollution impairs water quality in the lower basin. Observed pollution negatively affects the health of miners. Pollution also affects the health of communities downstream by contaminating fish that form an important part of their diet, and to a lesser extent by contaminating drinking water. Mercury exposure was a problem only in the downstream part of the basin, where methylation resumes, perhaps due to the lower cyanide levels. Children were showing evidence of chronic exposure and toxicity to manganese and lead. All communities may be experiencing substantial added lead exposure from cooking in metal pots that contain lead. The project contributed to the development of pollution-reducing municipal regulations and to processes for greater community engagement in decision making. The sharing and discussion of project results increased awareness among both miners and decision makers in all parts of the Puyango River basin.

Acknowledgments We thank the residents of Portovelo, Zaruma, Puyango Viejo, Gramadal, and Las Vegas along the Puyango River in Southern Ecuador and many other stakeholders for their long-term contributions. IDRC support was provided through projects 004291, 101415, and 105145.

References

Appleton, J., Williams, T., Orbea, H., and Carrasco, M. (2001). Fluvial Contamination Associated with Artisanal Gold Mining in the Ponce Enriquez, Portovelo-Zaruma and Nambija Areas, Ecuador. Water, Air and Soil Pollution, 131, 19–39.

ATSDR (Agency for Toxic Substances and Disease Registry). (2000). Toxicological Profile for Manganese. Available at: http://www.atsdr.cdc.gov/toxprofiles/tp151.html.

Barriga, R. (1991). Lista de Peces de Agua Dulce del Ecuador. Politécnica, 16(3), 7–88.
Betancourt, O., Narváez, A., and Roulet, M. (2005). Small-Scale Gold Mining in the Puyango River Basin, Southern Ecuador: A Study of Environmental Impacts and Human Exposures. EcoHealth, 2(4), 323–332.
Boischio, A., and Henshel, D. (2000). Fish Consumption, Fish Lore, and Mercury Pollution-Risk Communication for the Madeira River People. Environmental Research, Section A, 84, 108–126.
Cattell, R.B., and Cattell, A.K.S. (2001). Factor "g" Escalas 2 y 3 (10th Edition). TEA Ediciones S.A, Madrid.
Chincheros, J. (2007). Procedimientos para la Determinación de Mercurio, Plomo, Arsénico y Manganeso en Muestras Ambientales. Laboratorio de Calidad Ambiental (LCA), Universidad Mayor de San Andrés (UMSA), La Paz.
Franco, G. (1999). Ramazzini and Workers' Health. Lancet, 4, 858–861.
FUNSAD (Fundación Salud Ambiente y Desarrollo). (2007). Impactos en el Ambiente y la Salud por la Minería del Oro a Pequeña Escala en el Ecuador (Segunda Fase): Informe Técnico Final (Project 101415). Final Technical Report to IDRC (in Spanish). Available at http://idl-bnc.idrc.ca/dspace/handle/10625/36568.
Gray, J.E., Hines, M.E., Higueras, P.L., Adatto, I., and Lasorsa, B.K. (2004). Mercury Speciation and Microbial Transformations in Mine Wastes, Stream Sediments and Surface Waters at the Almaden Mining District, Spain. Environmental Science and Technology, 38(16), 4285–4292.
Guimarães, J.R.D., Malm, O., and Pfeiffer, W.C. (1995). A Simplified Radiochemical Technique for Measurement of Net Mercury Methylation Rates in Aquatic Systems Near Goldmining Areas, Amazon, Brazil. Science of the Total Environment, 175(2), 151–162.
Guimarães, J.R.D., Roulet, M., Lucotte, M., and Mergler, D. (2000). Mercury Methylation along a Lake-Forest Transect in the Tapajós River Floodplain, Brazilian Amazon: Seasonal and Vertical Variations. Science of the Total Environment, 261, 91–98.
Lebel, J. (2003). Health: An Ecosystem Approach. In Focus Series, International Development Research Centre (IDRC), Ottawa. Available at: http://www.idrc.ca/in_focus_health/.
Malm, O. (1998). Gold Mining as a Source of Mercury Exposure in the Brazilian Amazon. Environmental Research, Section A, 77, 73–78.
Miranda, M.R., Guimarães, J.R.D., and Coelho-Souza, A.S. (2007). [^{3}H]Leucine Incorporation Method as a Tool to Measure Secondary Production by Periphytic Bacteria Associated to the Roots of Floating Aquatic Macrophytes. Journal of Microbiological Methods, 71(1), 23–31.
Nriagu, J.O. (1996). A History of Global Metal Pollution. Science, 272(5259), 223–224.
OIT (Orgnización Internacional del Trabajo). (1999). Los Problemas Sociales y Laborales en las Explotaciones Mineras Pequeñas: Informe para el debate de la Reunión Tripartita Sobre los Problemas Sociales y Laborales en las Explotaciones Mineras Pequeñas. Orgnización Internacional del Trabajo, Geneva.
PRODEMINCA (Proyecto de Desarrollo Minero y Control Ambiental). (1998). Monitoreo Ambiental en las Áreas Mineras en el Sur del Ecuador. Swedish Environmental Systems and Ministerio de Energía y Minas, Quito.
Raven, J.C. (2003). Test de Matrices Progresivas Escala Avanzada (Second Edition). Ediciones Paidós, Buenos Aires.
Roulet, M., Lucotte, M., Guimarães, J.R.D., and Rheault, I. (2000). Methylmercury in Water, Seston, and Epiphyton of an Amazonian River and its Floodplain, Tapajós River, Brazil. Science of the Total Environment, 261, 43–59.
Sandoval, F. (2002). Small Scale Mining in Ecuador. Mining, Minerals and Sustainable Development, Number 75. International Institute for Environment and Development (IIED) and World Business Council for Sustainable Development (WBCSD). Available at: http://pubs.iied.org/pdfs/G00720.pdf.
Torres, V.H. (2001). La Participación Comunitaria y Vecinal en la Formulación, Seguimiento y Evaluación de Proyectos. Sistema de Desarrollo Local-SISDEL, (First Edition). Ediciones Abya-Yala and Comunidades y Desarrollo en el Ecuador (COMUNIDEC), Quito.

Van Wendel de Joode, B., Mergler, D., Wesseling, C., Henao, S., Amador, A., and Castillo, L. (2000). Manual de Pruebas Neuroconductuales. Organización Panamericana de la Salud (OPS), Organización Mundial de la Salud (OMS) (IRET-CINBIOSE-OPS/OMS-CEST). San José.

Velasquez-Lopez, P.C., Veiga, M.M., and Hall, K. (2010). Mercury Balance in Amalgamation in Artisanal and Small-Scale Gold Mining: Identifying Strategies for Reducing Environmental Pollution in Portovelo-Zaruma, Ecuador. Journal of Cleaner Production, 18, 226–232.

WHO (World Health Organization). (1990). Environmental Health Criteria Number 101: Methylmercury. World Health Organization, Geneva.

WHO (World Health Organization) (1995). Environmental Health Criteria Number 165: Inorganic Lead. World Health Organization, Geneva.

Part III
Poverty, Ecosystems, and Vector-Borne Diseases

Chapter 12
Introduction

Roberto Bazzani and Martin Wiese

Population growth and migration, environmental change and transformation of landscapes, and globalization of trade and economies have changed the kinds of health challenges faced by populations around the world. Wealthier nations, indeed, wealthier people in all nations, face a growing burden of diseases of affluent and sedentary lifestyles: cardiovascular disease; diabetes; cancer; and obesity. Despite substantial gains in social and economic development around the world, and enormous advances in sanitation and medical knowledge and technologies, many infectious diseases continue to pose an unacceptable burden to people who live in developing regions. A further threat is posed to people all around the world by the emergence of new infectious diseases, which appear to be on the rise (Jones et al. 2008).

While health systems face an increasing burden from chronic diseases, infectious diseases continue to be the single-most important contributor to the burden of disease in developing regions. Most of these diseases are linked to environment – whether vector-borne diseases like malaria and dengue or water-borne causes of diarrhoea (Weiss and McMichael 2004; Prüss-Üstün and Corvalán 2006). The prevailing approach to infectious disease control, a highly successful one, has rested on mass immunization or rapid diagnosis, isolation, and treatment for nonvaccine preventable diseases. The emergence late in the twentieth century of antimicrobial resistance, the difficulties in finding effective vaccines for diseases such as HIV and dengue, and the experiences from SARS and pandemic influenza have highlighted the need for alternative or complementary approaches that emphasize disease prevention, in addition to control.

Ecosystem approaches to health incorporate a systems view of the drivers of disease transmission and persistence and participatory community-based approaches to prevention and control. For vector-borne diseases like dengue, vector-eradication

R. Bazzani (✉) • M. Wiese
International Development Research Centre, Ottawa, ON, Canada
e-mail: ecohealth@idrc.ca

campaigns relying primarily on insecticides have remained the most widely used approach. Meanwhile, changes in the ecology of vectors have accelerated because of environmental and climate changes, changes in host ecology, and selection pressures from insecticides. Because microbes and vectors are living organisms, they will continue to adapt, change, and pose threats to human health.

Because of the importance of environmental conditions to the presence, reproduction, and survival of vectors, as well as the environmental context for vector–human contact, diseases transmitted by invertebrate vectors are emblematic of the close connection between human health, ecosystems, and social and economic activities (Campbell-Lendrum and Molyneux 2005). The four case studies in this part highlight how an ecosystem approach can lead to innovative strategies to prevent and control vector-borne diseases and enhance traditional public-health services.

An ecosystem approach to preventing infectious diseases considers drivers of risk in terms of the ecological, social, cultural, political, and economic underlying factors of transmission dynamics. Like other health problems, the ecology and transmission of most infectious diseases can be linked to interactions among several factors, for example, demographic changes, poverty, urbanization, deforestation, changes in agriculture models of production, changed relationships between people and animals, natural resources management, and gender differences and cultural patterns. An understanding of complex interactions among these factors, and their local manifestations and risk factors for disease, require research across current boundaries of scientific disciplines and sectors (Waltner-Toews 2001; Parkes et al. 2005; Spiegel et al. 2005; Boischio et al. 2009). The prevention of disease and the prevention of serious harm to livelihoods from disease depend on sound and resilient environments in ecological, social, and economic terms. It is therefore essential to build multisector policy options that target disease prevention.

The case studies presented in this chapter address three major vector-borne diseases: malaria, dengue, and Chagas disease. In 2008, 60 million cases of malaria were reported in Africa (WHO 2010). Dengue and dengue hemorrhagic fever (DHF) are acute viral diseases that have significantly increased in the past 25 years in Latin America and the Caribbean (LAC). More than one million cases of dengue were reported in 2010 (PAHO 2010). Chagas disease is a chronic infection and important cause of disease and mortality in Latin America, where more than ten million people are infected with *Trypanosoma cruzi*, the parasite responsible for the disease (WHO-TDR 2010).

In East Africa, a substantial proportion of malaria is believed to be closely associated with agricultural systems (Mutero et al. 2006). The paper devoted to malaria in Uganda and Tanzania integrates the results of three projects developed to explore links between agriculture and health as part of a regional network. Chagas disease, a zoonosis responsible for the highest burden of vector-borne disease in Latin America, is highlighted in a case study in Guatemala. Here the disease is associated with deforestation and with poor housing conditions (Briceño-León 2009; Bustamante et al. 2009; Petherick 2010). Two case studies focus on dengue and its control in Asian and Caribbean contexts. The ecohealth perspective to dengue seeks to overcome current limitations of vector control through community-driven environmental and social strategies, and the development of early warning systems based on environmental monitoring and coordinated case reporting.

These case studies on vector-borne diseases are good illustrations of ecohealth multistakeholder processes. Each case study reports how a range of multiple actors became involved in the early stages of the research process and to what effect. In some cases, as in Guatemala and Cuba, they remained engaged as part of the intersectoral response. In all case studies, multidisciplinary teams interacted with civil society organizations and decision makers to reduce risks of vector-borne disease. In Guatemala, researchers worked closely with municipalities, school authorities, village leaders, vector control staff, and the Ministry of Health. Community leaders adopted housing-improvement techniques and, with the endorsement of vector-control staff, trained community members in a new wall-plastering technique that prevented infestation by the vector of Chagas disease. In Cuba, in the Municipality of Cotorro in Havana City, a management group composed of members of the community and local institutions took part in the design and implementation of a participatory dengue-surveillance system. In both the African malaria and Asian dengue studies, stakeholders such as district health officers, village heads and local leaders of community groups, local nongovernmental organizations, and neighborhood committees were included from the outset and formal mechanisms put in place for their input. The early involvement of strategic stakeholders helps explain how these projects managed to influence vector-control policies.

In many of the case studies, research to better control vector-borne diseases was embedded in a community-development process, acting on the risk of disease while improving quality of life. In Guatemala, the Chagas study exposed potential benefits of improved poultry husbandry for vector control, while increasing women's income and household food security.

All four case studies highlight the importance and specificity of the local context for successful disease-control strategies. The projects show that while general control strategies may be developed, their application must be grounded in the understanding of specific ecosystem dynamics. Although some strategies and innovations acting on the vector are likely to be effective wherever vector ecology may be similar, the more highly variable human dimensions (social, cultural, and economic) complicate matters. Different vector species and ecologies, and changeable ecosystem conditions complicate things further. Thus, scaling-up of ecohealth disease-control strategies must be adaptable to the local context. The leadership of local stakeholders helps maintain this adaptability while higher levels of government provide the framework for uptake and application in other contexts. In Cuba, the participatory integrated surveillance system was successfully replicated in other municipalities of Havana City. In Guatemala, the Chagas ecosystem intervention is now approved as an official policy in Central America to control *Triatoma dimidiata*, an important Chagas vector in the region.

Multiple constituencies are currently contributing to build the field of ecohealth and infectious diseases in different regions. Although vector-borne diseases are an obvious set of problems worth approaching from an ecohealth perspective, researchers around the world are also seeking to better understand and prevent the emergence and spread of new zoonoses like avian influenza (H5N1) and old problems such as leishmaniasis. The persistence or reemergence of infectious diseases, and the emergence of some new ones, may be indicative of wider problems affecting the dynamics of social-ecological systems. Are there now more mosquitos and other disease vectors than before? Why are people increasingly infected with animal diseases? Why do some diseases like dengue appear to spread unchecked while others like SARS disappear? The answers to these questions may lie in gaining a better understanding of the upstream drivers of infectious disease emergence and spread: environmental, social, and economic change. Large groups of people continue to be vulnerable to infectious diseases due to poverty, absence of adequate infrastructure, lack of access to health services, and degraded living environments, and these need to be addressed to protect people. Further progress in controlling infectious diseases almost certainly requires some combination of biomedical technologies and social and environmental strategies coordinated among relevant sectors. These case studies highlight early successes and insights from applying an ecosystem approach to health to vector-borne disease prevention and control. There is still much more to be learned about controlling vector-borne diseases, and lessons that can be applied to zoonoses and other emerging infectious diseases.

References

Parkes, M.W., Bienen, L., Breilh, J., Hsu, L-N., McDonald, M., Patz, J.A., Rosenthal, J.P., Sahani, M., Sleigh, A., Waltner-Toews, D., and Yassi, A. (2005). All Hands on Deck: Transdisciplinary Approaches to Emerging Infectious Disease. EcoHealth, 2(4), 258–272.

Boischio, A., Sanchez, A., Orosz, Z., and Charron, D. (2009). Health and Sustainable Development: Challenges and Opportunities of Ecosystem Approaches in the Prevention and Control of Dengue and Chagas. Cadernos de Saúde Pública, 25 (Suppl 1), S149–154.

Briceño-León, R. (2009). Chagas disease in the Americas: an ecohealth perspective. Cadernos de Saúde Pública, 25 (Suppl. 1), S71–S82.

Bustamante, D.M., Monroy, C., Pineda, S., Rodas, A., Castro, X., Ayala, V., Quiñones, J., Moguel, B., and Trampe, R. (2009). Risk factors for intradomiciliary infestation by the Chagas disease vector *Triatoma dimidiata* in Jutiapa, Guatemala. Cadernos de Saúde Pública, 25 (Suppl. 1), S83–S92.

Campbell-Lendrum, D., and Molyneux, D. (2005). Ecosystems and Vector-Borne Disease Control. In: Epstein, P., Githeko, A., Rabinovich, J., and Weinstein, P. (Editors), Ecosystems and Human Well-Being: Policy Responses, Vol. 3., pp. 353–372, Island Press, Washington. Available at: http://www.maweb.org/documents/document.317.aspx.pdf.

Jones, K.E., Patel, N.G., Levy, M.A., Storeygard, A., Balk, D., Gittleman, J.L., and Daszak, P. (2008). Global Trends in Emerging Infectious Diseases. Nature, 451(7181), 990–993.

Mutero, C.M., McCartney, M., and Boelee, E. (2006). Understanding the Links Between Agriculture and Health: Agriculture, Malaria and Water-Associated Diseases. Brief 6. International Food Policy Research Institute, Washington.

PAHO (Pan American Health Organization). (2010). Data and Statistics. Number of Reported Cases of Dengue and Dengue Hemorrhagic Fever (DHF) in the Americas. PAHO, Washington. Available at: http://new.paho.org/hq/index.php?option=com_content&task=view&id=719&Itemid=1119.

Petherick, A. (2010). Chagas Disease in the Chaco. Nature, 465, S18–S20.

Spiegel, J., Bennett, S., Hattersley, L., Hayden, M.H., Kittayapong, P., Nalim, S., Wang, D.N.C., Zielinski-Gutiérrez, E., and Gubler, D. (2005). Barriers and Bridges to Prevention and Control of Dengue: The Need for a Social-Ecological Approach. EcoHealth, 2(4), 273–290.

Waltner-Toews, D. (2001). An Ecosystem Approach to Health and its Applications to Tropical and Emerging Diseases. Cadernos de Saúde Pública [online], 17, S07–S36. Available at: http://www.scielosp.org/scielo.php?pid=S0102-311X2001000700002&script=sci_arttext&tlng=en.

Weiss, R.A., and McMichael, A.J. (2004). Social and Environmental Risk Factors in the Emergence of Infectious Diseases. Nature Medicine, 10, S70–S76.

WHO (World Health Organization). (2010). World Health Statistics 2010. WHO, Geneva. Available at: http://www.who.int/whosis/whostat/2010/en/index.html.

WHO-TDR (World Health Organization, Special Programme for Research and Training in Tropical Diseases). (2010). Chagas Disease — *Trypanosomiasis americana*, Fact Sheet 340, June 2010. WHO, Geneva. Available at: http://www.who.int/mediacentre/factsheets/fs340/en/index.html.

Prüss-Üstün, A., and Corvalán, C. (2006). Preventing Disease Through Healthy Environments: Towards an Estimate of the Environmental Burden of Disease. World Health Organization, Geneva.

Chapter 13
Malaria Research and Management Need Rethinking: Uganda and Tanzania Case Studies

Joseph Okello-Onen, Leonard E.G. Mboera, and Samuel Mugisha

Of the estimated 247 million cases of malaria in the world annually, 86% (212 million cases) occur in Africa, and kill 800,000 people (mostly children). Most deaths occur among the poorest, who lack access to adequate prevention and treatment (Rowe et al. 2006). As elsewhere in sub-Saharan Africa, malaria is the leading public health problem in East Africa (Burundi, Kenya, Rwanda, Tanzania and Uganda). Of the 115 million people in the region, 73% are at risk of malaria infection. Malaria is responsible for more than one-third of deaths among children less than 5 years old, and one-fifth of deaths among pregnant women (WHO 2008). The disease situation is markedly diverse among the neighbouring countries, and even within localized geographical areas, in terms of the infectivity of mosquito vectors, the intensity of human–vector contacts and disease prevalence. In East Africa, the burden and persistence of malaria depend on anthropogenic, topographical and ecological factors.

Agriculture, for subsistence and for cash, is the main economic activity in rural East Africa. The development of agricultural and water resources affects the environment, which in turn affects human health. For example, agricultural practices and water-resource management may create new mosquito breeding sites, and facilitate the spread of malaria. This case study presents highlights from three interlinked projects that explored the impact of various agro-environmental factors and socio-cultural practices on malaria in East Africa (Mboera et al. 2007;

J. Okello-Onen (✉)
Department of Biology, Gulu University, Gulu, Uganda
e-mail: jonen65@hotmail.com

L.E.G. Mboera
National Institute for Medical Research, Dar es Salaam, Tanzania

S. Mugisha
Department of Zoology, Makerere University, Kampala, Uganda

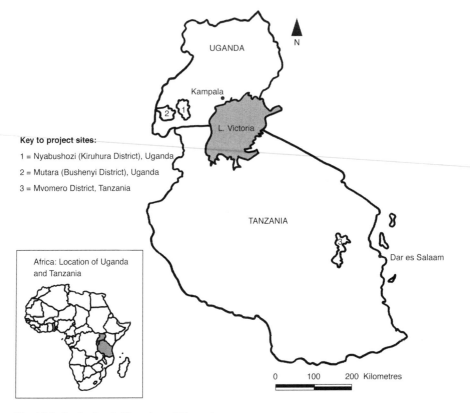

Fig. 13.1 Study sites in Uganda and Tanzania

Okello-Onen, unpublished). The three projects were conducted in Nyabushozi (Kiruhura District) and Mutara (Bushenyi District) in Uganda and Mvomero District, Tanzania (Fig. 13.1). Together they formed part of the IDRC and Consultative Group on International Agricultural Research (CGIAR) Systemwide Initiative on Malaria and Agriculture (SIMA), initiated in 2001 (Mutero et al. 2005). The highlights and cross-cutting findings of these projects underscore the importance of assessing epidemiological data for malaria in different agro-ecosystems and farming practices and taking into account the needs of different groups.

In Uganda and Tanzania, malaria is mostly a disease of rural communities where agriculture is the backbone of household economy, and poverty and illiteracy are common. In these countries, the tropical climate is suitable for the survival of vector and parasite populations, and consequently for high rates of malaria transmission. The disease is highly prevalent and constrains economic and social development.

Perceived Causes of Malaria

Southwestern Uganda is known for its early 1990s malaria epidemics with morbidity and mortality exceeding 75 and 25%, respectively, among outpatients and inpatients in the rural areas (UG-MOH 1998). Previous malariometric surveys had characterized the area as a low to moderate malaria-transmission zone. These new and large outbreaks were attributed to factors such as changes in weather, environmental management, patterns of population migration, inadequate knowledge of the disease, and a breakdown in health-service provision (Lindsay and Birley 1996).

Community members felt that the increased occurrence of malaria in these two districts in Uganda was due to a number of marked changes in livelihoods and agro-ecosystems. These included: introduction of fish-farming in traditional crop and animal-farming systems; increases in water storage (valley dams and tanks); a reduction in the burning of bush and dry cow dung in kraals (animal housing); changes in gender roles; changes in child-feeding habits (less reliance on milk); more exposure to outside ideas; changes in sources of income; disparities among men and women in access to and control of household and community resources; disparities in access to training; and gender variations in treatment-seeking patterns for different members of the households.

The changes in livelihoods and the agro-ecosystem were driven by government policies and the need to transform the rural economy. In Uganda, the government persuaded nomadic pastoralists to cease their migrations and to practice ranching or dairy farming. After some years they complained that their children were dying of malaria because of: changes in livestock-management practices; restructuring of the land to households (to minimize movement of pastoralists); the shift from pastoralism to sedentary agro-pastoralism; and the increased distance between human dwellings and animal housing. The investigators hypothesized that reductions in the size of the cattle herds, and housing of cattle in sheds (rather than closer to people), might have favoured increased human–vector contact with the most likely, zoophilic (preferring to feed on animals rather than people) anopheline vectors of malaria. As a result, the mosquitoes fed more on people than on cattle, and this greatly increased malaria transmission.

Malaria is endemic in almost all parts of Tanzania. However, the situation is not homogenous, and there are marked differences in the seasonality of malaria. As in Uganda, malaria is mostly a disease of rural communities, but little is known of the impact of agriculture on malaria. In Africa, where malaria reaches a peak at harvest time, a single bout of the disease costs an estimated equivalent of ten lost working days. As a result, affected families manage to work only 40% of the land available for crops, compared with healthy families who are able to work up to 75% of the available land (deBartolome and Vosti 1995). Agricultural and natural-resource management practices directly impact the risk and vulnerabilities for malaria.

Mvomero district in central Tanzania has perennial malaria transmission, with an overall prevalence of malarial infection of about 43% (Mboera et al. 2006). The district's economy depends heavily on agriculture, mainly crop production. More than 80% of adults in Mvomero earn their livelihood from subsistence agriculture. Monoculture, mixed cropping and multiple cropping are common in the district. However, the impact of different land-use patterns and agro-ecosystems on the malaria burden is little known. Knowledge of the interactions between agricultural production systems and malaria remains relatively weak. Reliable information on environmental risks to health is fundamental for the prevention and control of malaria, for evidence-based guidance of health policy and planning, and for the promotion of intersectoral action for the reduction of transmission.

Links Between Agriculture and Malaria

Farming systems influence malaria by changing the ecological conditions for mosquito vectors or by altering the contact between the human population and the vector mosquito (Ijumba and Lindsay 2001). Water-resource development results in the creation of new mosquito breeding sites, and is usually coupled with demographic changes. As a result, human–vector–parasite contact patterns are altered, and the spread of malaria and other vector-borne diseases facilitated (Keiser et al. 2005; Mutero et al. 2006). However, the impact of agriculture and natural-resource management practices on malaria has not been widely acknowledged or studied in Africa. For example, there is paucity of information on the negative impact of malaria on household food security, labour and general investment in agriculture. Furthermore, only limited scientific data exist on the links between farming systems, socio-cultural and economic factors and the risk of exposure to malaria. There is a need to examine malaria in an agricultural context, because improvements to agricultural practices present opportunities not only for improving health and livelihoods, but also for changing risk patterns of malaria in East Africa.

About the Three Projects

As part of the SIMA, the three projects sought to contribute new knowledge by identifying key determinants of malaria transmission in relation to agricultural, pastoral and aquacultural systems. They also wished to inform key opportunities for malaria control that could be scaled up and out for wider use by the communities and government authorities in the region.

The objectives of these studies were in line with the objectives and mandate of SIMA: to generate knowledge on the relations between malaria and agriculture (the complex links among socio-cultural, economic and environmental factors resulting from natural-resource management and production activities); and to contribute to a reduction of malaria, resulting in improved health and well-being, increased agricultural productivity and poverty alleviation.

Methodology

All three projects (in keeping with the idea of SIMA) adopted an ecosystem approach to human health (Lebel 2003). They each examined the links between agricultural and natural-resource management practices and malaria, so as to unravel the different elements of malaria epidemiology and socio-economic impacts. In Uganda (two projects), the thrusts were on the role of livestock and its management practices among Bahima communities, and the aquacultural practices in malaria epidemics in Mutara communities in the malaria-prone southwestern areas of the country. In Tanzania (one project), the focus was on malaria in relation to diverse agro-ecosystems, farming practices and socio-economic factors in a rural district of Mvomero in the central part of the country.

The geographic focus and characteristics of each study area differed between these projects. In Uganda, increased reports of malaria epidemics from the southwest attracted the focus of two projects. In Tanzania, it was the availability of different agro-ecosystems within short distances in a perennial malaria zone that guided the study location. These zones included sugarcane, flooding rice irrigation, non-flooding rice irrigation, wet savannah and dry savannah (Fig. 13.2).

Although each project was independently designed, each employed similar entomological (i.e., sampling of larvae and adult mosquitoes, determining breeding sites of mosquitoes, sporozoite rates), malariometric (prevalence and incidence of malaria) and sociological surveys. In addition, community diagnosis workshops were conducted for the farming communities and other stakeholders. The locations of the study areas, including farming systems, villages and health facilities, were geo-referenced using a Global Positioning System (GPS). All coordinates were translated into geographic information system (GIS) supported maps.

Each project was planned and designed by a separate multidisciplinary research team in collaboration with policymakers, local politicians and other stakeholders who were essential to the success of the project. The projects drew on the methods and experiences of scientists from diverse disciplines. These included experts in agriculture (agronomy, agribusiness, agricultural extension and education), sociology, public health (epidemiology, disease ecology, parasitology and entomology), GIS, environment and biometrics. In each project, teams synthesized their work into a common definition of the research problem and approach. Extensive contextual data were collected that proved crucial to reformulating the questions and pointing to the likely causes of the problems.

In keeping with a participatory approach, in each project, stakeholders were sensitized to the project objectives and planned research activities, and were involved in mobilizing their communities, guiding the research team during household surveys,

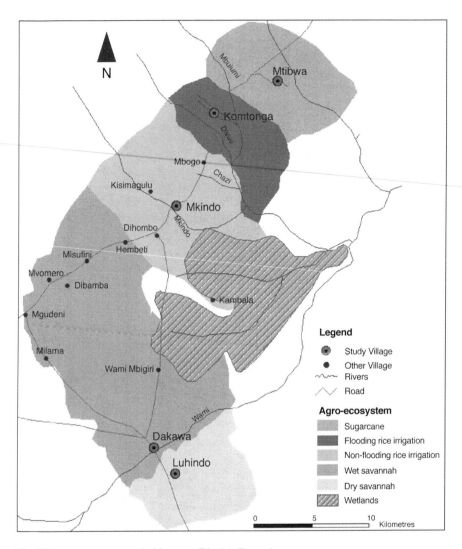

Fig. 13.2 Agro-ecosystems in Mvomero District, Tanzania

and collecting malaria-related data. Data were also collected on socio-economic and gender variables, and livelihood patterns. Regular meetings were held with the leadership of villages to review the research progress and plan the way forward. The data were synthesized in a participatory manner by the research teams and the communities, district officials and policymakers. The communities received feedback on the results and made invaluable contributions to knowledge translation and the identification of potential options for management of mosquitoes and malaria.

Research Findings

Uganda Livestock–Malaria Project

The livestock–malaria project in Uganda established that malaria incidence in the study area had greatly increased since the national entomological surveys of the 1960s (Bennett and Hall 1968; Okello-Onen, unpublished). In Uganda, land restructuring policies have changed the traditional relationships between cattle keepers and animals: there has been a gradual shift from living in traditional huts, sited close to animal night paddocks (kraals), to living in permanent structures sited far from the paddocks. A comparative project design aimed to compare malaria burden among traditional nomadic pastoralists, settled cattle farmers, and those in transition. During project implementation, far fewer truly nomadic pastoralists were found than expected. Expectations were for 150 nomadic households in the eight villages; but only 16 were found. This did not allow for effective comparisons to be made on key variables between the three categories of communities. Nonetheless, preliminary research findings were different from what had been anticipated. Although there was no substantial difference in malaria prevalence between the nomadic, transitional and settled people (Fig. 13.3), the transitional group was more likely to have an increased malaria burden than the other two groups.

The settled communities were expected to be at greater risk of exposure to malaria parasites than the transitional and nomadic pastoralists. This is because the traditional close contact of nomadic people with cattle would deflect malarial mosquitoes from feeding on humans (a phenomenon termed passive zooprophylaxis), and reduce the malaria problem. Instead, it was found that *Anopheles arabiensis*, which is widely known to be zoophilic and opportunistic in its feeding (on both cattle and humans), was absent from the study sites. One possible reason is that these mosquitoes were inhibited by the intensive application of synthetic pyrethroids on cattle to control ticks and tsetse flies. The project found that the most prevalent vector was the morphologically indistinguishable *Anopheles gambiae*, which may feed preferentially on people. Ancillary data collected on ecological, social, economic and cultural aspects of communities showed that land allocation and privatization at the time of settlement had not considered access to water. This led to massive local proliferation of water-storage ponds for animals and people, which increased vector-breeding sites with consequences for malaria transmission. Increased malaria prevalence was also attributed to a decline in the use of cultural and traditional preventive measures such as burning cow dung and using herbs to treat malaria; increased density of the human population; local manifestations of global climatic change; and increased anti-malarial drug resistance, which make clinical malaria much more apparent in the population.

The results also showed that the breeding of anopheline mosquitoes was influenced by floating vegetation and shade levels around water bodies. When floating vegetation (*Azolla filiculoides* or "water fern") covered more than 90% of the

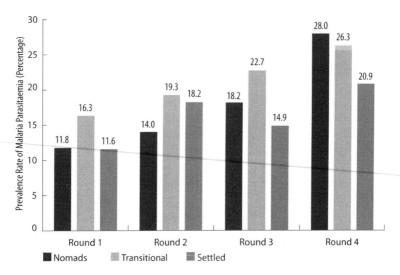

Fig. 13.3 The prevalence of malaria in the nomadic, transitional and settled communities in Nyabushozi County, Kiruhura District, Uganda

pond surface, no mosquito breeding sites could be detected and the anophelines were almost completely eliminated in these areas. The planting of trees provides a thick canopy that shades the ponds and reduces water temperature, thus creating unsuitable conditions for some mosquito species to lay eggs. Further, planting of trees around ponds and dams can deny access by animals to watering points, thus reducing the creation of hoof prints, which are fertile grounds for mosquito breeding.

Housing structures also had significant influence on the population density of adult anopheline mosquitoes. Grass-thatched structures had a higher number of mosquitoes when compared with permanent structures roofed with iron sheets or tiles. Provision of polythene ceilings in grass-thatched houses may reduce the number of resting places for mosquitoes indoors.

Uganda Aquaculture–Malaria Project

This SIMA project demonstrated that changing agricultural practices, especially introducing fish-farming to traditional farming of crops, aggravated the malaria problem among the Mutara community. High prevalence rates of malaria among humans, averaging 60%, and a sporozoite rate of 4% among mosquitoes were reported. This information on malaria was new to the communities and triggered a greater appreciation of malaria, and a higher priority was placed on control measures. Before the project, the communities misconstrued repeated infections and severe symptoms of malaria (such as convulsions and confusion) as the effects of witchcraft, which hampered efforts to recognize and control the disease. They were not aware that malaria is transmitted by female *Anopheles* mosquitoes that breed in stagnant water held in the leaf axils of some plants and in water ponds devoid of fish. They were also not aware of the behaviour and infectivity of adult mosquitoes, or how to interpret different levels of parasitemia. As a result of participating in the project and sharing in the learning from research results, the representatives of the communities gained valuable new knowledge on malaria and then demanded new malaria laboratory services. They developed drama (community theatre) as a vehicle to disseminate the knowledge to neighbouring communities. The changed behaviour of Mutara communities to seek more modern diagnostic services for malaria offers a counterbalance to traditional beliefs and superstitions concerning clinical manifestations of the severe form of malaria.

Tanzania Malaria in Agro-Ecosystems Project

The Tanzania study found that malaria transmission and disease burden vary from one agro-ecosystem to another. As expected, irrigated rice farming contributed more highly to malaria transmission intensity and prevalence than other agro-ecosystems. The prevalence of malaria and helminth co-infections was highest among school children from rice-farming systems. Only a third of sick individuals sought treatment from health facilities, and the formal health-care system reached fewer than half of the individuals who required treatment. There was a high frequency of consulting drugstores as the major alternative source of treatment. As a result, a precious part of already limited household income was spent on self-prescribed malaria treatment and prevention.

Although malaria was considered by the majority of farming communities as the most important public-health problem, the level of knowledge about malaria disease and its links with agricultural practices and control was generally low, but depended on the respondent's level of formal education (Mboera et al. 2009). Household economic losses as a result of malaria illness, and expenditures on its treatment and prevention, were observed to be substantial. High levels of malaria prevalence and transmission were observed among communities that practiced irrigated rice farming; especially those communities that used traditional subsistence rice-farming systems that are fragmented and often have faulty irrigation systems that create "pockets" of mosquito breeding sites. Late treatment-seeking behaviour was common among most households.

Participatory involvement of the community and sectors other than health (agriculture, water management) in developing malaria-control strategies at local level contributed to changing malaria control policy including changes in the current National Medium-Term Malaria Strategic Plan (2008–2013).The National Malaria Control Advisory Committee now includes members from outside the Health Sector, particularly from Agriculture and Irrigation (TZ-MOH 2008).

Outcomes

The two projects in Uganda registered some important outcomes and opportunities for malaria control that encompass the use of environmental-management strategies and cultural practices, backed up by enhancing community capacity for sustaining malaria control. The strategies include management of water bodies, cleaning and improving the design and construction of artificial ponds to capture rainwater, covering or filling areas that accumulate water (e.g., hoof prints and other depressions, and containers), introducing *Azolla* plant species into water bodies, using biocontrol agents such as *Bacillus thuriengiensis*, planting shade trees around water bodies, and improving the design and structures of houses to keep out mosquitoes.

Important reductions in mosquito breeding have been achieved by using *Azolla* on ponds used to water livestock as well as other methods that provide shade to bodies of water. These can be regarded as key agro-ecosystem interventions that can be exploited in other communities, in combination with other malaria-control strategies.

The Uganda malaria–livestock project findings were important in shifting prevailing beliefs that settled communities were at greatest risk of malaria, by linking the problem to the proliferation of mosquito larvae in water-storage ponds in the transitional communities. They also found that a particular species of mosquito (*A. gambiae*) predominated as the malaria vector. This might be linked to certain practices that occur during the settlement transition such as changes in structures of human housing, increased distances between human housing and cattle "kraal", engagement in crop farming, or lower youth-acquired immunity in this previously less-exposed population. A narrow initial approach would have simply led to an unproductive negative result.

The aquaculture–malaria project in Uganda provided evidence-based proof of the benefits of an ecohealth approach. As a result of generating and using scientific knowledge in a participatory manner with communities in Mutara village, they demanded and are currently using privately offered diagnostic laboratory services to ensure early detection of malaria before purchasing anti-malarial drugs from drugstores.

The lessons learned during the research process in Tanzania have reinforced the importance of involving communities and policymakers in identifying research problems and priorities. The most challenging issues encountered were the lack of links, and the presence of contradictions, in the kind of information and education that were provided by health and agricultural extension workers to the community about malaria control and farming practices. The project generated new knowledge on the links between agriculture and malaria, produced evidence of the relationships of parasitic co-infections with agricultural practices, and stimulated the willingness of communities to develop appropriate malaria-control strategies.

The findings showed that land use and water management in rice farming have an impact on mosquito productivity and the prevalence of malaria and other parasitic diseases. The use of appropriate agricultural practices, including intermittent irrigation practices, can be improved through community-education packages. Realizing the significant variation in malaria transmission between different agricultural production systems, interventions need to take into consideration the farming systems and practices and the local malaria prevalence and transmission patterns.

The knowledge-translation techniques used in Tanzania provided a crucial forum for sharing research findings with the community, decision makers and policymakers. Reaching out to the media was another important approach for conveying research results to the public and policymakers. This participatory research methodology was a new experience for communities and was highly commended by all stakeholders (Mlozi et al. 2006).

A strong model for capacity building in transdisciplinary research was created with these three projects. The studies promoted interactions and enhanced partnership between research teams, policymakers, practitioners, and the communities to understand each other's thinking and evolve a common understanding of the research questions being addressed. The projects established links between agricultural practices, livelihood patterns, and malaria risk, and the communities were able to identify potential malaria-control interventions that could be implemented. The projects also supported the training of postgraduate students.

Overall, the projects generated a better picture of the complex epidemiology of malaria and evidence of how the team's understanding of the communities, malaria, and agricultural environment developed. In Uganda, notable changes in perception, knowledge, and behaviour of the communities were noted with respect to relationships between mosquitoes and malaria. Several people began to clean their water points and homesteads to reduce mosquito breeding and resting spots. Many of them bought mosquito nets for their families; something which they never previously thought was a priority.

In both countries, the Ministries of Health provided financial support to the projects. In Uganda, the districts where the studies were undertaken promised to include strategies for environmental management of malaria in their strategic and development plans. In Tanzania, the Permanent Secretaries of the Ministries of Health, Agriculture, Environment, and Livestock Development willingly participated in the Research Translation Workshops. This participation was key to achieving greater integration between sectors in formulating Tanzanian malaria policy.

Conclusion

The three projects made progress towards achieving a practical understanding of environmental and other changes that led to more malaria. Without the ecohealth framework, some of these research projects might have ground to a halt after clear refutation of the initial hypothesis formulated by villagers and researchers. An ecohealth approach provided an adaptable framework to reframe the research to pursue further aspects of malaria (e.g., the Uganda team worked with settled former pastoralists on water-storage issues). The project generated new information on the determinants of malaria burdens, malarial transmission, and its complex and diverse determinants and impacts in rural settings. Prevailing assumptions about malaria were shown to reflect only one aspect of a complex reality, and stakeholders were engaged in thinking about a wider set of drivers of malaria, and of obstacles to control. Within the ecohealth context, new and more fruitful approaches emerged. The studies provided evidence for developing adaptable, rational, and feasible malaria-control strategies, and point to the need for more research to test and implement these strategies to achieve effective malaria control.

Acknowledgments This paper is the product of collaborative work that adopted a transdisciplinary research protocol in Uganda and Tanzania, and the culmination of the efforts of many communities and a broad range of stakeholders. We recognize Dr. Clifford Mutero for his valuable contributions as the coordinator of the Systemwide Initiative of Malaria and Agriculture (SIMA) and also thank Professor David Bradley, London School of Hygiene and Tropical Medicine and Dr. Anne Bernard for their insights. IDRC support was provided through the projects 100927-02 and 102155. The Dutch Ministry of Foreign Affairs (DGIS), through the International Water Management Institute, provided funding for research on agro-ecosystems and malaria in Tanzania.

References

Bennett, J., and Hall, S.H. (Editors) (1968). Uganda Atlas of Disease. Oxford University Press, Nairobi.
deBartolome, C.A., and Vosti, S.A. (1995). Choosing Between Public and Private Health Care: A Case Study of Malaria Treatment in Brazil. Journal of Health Economics, 14, 191–205.
Ijumba, J.N., and Lindsay, S.W. (2001). Impact of Irrigation on Malaria in Africa: Paddies Paradox. Medical and Veterinary Entomology, 15, 1–11.
Keiser, J., De Castro, M.C., Maltese, M.F., Bos, R., Tanner, M., Singer, B.H., and Utzinger, J. (2005). Effect of Irrigation and Large Dams on the Burden of Malaria on a Global and Regional Scale. American Journal of Tropical Medicine and Hygiene, 72, 392–406.
Lebel, J. (2003). Health: An Ecosystem Approach. In Focus Series. International Development Research Centre (IDRC), Ottawa. Available at: http://www.idrc.ca/in_focus_health/.
Lindsay, S.W., and Birley, M.H. (1996). Climate Change and Malaria Transmission. Annals of Tropical Medicine and Parasitology, 90, 573–588.
Mboera, L.E.G., Fanello, C.I., Malima, R.C., Talbert, A., Fogliati, P., Bobbio, F., and Molteni, F. (2006). Comparison of the Paracheck-Pf Test to Microscopy for Confirmation of *Plasmodium falciparum* Malaria in Tanzania. Annals of Tropical Medicine and Parasitology, 100(2), 115–122.
Mboera, L.E.G., Mlozi, M.R.S., Senkoro, K.P., Rwegoshora, R.T., Rumisha, S.F., Mayala, B.K., Shayo, E.H., Senkondo, E., Mutayoba, B., Mwingira, V., and Maerere, A. (2007). Malaria and Agriculture in Tanzania: Impact of Land Use and Agricultural Practices on Malaria Burden in Mvomero District. Policy Report, National Institute for Medical Research, Dar es Salaam.
Mboera, L.E.G., Shayo, E.H., Senkoro, K.P., Rumisha, S.F., Mlozi, M.R.S., and Mayala, B.K. (2009). Knowledge, Perceptions and Practices of Farming Communities on Linkages Between Malaria and Agriculture in Mvomero District, Tanzania. ActaTropica, 113(2), 139–144.
Mlozi, M.R.S, Shayo, E.H., Senkoro, E.H., Mayala, B.K., Rumisha, S.F., Mutayoba, B., Senkondo, E., Maerere, A., and Mboera, L.E.G. (2006). Participatory Involvement of Farming Communities and Public Sectors in Determining Malaria Control Strategies in Mvomero District, Tanzania. Tanzania Health Research Bulletin, 8(3), 134–140.

Mutero, C.M., Amerasinghe, F., Boelee, E., Konradsen, F., Van der Hoek, W., Nevondo T., and Rijsberman, F. (2005). Systemwide Initiative on Malaria and Agriculture: An Innovative Framework for Research and Capacity Building. EcoHealth, 2(1), 11–16.

Mutero, C.M., McCartney, M., and Boelee, E. (2006). Understanding the Links Between Agriculture and Health: Agriculture, Malaria and Water-Associated Diseases. Brief 6. International Food Policy Research Institute, Washington.

Rowe, A.K., Rowe, S.Y., Snow, R.W., Korenromp, E.L., Armstrong-Schellengberg, J.R.M., Stein, C., Nahlen, B.L., Bryce, J., Black, R.E., and Steele, R.W. (2006). The Burden of Malaria Mortality Among African Children in the Year 2000. International Journal of Epidemiology, 35, 691–704.

TZ-MOH (Tanzania Ministry of Health and Social Welfare). (2008). Medium Term Malaria Strategic Plan, 2008–2013. Ministry of Health and Social Welfare, Dar es Salaam, United Republic of Tanzania.

UG-MOH (Uganda Ministry of Health). (1998). UGANDA National Malaria Control Policy. Ministry of Health, Kampala, Republic of Uganda.

WHO (World Health Organization). (2008). World Malaria Report 2008. World Health Organization, Geneva.

Chapter 14
An Ecosystem Approach for the Prevention of Chagas Disease in Rural Guatemala

Carlota Monroy, Xochitl Castro, Dulce Maria Bustamante, Sandy Steffany Pineda, Antonieta Rodas, Barbara Moguel, Virgilio Ayala, and Javier Quiñonez

In Latin America, more than 10 million people carry a parasite that puts them at risk of developing Chagas disease. This chronic and debilitating illness is caused by a microscopic blood parasite called *Trypanosoma cruzi*. The parasite is transmitted by several species of insects called chinches, kissing bugs, or triatomines. People usually acquire the disease in childhood, experiencing flu-like symptoms; but later in life, about one-third of cases develop more severe symptoms affecting the digestive system and the heart. Chagas disease kills more than 10,000 people per year (WHO 2010).

The parasite naturally lives in more than 150 mammal species, including people. The parasite is mainly transmitted by contact with infected vector excrement. Many vector species have become particularly well-adapted to living in and around traditional mud-brick and thatch homes common to poorer areas. Some of these vectors are not native to the region and have no natural habitat or hosts. They can be controlled with improved hygiene and regular insecticidal spraying, because they cannot survive anywhere else. However, in Guatemala and other areas of Central America, some of the insect vectors are native and widespread, and can easily re-infest houses after insecticidal treatment.

One such insect, *Triatoma dimidiata*, is native to Central America and an important vector of Chagas disease. It lives in forests and other areas, but frequently invades houses and yards (peridomestic areas). In Guatemala, the parasite *T. cruzi* is

---

C. Monroy (✉) • X. Castro • D.M. Bustamante • A. Rodas • B. Moguel
Laboratorio de Entomología Aplicada y Parasitología (LENAP), Facultad de Ciencias Químicas y Farmacia, Universidad de San Carlos de Guatemala (USAC), Guatemala City, Guatemala
e-mail: mcarlotamonroy@gmail.com

S.S. Pineda
Institute for Molecular Biosciences, The University of Queensland, St Lucia, QLD, Australia

V. Ayala • J. Quiñonez
Facultad de Ingeniería, Universidad de San Carlos de Guatemala (USAC), Guatemala City, Guatemala

present in wild populations of *T. dimidiata* and its wild animal hosts, including opossum and armadillo (Calderon et al. 2004; Monroy et al. 2003). Thus, the highest infestation rates in houses occur in the most deforested areas in the eastern part of the country (Tabaru et al. 1999). The control of house infestations using insecticide sprayings is challenging (Nakagawa et al. 2003a, b) and re-infestations are frequent (Hashimoto et al. 2006). Re-infestation is a consequence of the capability of these insects to inhabit many different environments (domestic, peridomestic and wild) and to migrate among them (Dumonteil et al. 2004; Monroy et al. 2003).

Vector distribution and Chagas risk in Guatemala are closely related to poor socioeconomic conditions, certain cultural characteristics, adobe housing construction and poor hygienic conditions (Bustamante et al. 2009). It is impossible to improve the problem of Chagas disease without involving people in the management of their environment. An ecosystem approach promotes a pragmatic response to this complex reality.

Simply keeping the house tidy (i.e., removing clutter that could serve as hiding places for the bugs) can help prevent infestation (Zeledón and Rojas 2006), although in many contexts this is not sufficient. Achieving the substantial and permanent behavioural changes required to control Chagas at the level of the community is not simple. It requires that changes be put into practice and incorporated into people's daily lives, and passed on to future generations.

In this study, an ecosystem approach to human health research (Lebel 2003) was adopted to address *T. dimidiata* infestation in rural Guatemala. A team of scientists from different disciplines (medical entomology, anthropology, microbiology, architecture and civil engineering) worked together to understand the different facets of the problem and to devise practical interventions to reduce Chagas risk and improve community well-being. The research was undertaken in the Department of Jutiapa in collaboration with the local public health officers (vector-borne diseases unit – ETV), village authorities and ordinary citizens who provided input to adapt the project's recommendations. The final intervention was based on the research findings, incorporated local knowledge and practices, and was realistic in both economic and cultural senses, and thus more easily implemented by the local people.

This paper summarizes the experiences in adopting an ecosystem approach, the scientific findings, and the challenges encountered while coordinating efforts among different stakeholders. The results of a knowledge, attitudes and practices (KAP) survey, a socioeconomic survey, and the efforts to engage the community in environmental management activities are presented.

Implementation of the Approach

Four villages in the Department of Jutiapa participated in this study: El Tule and La Brea in the Municipality of Quesada; and El Sillon and La Perla in the Municipality of Yupiltepeque (Fig. 14.1). These municipalities share similar socioeconomic levels and are typical of this region (Bustamante et al. 2009). The study sites, the risk factors

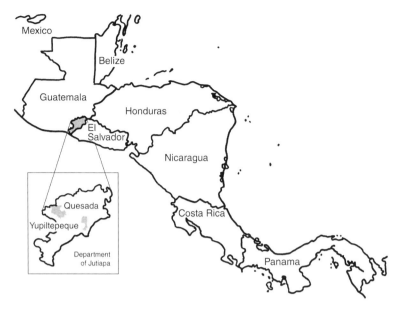

Fig. 14.1 Location of the study site: Municipalities of Quesada and Yupiltepeque, Department of Jutiapa, Guatemala, Central America

for house infestation (Bustamante et al. 2009), house conditions, entomological surveys, insecticide sprayings and the development of an improved plaster mixture for adobe walls have been described in detail elsewhere (Monroy et al. 2009).

After initial surveys on housing conditions and *T. dimidiata* infestation, houses in the four villages were classified into three categories (A, B, or C) depending mostly on the conditions of house walls and hygiene (Monroy et al. 2009). Houses of type A had the best conditions and type C had the worst conditions, meaning incomplete or absent wall plastering of adobe walls, and overall poor hygiene.

Data from these initial surveys were analyzed to identify risk factors for house infestation by *T. dimidiata* (Bustamante et al. 2009). Regardless of the age of the house, bug infestation was less likely in adobe houses that had walls with complete plastering and houses with better hygiene (i.e., overall cleanliness, no animals inside the house, clean floors). Most of the houses with these conditions were houses in category A and some in category B, thus most of the high-risk houses corresponded to those in the category C. Rudimentary wall plastering was the most common house improvement that villagers already implemented in their houses, which make it culturally appropriate to suggest the introduction of improved wall plasterings as part of the intervention to improve homes and prevent *T. dimidita* infestation (Monroy et al. 2009). Interviews were used to identify the traditional methods of wall plastering, finding that it is not men but usually women and children who plaster the walls using their bare hands (Monroy et al. 2009).

"Ecosystem Intervention" and "Traditional Intervention"

The villages of El Tule and La Brea participated in the "Ecosystem Intervention" (EI), and El Sillon and La Perla participated in the "Traditional Intervention" (TI). Both types of intervention (EI and TI) included initial application with insecticide (Deltamethrin 5%) in almost all houses to reduce the intradomestic vector population to near zero. Serological assessment [enzyme-linked immunosorbent assay (ELISA) screening plus confirmatory immunofluorescence] was done in children less than 15 years old with parental consent. All children found to test positive for *T. cruzi* were treated with benznidazole by the Ministry of Health. The EI villages also received housing improvement activities. The research group spent 6 months conducting workshops and promoting community participation to improve the plastering of walls using local materials. Various other activities described in the next section were conducted in the EI villages that supplemented house improvement and substantially helped in the success of this type of intervention.

Two types of surveys were conducted in both EI and TI villages before and after the interventions to evaluate their efficacy. Entomological surveys were conducted in 100% of the houses in each village, using the person/hour method to search for bugs. A KAP survey was applied in 30% of the households in each village to investigate the perceptions of the population about bug infestation and Chagas disease.

Research Team and Community Participation

Effective collaboration among researchers from different disciplines was not easily achieved. Frequent meetings among the scientists were essential. It was important to set the common goals and actions according to the findings, and to discuss how to best interact with the people in the villages to introduce the interventions, while always respecting local traditions. A house was rented in a nearby village (Jutiapa), and this became the centre of operations and a meeting point, which facilitated interactions among stakeholders. Working with the local public health officers of the ETV was essential to gain access to the communities. The periodic monitoring of project progress by public health officers from the Central Government and the Department of Jutiapa was important to help change their perception that insecticide application alone was the best strategy for the control of Chagas vectors.

The principles of participation and equity were sought through various activities. Emphasis was placed on taking local traditions into consideration during the implementation of the intervention. Interviews with local community leaders helped project staff understand the cultural practices related to wall plastering (Monroy et al. 2009).

The anthropologist in the team lived near the EI villages, which facilitated daily visits, and she became well known and trusted among the local people. She visited with women and spoke to them about house hygiene and how the whole family can contribute to house cleanliness. The engineers and architects also worked with villagers, seeking the best local materials to formulate the wall plaster. All the materials were then subjected to laboratory assays at the university to find an end-product that would be durable and suitable for application with bare hands, as is traditionally done.

River sand was required as one of the ingredients of the new wall plastering, but it was scarce at the EI villages. The research team and community leaders worked together to negotiate the procurement of sand from the municipal authorities of Quesada, who donated ten loads of sand. The experience helped the village leaders develop negotiation skills. Because agriculture is the main occupation in the villages, it was essential to consider the seeding and harvesting times when planning the timing of interventions (i.e., sprayings and wall plastering).

The project employed a local community self-development approach, rather than focusing only on disease control. This approach motivated the communities to participate in the study, because Chagas disease was not recognized as a community priority in the beginning. In response to community interests raised during village discussions, other activities were promoted to aid community development. Greenhouses of both native and fruit trees were established, free seedlings were provided by the Quesada Municipality nursery, and reforestation was promoted among the population. Workshops were also conducted on the management of stingless "Maya bees", which are local insect pollinators also appreciated by rural people for the medicinal qualities of their honey. The number of beehives has increased in the region, and some families are generating income from selling the hives to other villagers. Many other activities were conducted that helped the project to be perceived by the locals as more than a Chagas-control project but as a community development project.

The project recruited research assistance from two local young men from the EI villages. They helped the project to understand the needs and circumstances of each family in the villages. They were the first to pilot the house improvements developed by the project. They became examples for the entire community, who witnessed their changes in attitudes and the conditions of their houses. One of them was later elected to the local village authority.

Socio-Anthropological Survey and Data Analysis

The KAP survey collected information on knowledge about Triatomine bugs and Chagas disease, attitudes towards the disease, hygienic practices in the household and economic conditions (including source of income). At each house, a person older than 10 years of age (with informed consent or informed parental consent) was interviewed using a 45-item questionnaire. The survey was pretested in other villages from the same department with similar socio-economic and cultural conditions. The people who conducted the survey also evaluated the overall hygienic conditions of the house during household visits for the interviews. Statistical differences between pre- and post-intervention responses, categorized by intervention type, were assessed using a test of the difference between two proportions.

Results

After intervention in the EI villages, the peridomestic (yard) environment became the more preferred habitat for bugs, holding 82.5% of them. This was a significant change from before the intervention (47.6%). Because the vector primarily feeds on humans in houses at night, these changes of spatial distribution of the bugs are likely to reduce human–vector contact, and thus disease transmission.

Improvements were not achieved in all houses of the EI villages. *T. dimidiata* was eliminated from all improved houses in those villages, but the bugs remained inside the houses that were not improved. In the EI village of La Brea, 39.3% of homes were positively improved (changed from category C to B). There was not the same level of interest in house improvements in the other EI village of El Tule, where there was only a 19.8% decrease in category C houses. No differences in the proportion of house category were found in the TI villages. More detailed results can be found in Monroy et al. (2009).

Sustainability of the project interventions depended on ongoing community participation. A culturally sensitive approach resulted in the implementation of an improved plastering technique informed by local traditions. Communities continued basically doing what they knew how to do, but with the new mixture of local materials. This is a major contribution toward sustainability of this change: they are likely to continue using this technique, leading to substantial reduction in exposure to the Chagas vector. In fact, the model is so similar to what they traditionally do, that the neighbouring villagers copied the "sand–mud" improvement technique. The communities and homes now look much better, and home-owners expressed pride in their improved houses.

Results from the KAP survey pre- and post-interventions (Table 14.1) show that some knowledge about Chagas disease and its vector existed among the population, but knowledge was incomplete and poor hygiene practices favoured infestation. Some of the unhygienic practices were changed for the better after the interventions in both study groups (EI and TI), but especially in the EI villages. For example, there was significantly less clutter on the walls (reducing hiding places for bugs), and there was better cleanliness of the beds and floors.

Other practices remained unchanged. The percentage of people keeping the beds away from walls (where bugs are known to hide, facilitating night feeding on sleeping people), or keeping animals inside (a source of blood for the bugs) did not show significant changes in either group. Women remained responsible for bed cleaning (which might increase their exposure to infective bug excrements) in all villages.

Socio-economic characteristics are related to Chagas risk. The sources of income are constantly changing in these villages. In 2006, changes in prerequisites of the National Civil Police coincidentally reduced employment for inhabitants of TI villages, who had previously counted on some employment with the police. An increase in commercial (retail) activities was observed. Out-migration for work is a factor in this region. In both TI and EI villages, there was an increase in sales of agricultural land by villages to pay for travel to attempt to immigrate to the United States. Some

Table 14.1 Findings of KAP surveys conducted before and after Traditional (TI) and Ecosystem Interventions (EI)

	TI Villages			EI Villages		
	Before	After	SS[a]	Before	After	SS
Knowledge						
Identification of *T. dimidiata*	91	94	ND	89	97	SD
Association of *T. dimidiata* with Chagas disease	5	24	SD	16	84	SD
Identification of *T. dimidiata* faecal streaks	1	21	SD	16	79	SD
Association of *T. dimidiata* fecal streaks with Chagas disease	1	18	SD	6	80	SD
Attitudes and practices						
Good hygiene of floors	69	76	SD	66	96	SD
No cluttering under the beds	44	61	SD	60	83	SD
No cluttering on walls	27	28	ND	18	37	SD
Good hygiene of the beds	69	71	ND	66	89	SD

Results are percentages of respondents for each characteristic
[a]*SS* statistical significance: SD indicates a significant difference ($p<0.05$) using a test of the difference between the two proportions

of these attempts led to deportation back to Guatemala, with substantial economic losses for some families due to the costs of preparing for an emigration attempt, combined with the loss of remittances from abroad. Poverty increased in these households because land could not be reacquired, and suitable land for agriculture is essential for the survival of households in these villages.

The "EI" plastering technique was successful because (1) local knowledge was directly incorporated into both how to do the plastering and the technological improvements in the technique; (2) workshops were held to promote the new technique; (3) the regular and frequent presence of the research team at the EI villages garnered the trust of the local people; (4) the house-to-house visits by the team anthropologist meant she became well known and trusted; (5) the local vector-borne disease control officers participated; (6) the municipal authorities donated river sand; (7) local people were hired to work on the project; and (8) the planned interventions considered the local agricultural calendar.

Discussion

Generally, poverty may not be improving as much as hoped in Guatemala. The project found that, in our study villages, it may be increasing. The economic struggle of the people in the participating villages is evident. Agriculture produces low income and people tend to migrate to bigger cities or abroad in search of a better future, a phenomenon shared with other Latin American countries where Chagas disease is endemic (Avila et al. 1998; Briceño-León 1994). This can have implications for spread of Chagas disease, where there are competent vectors.

The survey revealed that many unhygienic practices were improved in the EI villages, but others remained unchanged. Change is difficult, and long-standing practices associated with a particular culture can be particularly resistant to change. This worked in the project's favour for adoption of the new plastering technique, but hampered other desirable changes, like the keeping of animals in the house.

The full participation of local public-health officers with the research team was essential to demonstrate interinstitutional coordination, not only to the community but to the municipality and other local organizations. The public-health officers had the opportunity and challenge to work with the different experts (anthropologist, architects, engineers, biologists and entomologist) involved in the project. Public-health officers from the central and local governments witnessed the progress of the activities in the EI communities and were able to observe the differences with the TI villages. The tangible results influenced the formal adoption of the project's plastering technique using local materials as part of an official strategy for the control of *T. dimidiata* in Guatemala.

Motivating community participation and engagement in tackling Chagas disease and its risk factors is a challenging enterprise. To achieve this, it was important to avoid a "single-minded" approach (focusing only on bug control) and to appropriately consider community perceptions and views of desirable community development. The research team promoted several activities: reforestation with native trees; a school library; a coffee-plants nursery; vaccination of chickens against the most common diseases; establishment of Maya beehives; construction of chicken coops; sterilization of dogs; planting of fruit trees in yards; and meetings with municipality authorities and villages leader. Many of these activities emerged from the needs expressed by the communities. The inclusion of activities that were aimed at creating a better quality of life facilitated participation and engagement in Chagas-control activities.

Conclusion

After 2 years of intensive team work, the exposure of humans to Chagas-disease vectors in the EI villages was reduced by decreasing the level of human–vector contact (Monroy et al. 2009). House improvements led to a decrease of domestic infestation rates in the EI villages, which demonstrated the protective ability of the plastering and improved hygiene interventions. By respecting local culture, improving housing conditions, and responding to community needs, the EI prevented vector re-infestation and simultaneously improved the quality of life in two rural communities of Guatemala.

This experience produced methodologies and techniques that have already been introduced to other villages in eastern Guatemala. The involvement of the local municipalities in the resolution of health problems (donation of sand to improve houses), and the leadership of the public-health workers, made this municipal-scale application feasible. Honduras and El Salvador have similar cultural backgrounds, and these technologies may be applicable in these countries. This innovative procedure for house improvements has been transferred to personnel from the Ministries of Health in El Salvador and Honduras thanks to the collaboration of the Japan International Cooperation Agency.

Acknowledgments We thank the village investigators Belter Alcántara, Oscar Villanueva, Ranferi Trampe and Leonicio Revolorio (ETV), and the villagers of La Brea, El Tule, La Perla, and El Sillon for their participation. IDRC support was provided through projects 101812 and 103696-005.

References

Avila, G., Martínez, M., Ponce, C., Ponce, E., and Soto, R. (1998). La Enfermedad de Chagas en la Zona Central de Honduras: Conocimientos Creencias y Prácticas. Revista Panamericana de Salud Pública, 3(3), 158–163.

Briceño-León, R. (1994). Retos y Problemas Para Alcanzar la Participación Comunitaria en el Control de las Enfermedades Tropicales. Fermentum, 8–9, 165–176.

Bustamante, D.M., Monroy, C., Pineda, S., Rodas, A., Castro, X., Ayala, V., Quiñonez, J., Moguel, B., and Trampe, R. (2009). Risk Factors for Intradomiciliary Infestation by the Chagas Disease Vector *Triatoma dimidiata* in Jutiapa, Guatemala. Cadernos de Saude Publica, 25 (Suppl. 1), S83–S92.

Calderon, C., Dorn, P., Melgar, S., Chavez, J., Rodas, A., Rosales, R., and Monroy, C. (2004). A Preliminary Assessment of Genetic Diferentiation of *Triatoma dimidiata* (Hemiptera: Reduviidae) in Guatemala by RAPPD-PCR. Journal of Medical Entomology, 41, 882–887.

Dumonteil, E., Ruiz-Piña, H., Rodríguez-Félix, E., Barrera-Perez, M., Ramirez-Sierra, M.J., Rabinovich, J.E., and Menu, F. (2004). Re-Infestation of Houses by *Triatoma dimidiata* after Intradomicile Insecticide Application in the Yucatán Peninsula, Mexico. Memórias do Instituto Oswaldo Cruz, 99, 253–256.

Hashimoto, K., Cordon-Rosales, C., Trampe, A., and Kawabata, M. (2006). Impact of Single and Multiple Residual Sprayings of Pyrethroid Insecticides Against *Triatoma dimidiata* (Reduviidae; Triatominae), the Principal Vector of Chagas Disease in Jutiapa, Guatemala. American Journal of Tropical Medicine and Hygiene, 75, 226–230.

Lebel, J. (2003). Health: An Ecosystem Approach. In Focus Series. International Development Research Centre (IDRC), Ottawa, Canada. Available at: http://www.idrc.ca/in_focus_health/.

Monroy, C., Bustamante, D.M., Rodas, A., Enriquez, E., and Rosales, R. (2003). Habitats, Dispersion and Invasion of Sylvatic *Triatoma dimidiata* (Hemiptera: Reduviidae Triatominae) in Peten, Guatemala. Journal of Medical Entomology, 40(6), 800–806.

Monroy, C., Bustamante, D.M., Pineda, S., Rodas, A., Castro, X., Ayala, V., Quiñonez, J., and Moguel B. (2009). House Improvements and Community Participation in the Control of *Triatoma dimidiata* Re-Infestation in Jutiapa, Guatemala. Cadernos de Saúde Pública, 25 (Suppl. 1), S168–S178.

Nakagawa, J., Cordón-Rosales, C., Juárez, J., Itzep, C., and Nonami, T. (2003a). Impact of Residual Spraying on *Rhodnius prolixus* and *Triatoma dimidiata* in the Department of Zacapa in Guatemala. Memórias do Instituto. Oswaldo Cruz, 98, 277–281.

Nakagawa, J., Hashimoto, K., Cordon-Rosales, C., Juarez, J.A., Trampe, A., and Marroquin, L. (2003b). The Impact of Vector Control on *Triatoma dimidiata* in the Guatemalan Department of Jutiapa. Annals of Tropical Medicine and Parasitology, 97(3), 289–298.

Tabaru, Y., Monroy, C., Rodas, A., Mejia, M., and Rosales, R. (1999). The Geographical Distribution of Vectors of Chagas Disease and Populations at Risk of Infection in Guatemala. Medical Entomology and Zoology, 50, 9–17.

WHO-TDR (World Health Organization, Special Programme for Research and Training in Tropical Diseases). (2010). Chagas Disease — *Trypanosomiasis americana*, Fact Sheet 340, June 2010. WHO, Geneva, Switzerland. Available at: http://www.who.int/mediacentre/factsheets/fs340/en/index.html.

Zeledón, R., and Rojas, J. (2006). Environmental Management for the Control of *Triatoma dimidiata* (Latreille 1811), (Hemiptera: Reduviidae) in Costa Rica: A Pilot Project. Memórias do Instituto Oswaldo Cruz, 101(4), 379–386.

Chapter 15
Preventing Dengue at the Local Level in Havana City

Cristina Díaz

Dengue is one of the most important viral vector-borne diseases in the world. It is endemic in several continents; almost half of the world's population lives in risk areas. The cost of the disease is estimated to be in the billions (Suaya et al. 2009). The four serotypes of this disease are widespread and their incidence continues to escalate (Hales et al. 2002; Guzmán et al. 2004). It is transmitted by several mosquito species, particularly the tiger mosquito, *Aedes aegypti*. Cases present a variety of symptoms including fever, fatigue, and headache. The disease in its mild form in adults generally resolves after several days of discomfort. In its serious form, dengue haemorrhagic fever (DHF), the disease is life-threatening. The persistence of this disease around the world is assisted by the widespread presence of competent vector mosquitoes well adapted to human environments, clinical manifestations that in the mild form resemble many other diseases, lack of permanent immunity in recovered cases, and lack of effective vaccination or specific treatment. Climate, land-use change, and urbanization are contributing factors. As a result, the disease continues to spread and re-emerge in all tropical and subtropical zones, and poses a persistent challenge to public health authorities and communities.

In spite of recognized efforts (Halstead 2000) and resources dedicated to preventing dengue transmission, dengue outbreaks in Cuba continue to be cyclical and are occurring more frequently (Kourí et al. 1998). This is primarily due to the presence of *A. aegypti* and the high incidence of the disease in countries where an active exchange is maintained (Díaz et al. 2009). Outbreaks in Cuba have been controlled because of the existence of political will, a community with high educational levels, highly qualified human resources, a universal health care system, and legislation

---

C. Díaz (✉)
Formerly of Pedro Kourí Tropical Medicine Institute, La Habana, Cuba
e-mail: cristydiazpantoja2003@yahoo.es

(Cuban Law 91, 2000) that endorses community and intersectoral participation in the social production of health. Cuba has much experience fighting dengue. In 1987, a severe outbreak of DHF swept the country. Paradoxically, part of the reason this outbreak was so severe was that the country had effectively controlled the disease for nearly 20 years (Vaughn 2000).

During dengue epidemics, a strong relationship is established among stakeholders from the Cuban government, local communities, public health organizations, and other sectors. Through this relationship, community messages are generated and actions taken to better control *A. aegypti* breeding sites, eliminate vectors in any life stage, and strengthen detection of active cases. However, these actions weaken after epidemics are controlled (Martínez et al. 2004).

The health sector alone continues to assume responsibility for dengue control between epidemics when political and community attention to the issue wane, although solutions still rely on many sectors working together.

Developing an Integrated Dengue-Surveillance System

The integration of epidemiological surveillance into dengue prevention continues to be a challenge. In 2002, a multidisciplinary commission of representatives from the Cuban Government, the Ministry of Public Health, and several research institutions such as the Pedro Kourí Tropical Medicine Institute and the National Institute of Hygiene, Epidemiology, and Microbiology, proposed the development of an integrated dengue-surveillance system (IDSS).

Besides the integration of the environmental analysis (the vector and the host in a single surveillance system), the IDSS was to include community participation and the integration of the government, NGOs, and other sectors. This paper presents the results of a project designed to develop an intersectoral, continuous, and participatory IDSS at the local level using an ecosystem approach to human health.

The project was conducted from 2003 to 2006 in the Municipality of Cotorro in southeastern Havana. Characteristics of both rural and urban zones were present. Health services are organized in three areas under the direction of their respective polyclinics. The municipality is divided into six popular council areas, and each popular council oversees an area further divided into a number of circumscriptions (municipal wards).

Health Councils are created by the Cuban Ministry of Public Health and operate at the national and local levels. As part of their intersectoral mission, the Health Councils contribute to the improvement of the quality of life of the population, by formulating and implementing intersectoral public health plans.

San Pedro-Centro Cotorro (Popular Council No. 1) was chosen for this investigation because of its high level of urbanization, its population density (17,030 inhabitants, 51% of whom were women living in 4,215 houses), and its real and potential sites for vector breeding. Just before this study began, an epidemic of dengue (serotype 3) occurred in Havana City from June 2001 to March 2002, in which 12,889 cases were reported (González et al. 2005). The Municipality of Cotorro, Popular Council No. 1 reported the highest number of dengue cases.

The investigation included two stages – strategy design and implementation. Both entomological and clinical–epidemiological surveillance were implemented. The introduction of environmental surveillance represented a fundamental change in vector-borne disease surveillance in Cuba, which had relied on extensive entomological surveillance in the past. A process of community and intersectoral participation was also incorporated to improve the relationship of neighbourhood groups (NGs) with the government, public health organizations, and other sectors, and to help manage the environment to improve health. All information was collected and managed at the level of the San Pedro-Cotorro Centro Health Council (Díaz et al. 2009).

An Ecosystem Approach: Transdisciplinarity and Participatory Action Research

An ecosystem approach to health had previously been applied in Cuba (Castell-Florit 2008; Funes 2007; Ibarra et al. 2007; Mugica et al. 2007; and CoPEC-LAC).[1] An urban ecosystem approach was implemented in Havana City to evaluate the impact on human health of changes made in the urban environment as a result of government interventions in Cayo Hueso in the Municipality of Centro Habana in Havana City (Spiegel et al. 2003, 2004). Our project in Cotorro used this innovative approach to address the interactions among environmental, health, economic, social, and cultural factors. The aim of the integrated and sustained environmental management system was to control factors that favoured vector appearance and thereby prevent dengue transmission.

At the beginning of the project, a multidisciplinary research team was created that included 19 researchers from biology, entomology, medicine, virology, psychology, sociology, pedagogy, communications, geography, and biostatistics. The researchers possessed a variety of specializations, including environmental health, ecology and control of vector-borne diseases, clinical care and control of dengue epidemics, health promotion and education, and software development and database management. Two-thirds of the researchers were women. The team was modified during the project as members from both the community and the local-level institutions joined. This resulted in the creation of a management group (MG).

Participatory action research (Hernández et al. 1999) and scientific research methodologies were used to understand the characteristics of dengue in the study area and to carry out the interventions. The initial tasks included: characterizing the study area, developing links with the community, and explaining to the stakeholders the interactions between the ecosystem and health.

---

[1] See http://www.unites.uqam.ca/copeh/design/English/Home%20CoPEH-TLAC.htm.

Situational Analysis

The project found the factors associated with *A. aegypti* infestation to be: irregular water supplies, water being stored for a long time, unprotected water storage containers, inadequate control of mosquito breeding sites both near homes and workplaces, problems with environmental sanitation and insufficient knowledge of waste management and sanitation, incorrect public perceptions of the risk of contracting dengue, and some deficiencies in the National Program for the Eradication of *A. aegypti*.

Initially, surveillance efforts of the project were found to be inadequate. Environmental, entomological, and epidemiological data were neither uniform nor comparable because different units of time and space were used in relation to the risks detected. An integrated analysis of the information provided by the different surveillance components was not possible. The MG reformulated its approach to achieve adequate intersectoral and community integration; to adapt and improve the activities of the vector control program, the statistical information system, and the environmental health system; and to identify capacity-building needs and integrate them into the new system.

Participatory Development of Technical Tools

The IDSS and the tools used for its implementation were designed and adjusted in a participatory way (Álvarez et al. 2007). The existing daily work forms for collecting the primary entomological data were improved with the inclusion of: environmental information, better entomological characterization, clinical–epidemiological and laboratory information on reported cases of nonspecific fever (a proxy for dengue), cases of positive dengue serological results, confirmed reports of imported and autochthonous dengue cases, and information on travellers with dengue symptoms. The IDSS was designed to integrate data from three components – environmental, entomological, and epidemiological systems (including clinical and laboratory data). By the end of the cycle, information was integrated by block ("manzana") and available for analysis at different levels of the municipality.

To integrate the information from the different components, a digital database and the Dengue 537 software were designed based on inputs from the different disciplines (Suárez 2007). The software produced maps and graphs of the various environmental, epidemiological, and entomological data (see García et al. 2007 for environmental example). With this tool, the project team was able to track the appearance or reappearance of the vector, and to focus on a 300-m range (presumed range of the mosquito) around an area positive for the vector to better target remedial actions that could improve the environment. The participatory planning process kept the new integrated information system relevant, and the new data strengthened the capacity for participatory management and engagement.

Over the course of the study, the IDSS identified 666 environmental problems favouring *A. aegypti* proliferation (188 outdoors and 478 indoors). As well, 6,031 water storage containers were detected (low and elevated tanks, barrels, and cisterns and wells), of which 2,138 (35%) stored water. Of those, 341 (16%) were unprotected, mainly because of deficient covers (213, 62%). Inaccessibility of elevated tanks was diagnosed as a frequent problem for monitoring these containers (108 tanks could not be accessed). These results helped stakeholders to prioritize existing resources to control environmental factors more effectively, and thus to better control mosquito breeding sites.

As part of the integrated dengue management protocol, new tools for biological control were assessed. Two copepod species, *Mesocyclops pehpeiensis* (Menéndez et al. 2006) and *Macrocyclops albidus,* were tested as biological controls for *Aedes* spp. *M. albidus* is an active predator of *Aedes* spp. larvae in the laboratory. In this project, the application of *M. pehpeiensis* in natural breeding sites showed a gradual decrease in larval density until vector elimination (unpublished data), indicating some promise for use of this species for controlling the dengue vector.

Community Participation and Social Process: The Neighbourhood Groups

In Cuba, community participation is generally well organized and easily achieved in programs like dengue control, although performance is variable (Bru-Martín and Basagoiti 2003). In this project, the MG promoted participatory environmental management through social and cultural activities targeted at different age groups (children, young people, and adults) (Díaz et al. 2009). With the collaboration of the municipal health team, neighbours organized themselves voluntarily into 17 NGs with a total of 230 members. Each NG nominated a president who also participated in the MG.

The NGs were the main agents of change for environmental management. They conducted surveillance and control of environmental conditions; education and promotion of ecohealth; and organization, mobilization, and advocacy needed to solve problems that were beyond their reach. In the beginning, the majority of NGs were comprised of women, who successfully recruited some men to help take on responsibilities. NGs received various types of training to be better able to execute their responsibilities.

The MG agreed that the information collected both from the IDSS and the complementary surveillance data provided by the NGs should be integrated by the local Health Councils. To facilitate this integration, and to promote action, NG presidents joined the local Health Council. This process, according to an evaluation by the beneficiary population, turned out to be more coordinated and efficient because it increased detection, helped control the risks of vector proliferation, and therefore reduced dengue transmission. The detection and description of vector breeding sites, and the needs felt by the community, were the basis of the NG's environmental risk-assessment strategy. Standard strategies to control *A. aegypti* generally start with the control of vector breeding inside houses and then move to control measures outside the houses. However, the population identified the greatest number of environmental problems to be outside their houses, and the entomological characterization showed that the highest percentage of *A. aegypti*-positive breeding sites were found in low tanks, mainly around the house, and in particular in the backyards.

The Rifkin spider anagram process (Rifkin et al. 1988) was used to assess the success of the participatory process and to allow the stakeholders and the representatives of the different sectors and the community to participate jointly in the evaluation of the results of their work. Five elements were analysed: leadership, resources mobilization, identification of needs, management, and organization. Progress was observed in all elements.

In the beginning, leadership resided mostly with the health sector, which included the local health representatives and the external research team. Later, leadership was shared between governmental representatives and the NGs. Relative improvements were obtained in resource mobilization at the local level by stakeholders and by the community, although many of the problems identified could not be solved because they required financing that had to be assigned by higher levels. The impact achieved by the management training given to representatives from the different sectors, the government, and the community was evident.

Indicators of Success

With the creation and modification of the primary data collection forms, and the design of computer software for data analysis, surveillance quality was improved. The introduction of the environmental component into the IDSS led to action on the existing hazards in the environment that favoured vector proliferation. By making spaces for more dynamic stakeholder participation, the decision-making process became stronger. Several indicators of this success were recorded (Díaz et al. 2009).

After 4 years, the number of unprotected water sources, mainly low tanks without covers or with bad covers, diminished from 62 to 8%. As well, the proportion of unhygienic backyards had dropped from 16 to less than 1%. Environmental improvements to public spaces were observed, including planting of orchards, creation of parks and sport fields, and transformation of common areas near houses into gardens.

The detection of febrile cases was higher in the Popular Council under study because of the new links between the health network and the NGs. *A. aegypti*-positive sites diminished during and after the project. In 2003, 85% of these positive sites were found in low tanks placed in the backyards of houses. In 2007, 2 years after the project was finished, no positive sites were found in the houses. The percentage of households that refused to allow inspections by vector-control workers diminished by 30%.

The community's capacity to solve its environmental problems with its own resources increased from 16 to 47% and, together with the other agencies they solved 74% of the environmental problems that were identified by the participatory surveillance system. The number of agencies actively participating in the Health Council and in ecosystem management increased from two at the beginning of the investigation (Garbage Collecting Services, and Water Supply and Sewer System) to five at the end of the 2005.

The relationships among researchers, community members, and sectoral representatives favoured knowledge integration and the adoption of a common language, and helped create a synergy that favoured the transformation toward a transdisciplinary team.

Replicating Results, Knowledge Transfer, and Scientific Recognition

The strategy in Cotorro was sustainable despite government changes and a reorganization of health services 2 years after the project finished. It was replicated in three Popular Councils in the Municipality of Centro Habana with satisfactory results (Bonet et al. 2007). This impact was aided by dissemination of results in 24 scientific events and registration of the "Dengue 537" software (Register Number 2443–2005). The work was nationally recognized in 2008 (National Award of the Academy of Sciences of Cuba and Most Relevant Result at the Provincial Forum of Science and Technology). The results were also endorsed by four National Units of the Ministry of Health, the Presidency of the Municipal Assembly of the People, and the Municipal Unit of Health.

Conclusion

Two elements contributed to the success of this project: (1) the training process implemented at the local level from the beginning of the project empowered the community and stakeholders in the use of the system tools; and (2) the collaborative process used by the health researchers, who gradually withdrew but stayed on as advisors to the MG.

The application of an ecosystem approach to health in this project led to better understanding of risk factors for dengue, the development of a participatory surveillance system, improved control of mosquito breeding sites, and improved and sustained community participation in dengue control. It also enabled them to approach the surveillance of the environment, the vector, and the disease in an integrated and participative way, and placed the team a step ahead in the prevention of dengue.

Acknowledgements This paper reflects the commitment and contribution of the people of Cotorro and the many stakeholders who were involved in the project. The guidance of Dr. Gustavo Kourí, Pedro Kourí Tropical Medicine Institute and the contribution of Armando Martínez are gratefully acknowledged. IDRC support was provided through project 101545. The Ministry of Public Health, Republic of Cuba, also provided support through the Communicable Diseases Program.

References

Álvarez, A.M., Díaz, C., García, M., Piquero, M.E., Alfonso, L., Torres, Y., Mariné, M.A., Cuellar, L., Fuentes, O., and de la Cruz, A.M. (2007). Sistema Integrado de Vigilancia para la Prevención del Dengue. Municipio Cotorro. La Habana. Cuba. Revista Cubana Medicina Tropical, 59(3), 193–201.

Bonet, M., Spiegel, J., Ibarra, A.M., Kourí, G., Pintre, A., and Yassí, A. (2007). An Integrated Ecosystem Approach for Sustainable Prevention and Control of Dengue in Central Havana. International Journal of Occupational and Environmental Health, 13(2), 188–194.

Bru-Martín, P., and Basagoiti, M. (2003). La Investigación-Acción Participativa Como Metodología de Mediación e Integración Socio-Comunitaria. Publicación periódica del Programa de Actividades Comunitarias en Atención Primaria n.6°, Barcelona cemFYC. Available at: http://www.pacap.net/es/publicaciones/pdf/comunidad/6/documentos_investigacion.pdf.

Castell-Florit, P. (2008). Intersectorialidad en Cuba, su Expresión a Nivel Global y Local. Editorial Ciencias Médicas, Ciudad de la Habana, Cuba. 63 p.

Díaz, C., Torres, Y., de la Cruz, A.M., Álvarez, A.M., Piquero, M.E., Valero, A., and Fuentes, O. (2009). Estrategia Intersectorial y Participativa para la Prevención de la Transmisión de Dengue en el Nivel Local. Cądernos de Saúde Pública, 25 (Suppl. 1), 59–69.

Funes, F. (2007). Alimentación, Medio Ambiente y Salud: Integrando Conceptos. LEISA Revista de Agroecología, 23(3), 12–15.

García, M., Mariné, M.A., Díaz, C., Concepción, M., and Valdés, I. (2007). El Componente Ambiental de la Vigilancia Integrada para el Control y la Prevención del Dengue. Revista Cubana Higiene y Epidemiología, 45(1). Available at: http://bvs.sld.cu/revistas/hie/vol45_1_07/hie07107.htm.

González, D., Castro, O., Kourí, G., Pérez, J., Martínez, E., and Vázquez, S. (2005). Classical Dengue Hemorrhagic Fever Resulting from Two Dengue Infections Spaced 20 Years or More Apart: Havana, Dengue 3 Epidemic, 2001–2002. International Journal Infection Disease, 9, 280–285.

Guzmán, M.G., Kourí, G., Díaz, M., Llop, A., Vázquez, S., González, D., Castro, O., Álvarez, A.M., Fuentes, O., Montada, D., Padmanabha, H., Sierra, B., Pérez, A.B., Rosario, D., Pupo, M., Díaz, C., and Sánchez, L. 2004. Dengue, one of the Great Emerging Health Challenges of the 21st Century. Expert Review Vaccines, 3(5), 89–98.

Hales, S., de Wet, N., Maindonald, J., Woodwar, A. (2002). Potential Effect of Population and Climate Changes on Global Distribution of Dengue Fever: An Empirical Model. Lancet, 360, 830–834.

Halstead, S.B. (2000). Successes and Failures in Dengue Control — Global Experience. Dengue Bulletin Volume 24. World Health Organization, Geneva, Switzerland.

Hernández, I., García, M., Moreno, L., and González, N. (1999). Selección de Lecturas Sobre Investigación-Acción Participativa. CIE "Graciela Bustillos," Asociación de Pedagogos de Cuba. La Habana, Cuba. 91 p.

Ibarra, E.J., Mugica, J.P., Jaime, A., González, R.M., Gravalosa, A.J., Menéndez, E., Guevara, M.E., and Cabrera, C. (2007). Valores de Referencia de Plomo en Sangre en la Población en Edad Laboral de la Ciudad de la Habana. Paper presented at Congreso de Salud y Trabajo Cuba

2007, 12–16 March 2007. La Habana, Cuba. Instituto Nacional de Salud de los Trabajadores, La Habana, Cuba.

Kourí, G., Guzmán, M.G., Valdés, L., Carbonell, I., del Rosario, D., and Vázquez, S. (1998). Reemergence of Dengue in Cuba: A 1997 Epidemic in Santiago de Cuba. Emerging Infectious Disease, 4, 89–92.

Martínez, S., Caraballoso, M., Astraín, M.E., Pría, M.C., Perdomo, V.I., and Arocha, C. (2004). Análisis de la Situación de Salud. Editorial Ciencias Médicas, La Habana, Cuba. 122 p.

Menéndez, Z., Reid, J.W., Guerra, I.C., and Ramos, I.V. (2006). A New Record of Mesocyclops pehpeiensis Hu, 1943 (Copepoda: Cyclopoida) for Cuba. Journal of Vector Ecology, 31(1), 193–195.

Mugica, J.P., Suárez, R., Díaz, H., Ibarra, E.J., and Baqués, R. (2007). Higiene del Trabajo. Métodos de Evaluación de Factores de Riesgo en el Ambiente de Trabajo. Paper presented at Congreso de Salud y Trabajo Cuba 2007, 12–16 March 2007. La Habana, Cuba. Instituto Nacional de Salud de los Trabajadores, La Habana, Cuba.

Rifkin, S., Muller, F., and Bichman, W. (1988). Primary Health Care: On Measuring Participation. Social Science and Medicine, 26(9), 931–940.

Spiegel, J.M., Bonet, M., Tate, G.M., Ibarra, A.M., Tate, B., and Yassi, A. (2004). Building Capacity in Central Havana to Sustainably Manage Environmental Health Risk in an Urban Ecosystem. EcoHealth, 1 (Suppl. 2), 120–130.

Spiegel, J.M., Bonet, M., Yassi, A., Tate, R., Concepción, M. and Cañizares, M. (2003). Evaluating the Effectiveness of a Multi-Component Intervention to Improve Health in an Inner City Havana Community. International Journal of Occupational and Environmental Health, 9(2), 118–127.

Suárez, R. (2007). Software para el Sistema Integrado de Vigilancia de Dengue "Dengue 537". Paper presented at the Convención Informática en Salud 2007. VI Congreso Internacional de Informática en Salud, April 2007. La Habana, Cuba. Ministerio de Salud Pública, La Habana, Cuba.

Suaya, J.A., Shepard, D.S., Siqueira, J.B., Martelli, C.T., Lum, L.C., Tan, L.H., Kongsin, S., Jiamton, S., Garrido, F., Montoya, R., Armien, B., Huy, R., Castillo, L., Caram, M., Sah, B.K., Sughayyar, R., Tyo, K.R., and Halstead, S.B. (2009). Cost of Dengue Cases in Eight Countries in the Americas and Asia: A Prospective Study. American Journal of Tropical Medicine and Hygiene, 80(5), 846–855.

Vaughn, D.W. (2000). Invited Commentary: Dengue Lessons from Cuba. American Journal of Epidemiology, 152(9), 800–803.

Chapter 16
Eco-Bio-Social Research on Dengue in Asia: General Principles and a Case Study from Indonesia

S. Tana, W. Abeyewickreme, N. Arunachalam, F. Espino, P. Kittayapong, K.T. Wai, O. Horstick, and J. Sommerfeld

Dengue fever, a viral disease primarily transmitted by the mosquito *Aedes aegypti*, is an emergent public health problem in Asia (WHO 2007). Numerous "top–down" approaches to vector control have not been able to achieve a sustainable reduction in vector densities or to prevent dengue transmission (Gubler 1989). The relative contribution of mosquitoes and humans (or both) to the distribution of dengue viruses is poorly understood. The continuous changes in the human population, vector population, and the virus along with and in response to the environmental changes present an ongoing challenge to control. This variation is associated with transmission

---

S. Tana (✉)
Centre for Health Policy and Social Change, Yogyakarta, Indonesia
e-mail: tanasusilowati@gmail.com

W. Abeyewickreme
Faculty of Medicine, University of Kelaniya, Kelaniya, Sri Lanka

N. Arunachalam
Centre for Research in Medical Entomology, Indian Council of Medical Research, Madurai, India

F. Espino
Research Institute for Tropical Medicine, Alabang, Muntinlupa City, Philippines

P. Kittayapong
Center of Excellence for Vectors and Vector-Borne Diseases, Faculty of Science, Mahidol University at Salaya, Nakhon Pathom, Thailand

K.T. Wai
Department of Medical Research (Lower Myanmar), Yangon, Myanmar

O. Horstick
Formerly of Special Programme for Research and Training in Tropical Diseases (TDR), World Health Organization, Geneva, Switzerland, currently of Deutsche Gesellschaft für Internationale Zusammenarbeit (GIZ) GmbH

J. Sommerfeld
Special Programme for Research and Training in Tropical Diseases (TDR), World Health Organization, Geneva, Switzerland

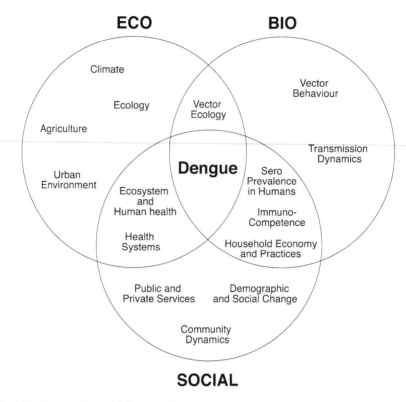

Fig. 16.1 The eco-bio-social framework

dynamics that produce a range of challenges from silent epidemics of dengue infection to overt outbreak of disease.

There is a gap in understanding how the transmission and dynamics of disease vary with ecological, biological, and social factors (Ellis and Wilcox 2009). Therefore, more insights into transmission dynamics and the possibility of intersectoral ecosystem management programs for dengue prevention and control are urgently needed. An ecosystem approach to successful reduction of vector-borne disease burden (Campbell-Lendrum and Molyneux 2005) can lead to considerable health gains.

The need for an approach that considers the multiple layers of the social and ecological determinants that influence dengue has been established (Spiegel et al. 2005), and attempts have been made to create conceptual frameworks that describe some of these factors (Pongsumpun et al. 2008). These insights play a crucial function in defining locally relevant and appropriate interventions with the prospects for sustainable control of vector populations.

Dengue can be described as "eco-bio-social" in nature (Fig. 16.1):

- Ecological factors refer to climate (e.g. rainfall, humidity, and temperature) and the natural and anthropogenic ecological setting (including the urban and agricultural environment).

- Biological factors relate to the behaviour of the main vector, *A. aegypti*, and transmission dynamics of the disease. The ecological and biological domains are linked by the ecology of the vector population.
- Social factors incorporate a series of factors that include: vector control and health services, and their political context (e.g. health sector reforms); public and private services such as sanitation and sewage, garbage collection, and water supply; macro-social events such as demographic growth and urbanization; and community- and household-based knowledge, attitudes, and practices and how these are shaped by large-scale forces such as poverty, social inequality, and community dynamics.

Such a generic conceptual framework can inform the development of ecosystem-specific frameworks that then can be investigated through research, and substantiated in models originating from that research. The characteristics of the eco-bio-social research initiative (a research initiative of the Special Programme for Research and Training in Tropical Diseases (TDR) and the International Development Research Centre (IDRC), referred to as the TDR/IDRC research initiative) with its multifactorial approach, ecosystem-specific framework, and multidisciplinarity are closely related to the principles of the ecohealth approach, described in other chapters of this book.

Situation Analysis for Vector-Borne Disease Control

Ecosystem and site-specific public health interventions necessitate a detailed multidisciplinary analysis of a disease in its eco-bio-social context. Such a situation analysis needs combined expertise from a multidisciplinary group that includes ecologists, entomologists, and social scientists. The application of the eco-bio-social framework to key outcome variables in an analytical research framework is critical in this regard (Fig. 16.2). In the case of the TDR/IDRC research initiative, vector density (measured by the pupae per person index and related indices) was determined as the core domain of interest and the dependent variable. Vector density is seen as dependent on the vector ecology and the social, ecological, and vector-control context. Vector density is considered as one of the proxy indicators for dengue transmission (Focks and Alexander 2006).

Applying this analytical framework with eco-bio-social dimensions, the TDR/IDRC research initiative was piloted with two studies on dengue in Latin America (Caprara et al. 2009; Quintero et al. 2009) and extended with six studies on dengue in Asia. This paper describes the initial findings of the six studies on dengue in Asia and presents a more detailed analysis of the situation in Yogyakarta, Indonesia. In 2009, the research initiative continued with nine studies on dengue and Chagas disease in Latin America and the Caribbean.

The main objective of the research initiative in Asia was *to strategise, and contribute to improved dengue prevention by better understanding its ecological, biological, and social ("eco-bio-social") determinants and to develop and evaluate a community-centred ecosystem management intervention directed at reducing dengue vector larval habitats, embracing intersectoral actions.* The research

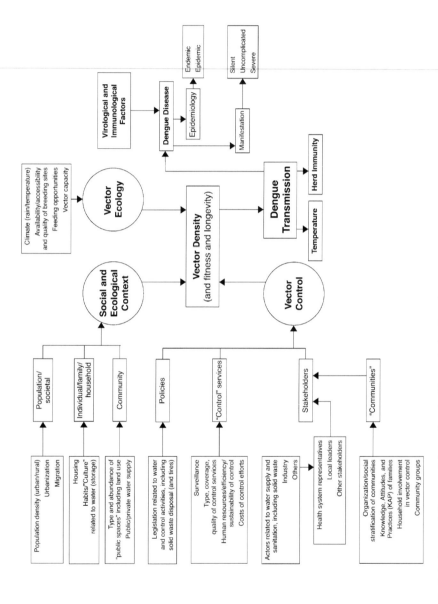

Fig. 16.2 Analytical research framework (reproduced with permission from Arunachalam et al. (2010):174)

Table 16.1 The sites of the dengue studies in Asia

Country	Study sites	Population and population density
India	Chennai City	5.9 million (26,000/km²)
Indonesia	Yogyakarta City	0.5 million (15,244/km²)
Myanmar	28 urban townships, Yangon Division	4 million (6,944/km²)
Philippines	Muntinlupa City	0.6 million (11,334/km²)
Sri Lanka	Gampaha District	2 million (1,336/km²)
Thailand	Urban and semi-urban areas of Chachoengsao Province	0.6 million (120/km²)

initiative consisted of two phases. Based on a situational analysis in phase 1, community-centred ecosystem management interventions appropriate for the ecosystem under study were implemented and evaluated in phase 2. The study was carried out simultaneously in six sites (Table 16.1).

Phase 1 entailed an analysis of existing routine data, an ecosystem characterization, cross-sectional entomological and household surveys, a participatory stakeholder analysis and problem assessment, and other qualitative social research methods.

The survey research was conducted in predefined clusters randomly selected from known clusters of high and low vector density areas. When vector density data were not available, high and low incidences of dengue were determined by using recent data or a history of reported dengue cases (hospital data) to define the clusters.

The phase 1 study employed both qualitative and quantitative approaches to data collection. The analysis specifically aimed to triangulate sources of evidence. The situation analysis was designed to inform the definition and development of a locally relevant ecosystem management intervention (in phase 2) with active participation of all stakeholders – communities, their governing structures (e.g. policymakers and decision makers), and related services (e.g. water supply and waste management). Interventions considered existing tools for dengue prevention and their employment in viable partnership models between community members and service providers within the framework of community development and environmental management. Once clearly defined and described, the interventions were implemented. Specific process and outcome indicators were defined beforehand and used as a baseline post-intervention to measure the impact and efficiency of the intervention. Phase 2 characterized human eco-management activities for a single transmission period during a total period of 12 months. Qualitative evaluation methods were employed to describe the process of the intervention.

In phase 1, data were collected on: (1) human ecology, which includes both environmental and social factors (descriptive and comparative data on demography, socio-economic data, water and sanitation, and public and private spaces); (2) vector ecology (descriptive and comparative data on preferred larval habitats, stratified by public and private spaces, and relation to rainfall); (3) the knowledge and practices of people related to dengue and dengue prevention; (4) vector control activities; and (5) stakeholders.

Details of the situational analysis have been published (Arunachalam et al. 2010) and are summarized here. Although the majority of data are site specific, some results are common to most sites. The characteristics of different neighbourhoods (clusters) were related to their total production of pupae per hectare of land (PPH), both in private and public spaces, and revealed the following pattern:

- Neighbourhoods with a high population density (inhabitants per hectare) had a much higher PPH than neighbourhoods with a lower population density – PPH 74.6 (CI 46.3–102.9) vs. PPH 11.0 (CI 7.8–14.1) (CI: confidence intervals).
- Neighbourhoods with schools had a higher PPH than neighbourhoods without schools – PPH42.7 (CI 25.21–60.3) vs. PPH 14.4 (CI 7.7–21.2).
- Neighbourhoods with religious sites had a higher PPH than neighbourhoods without religious sites – PPH 38.4 (CI 23.8–52.9) vs. PPH 11.8 (CI 3.2–20.4).
- Houses that were more than 4 m from each other had a higher PPH than houses that were close together – PPH 35.4 (CI 19.7–51.1) vs. PPH 11.6 (CI 5.4–17.8).
- People's knowledge of dengue vectors across all study sites was negatively correlated with PPH.

Other variables associated with increased PPH (although with a p value greater than 0.05) were middle and lower socio-economic strata, poor housing, houses with gardens, residential area (compared with commercial area), cemeteries or garbage dumps in the neighbourhood, abundant water through piped water (only in Myanmar was the absence of piped water associated with elevated PPH), and absence of recent interventions by vector control services.

These results across all sites are valuable to define interventions at the local level, but also to describe general patterns of dengue transmission across all sites. The advantage of a thorough analysis of eco-bio-social factors to describe the ecosystem becomes clear, because different interventions to prevent and control dengue may be appropriate according to local circumstances.

This may be one of the eventual overall conclusions of the study: dengue prevention and control is impossible without taking into account the factors that determine the disease within its eco-bio-social framework. The following section presents the analytical research framework for the situation analysis of dengue in urban Yogyakarta, Indonesia, and preliminary results of the intervention phase.

A Case Study from Indonesia

The City of Yogyakarta is the capital of the province with the same name, and covers 32.8 km^2. The city is about 100 m above sea level and has an average rainfall of 2,156 mm, which peaks in January (180.4 mm) and is lowest in November (0 mm). Humidity is between 72 and 87%, and average temperature is 27.8°C. These conditions make Yogyakarta an ideal breeding habitat for both *A. aegypti* and *Aedes albopictus* (WHO 2004). The population of Yogyakarta City is 435,236 (49% male and 51% female) (BPS 2005). Yogyakarta is the centre for business and tourism in the region, and a substantial part of the population commutes in and out of the districts surrounding the city. The city has urban and suburban areas, a very mixed population of migrants and indigenous community groups, different social strata, and a unique culture.

Dengue is endemic in Yogyakarta, with all 45 *Kelurahan* reporting cases (Kelurahan is a level of community organization in urban areas in Indonesia, equal to a village in rural areas). The Yogyakarta City District Health Office reported 705 cases of dengue haemorrhagic fever (DHF) in 2005 (incidence rate 16.2/10,000), 894 cases in 2006 (incidence rate 20.5/10,000), and 600 cases until mid-2007 (latest data available) (incidence rate 13.8/10,000).

In terms of entomological surveillance, the average house index [HI = (infested houses × 100)/houses inspected] fluctuated between 19% in the dry season and 33% in the rainy season. The average pupa per person index was 0.39 in the dry season and 0.48 in the rainy season. Water storage practices are important in this context, and keeping water for all daily family necessities in various water containers is a very common practice in all social strata.

For phase 1 (situational analysis) of this study, the research methods included household surveys (mostly based on knowledge, attitudes, and practices surveys), cluster background and entomology surveys, ecotope mapping, and multilevel stakeholder analysis, including mapping and gender analysis. The multilevel stakeholder analysis was carried out with different decision-making groups that included both genders (district health offices, village heads and local leaders of community groups, local nongovernmental organizations, and neighbourhood boards).

The household surveys revealed seasonal difference in various mosquito indexes. For example, the house index score was lower in the dry (19.4%) than the wet season (33.1%); the container index behaved similarly (dry season 6.1% vs. wet season 10.6%). In the dry season, 88.6% of pupae were produced in the household bathwater container. This main water container contributed 46.8% of the pupae found in the wet season, and the rest were contributed by other water containers. There was little community awareness or community action on dengue prevention and control. Despite a decade-long government campaign, there appeared to be little knowledge about the importance of community involvement in dengue prevention and control. Or perhaps there was some knowledge, but no obvious dengue control actions could be observed.

The government campaign, also known as "3M Plus" (Menguras, Menutup, Mengubur, Menjual – clean up, close, bury, and sell reusable solid waste) was widely known by households. People had some knowledge of dengue. For example, they knew that dengue virus is the cause of DHF, that dengue is transmitted by mosquitoes, and that dengue transmission can be prevented by 3M Plus.

But this knowledge did not translate into attitudes and behaviours that led to community participation in dengue prevention and control. There was a lack of knowledge about ecosystem-based vector control interventions. Dengue prevention and control was considered solely as a government responsibility. As well, wide-scale insecticidal "fogging" was reported to be the preferred dengue control intervention among community groups. Busy urban communities said they did not have time for clean-up activities, although this intervention was considered as important as other available interventions. Dengue was not considered to be a priority, when comparing with health in general, or with other daily issues like poverty. The issue of poverty and inequality was particularly important in the context of this community.

Table 16.2 The stakeholders and their potential roles in defining the components of the partnerships required to implement interventions

Stakeholders	Roles
Community groups	Partner for entering and organizing the community; they are decision makers for phase 2 interventions at the group level. Community educators will inform community groups about dengue-ecosystem management and facilitate group decision-making
Women	Partner in educating other family members; they are decision makers for phase 2 at the family level, particularly in relation to women's roles in the family, including water management
Men	Men should know the importance of dengue control and the interventions for phase 2 so they can participate in the family and community
Community leaders and cadres	Partners from the community who are trained and empowered to educate and work with communities to develop innovative dengue-ecosystem management interventions
School principals and managers	Decision maker for phase 2 interventions in schools
School children	Partners who are empowered to understand and assist phase 2 interventions in schools and families
RukunTetangga (RT)[a] Board	Decision makers and partners in developing dengue-ecosystem management interventions at the neighbourhood level
Communicable disease officers at district health office	Partners and decision makers of dengue control policy at city level
Other nongovernmental organizations	Partners working for the same community goal to prevent and control dengue
Scientists	Facilitators and consultants for community

[a]RukunTetangga (RT) is the lowest governmental organizational structure in Indonesia. Equal to a neighbourhood, each RT consist of about 50–100 households

In addition, knowledge about dengue prevention and control was lower among migrant workers when compared with resident (non-migrant) workers.

There was evidence of weak social cohesion among different community groups, especially between migrant and non-migrant groups, but also among different social strata. This may help explain the very weak participation of communities in health-related prevention and control activities.

When analyzing current vector control activities, it was evident that most educational efforts on larvae control (and multiplying these efforts by community members) remained with vector control personnel. Vector control staff were often strong in entomological methods, but had insufficient technical capacity to involve communities. Vector control staff perceived barriers to entering houses to execute their work, especially in wealthier areas of the city, and this, along with financial constraints, limited vector control operations.

The stakeholder analysis revealed ten important stakeholders. These stakeholders and their potential roles define the components of the partnerships required to implement the interventions in phase 2 (Table 16.2).

Phase 2 aimed to develop a community-centred ecosystem management intervention directed at reducing dengue larvae habitats and embracing community-based actions. Partnerships were developed between scientists and communities, and these partnerships gave equal weight to all members, whether lay person, technical expert, or political leader.

The scientists shared the results of phase I and some options for environmental management for dengue control with other stakeholders and members of the community partnership. This facilitated decision making and the development of an innovative dengue-environmental management ecosystem intervention for phase 2. The composition of these partnerships was defined by the stakeholder analysis presented in Table 16.2, and included community groups, men and women, community leaders, representatives of the local government, technical experts, and many others.

A participatory approach, reflecting gender issues, encouraged effective decision making within the community forum. The forum was organized to achieve better understanding of the eco-bio-social problems that contribute to dengue transmission and to find the most suitable and sustainable ecosystem-based interventions for primary prevention of dengue.

Although preliminary, there are early results of the intervention phase. The intervention activities consisted of a range of different dengue-ecosystem intervention options selected by the community research partnership, based on evidence from the situation analysis (phase 1). The contribution of different water containers, undegradable solid waste disposal, and wells to the production of mosquito pupae were the basis of the selection of control activities. Some additional considerations were included, such as the potential impact of the interventions on community livelihoods and environment, and how easily such interventions could be sustained over time. The options preferred by the community included: managing household waste (trash containers and other solid waste) and recycling for compost and handicrafts; covering wells and water tanks; repelling mosquitoes using zodia plantation (*Evodia suaveolens*), stocking bathwater containers with fish that eat mosquito larvae, and continued use of pyriproxyfen in water containers that have high levels of mosquito breeding. The community forum enabled problem identification and program development as well as monitoring and evaluation by community members. This forum also provided opportunities to discuss other community improvements. A local community leader had an important role in planning and leading the intersector thematic community self-help program, which linked to environment and economy.

The success of the program will be measured by (1) factors to prevent dengue, (2) positive changes in targeted behaviour, and (3) the likelihood of program sustainability. One measure of success was community response to an open-ended question on how to prevent DHF. After intervention, the answers given by community members were much more varied than before, including maintaining good water storage and improved environmental conditions (including sorting of solid household waste for disposal). This revealed a broader understanding of dengue prevention than before the project. One of the positive changes in targeted behaviour was the percentage of community members who participated in the collective dengue prevention program within the year of intervention (before intervention = 19.4%; after = 72.8%).

Sustainability of a community-based program cannot be achieved without the community gaining some benefit, either individually or as a member of the community.

In addition to tangible perceived benefits, such as potential economic benefit from selling of compost or handicrafts, or simply pride in a cleaner environment, the satisfaction level toward dengue prevention practices is also important for sustainable changes in behaviour. This was measured by asking, without being specific, about perceptions of the effectiveness of current dengue prevention practices. After the project interventions, most community members believed that the control measures had reduced the number of mosquitoes and would help to control dengue and other diseases caused by mosquito-borne viruses. Most people interviewed before the intervention believed that the dengue control measures would be effective; however, after the intervention all respondents perceived benefit from these practices. Furthermore, after the intervention, the proportion of community members who believed that mosquito control practices were solely a government responsibility was decreased.

Creating a supportive "environment" is essential for more sustainable changes in community attitudes and practices for dengue control. This environment includes broader systems that influence change at the community level, such as public services and infrastructures, policy and programming, and resources. The community forum, supported by the research team, is working hard to link this project to existing health and environmental systems and resources. Greater impact may also be produced by increasing resources through collaboration and financial sharing.

Although elimination of the traditional bath container and replacement by showers could substantially reduce mosquito indexes and risk of dengue in Yogyakarta, this will take time to implement. As well, many other changes are required, including improvement in local water supply and electricity services, and changes to the long-standing bathing habits of the community. However, eliminating bath-water containers in public spaces may be an important early step that could be more easily achieved. Discussions are underway to achieve this, and municipal officials including the Mayor and City Health Office are receptive to the idea.

Conclusion

A multidimensional research framework that focused on the eco-bio-social aspects of dengue led to a practical, public health-focused concept for situation analysis that could be applied to various vector-borne diseases. It would be interesting to apply this analytical framework to other diseases in future research and to assess its implications for human health.

By applying the framework in the context of dengue transmission, and focusing on relevant eco-bio-social variables, the project team developed an approach for a multicountry research initiative in very different ecological settings in Asia.

During phase 1 of the research initiative, it became clear that this approach poses several challenges because the teams must be multidisciplinary. But at the same time, there are enormous benefits. The teams are now in a position to decide together about appropriate interventions in the local context, have local ownership, and work toward sustainability of the expected satisfaction and participation gains flowing from phase 2. In this context, capacity building was important, notably in creating and sustaining multidisciplinary teams. The research team found that sustainable changes for dengue control needed both a holistic approach that considered the eco-bio-social factors and the active participation of the community. For this to occur, the community must be involved from the beginning of the project.

The applied eco-bio-social framework is a promising contribution to multidisciplinary–multistakeholder approaches to understanding the local dynamics of vector-borne disease and determining appropriate ecosystem management interventions for improved public health efforts directed at the control of vector-borne disease.

Acknowledgements This research initiative received financial support from IDRC through project 102741. The sub-study in Myanmar was supported by the Special Programme for Research and Training in Tropical Diseases at the World Health Organisation (WHO-TDR).

References

Arunachalam, N., Tana, S., Espino, F., Kittayapong, P., Abeyewickreme, W., Wai, K.T., Tyagi, B.K., Kroeger, A., Sommerfeld, J., and Petzold, M. (2010). Eco-Bio-Social Determinants of Dengue Vector Breeding: A Multicountry Study in Urban and Periurban Asia. Bulletin of the World Health Organization, 88, 173–184. Available at: http://www.who.int/bulletin/volumes/88/3/09-067892/en/.

BPS (BadanPusatStatistik — Indonesian Board of Statistics). (2005). BPS Statistics for Yogyakarta City, Population in Yogyakarta City. Indonesian Board of Statistics, Jakarta, Indonesia, p. 274.

Campbell-Lendrum, D., and Molyneux, D. (2005). Ecosystems and Vector-Borne Disease Control. Chapter 12. In: Epstein, P., Githeko, A., Rabinovich, J., and Weinstein, P. (Editors), Ecosystems and Human Well-Being: Policy Responses, Volume 3, pp. 353–372. Island Press, Washington DC, USA. Available at:http://www.maweb.org/documents/document.317.aspx.pdf.

Caprara, A., Lima, J.W.O., Marinho, A.C.P., Calvasina, P.G., Landim, L.P., and Sommerfeld, J. (2009). Irregular Water Supply, Household Usage and Dengue: A Bio-Social Study in the Brazilian Northeast. Cadernos de Saúde Pública, 25 (Suppl. 1), S125–S136.

Ellis, B.R., and Wilcox, B.A. (2009). The Ecological Dimensions of Vector-Borne Disease Research and Control. Cadernos de Saúde Pública, 25 (Suppl. 1), S155–S167.

Focks, D.A., and Alexander, N. (2006). Multicountry Study of *Aedes aegypti* Pupal Productivity Survey Methodology: Findings and Recommendations. Special Programme for Research and Training in Tropical Diseases, Document TDR/IRM/Den/06.1, World Health Organization, Geneva, Switzerland.

Gubler, D.J. (1989). *Aedes aegypti* and *Aedes aegypti*-borne Disease Control in the 1990s: Top Down or Bottom Up? American Journal of Tropical Medicine and Hygiene, 40, 571–578.

Pongsumpun, P., Garcia Lopez, D., Favier, C., Torres, L., Llosa, J., and Dubois, M.A. (2008). Dynamics of Dengue Epidemics in Urban Contexts. Tropical Medicine and International Health, 13(9), 1180–1187.

Quintero, J., Carrasquilla, G., Suarez, R., Gonzalez, C., and Olano, V. (2009). An Ecosystem Approach to Evaluating Ecological, Socioeconomic and Group Dynamics Affecting the Prevalence of *Aedes aegypti* in Two Colombian Towns. Cadernos de Saúde Pública, 25 (Suppl. 1), S93–S103.

Spiegel, J., Bennett, S., Hattersley, L., Hayden, M.H., Kittayapong, P., Nalim, S., Wang, D.N.C., Zielinski-Gutiérrez, E., and Gubler, D. (2005). Barriers and Bridges to Prevention and Control of Dengue: The Need for a Social-Ecological Approach. EcoHealth, 2(4), 273–290.

WHO (World Health Organization) (2004). WHO Regional Office for South East Asia 2004. Press Releases, SEA/PR/1372, 30 July 2004. As Dengue Spread to New Areas, WHO Alerts Member Countries. Available at: http://www.searo.who.int/EN/Section316/Section503/Section1549_6910.htm.

WHO (World Health Organization) (2007). WHO Regional Office for South East Asia 2007, Dengue/DHF. Press Releases, WHO Initiates Bi-Regional Approach to Tackle Dengue Fever in Asia Pacific. Available at: http://www.searo.who.int/en/Section10/Section332/Section2386_13110.htm.

Part IV
Building Community Health into City Living

Chapter 17
Introduction

Andrés Sánchez

People are moving into cities at unprecedented rates. Human beings became a predominantly urban species in 2006, and by 2030, two-thirds of the world's population are projected to be living in towns or cities, with seven of every ten urban dwellers located in Africa and Asia (Population Bulletin 2007). Urban growth will be accompanied by an increase in the ecological footprint of cities from higher use and extraction of resources, and from associated higher release of wastes (solid, liquid, and air pollution). If not well managed, larger and more crowded urban areas could exert excessive pressures on their surroundings and more distant ecosystems. Large urban centres present particular public health challenges, such as inadequate water quality and sanitation, vector-borne and transmissible infectious diseases control, air quality and shelter from extreme weather, and adequate provision of primary health-care services.

But with these challenges come opportunities. Given economies of scale and their compact configuration, cities offer great opportunities to reduce energy demand and the pressure on surrounding natural resources, including land. If cities can harness the inherent advantages that urbanization provides, they can become part of the solution to global environmental change. There are opportunities to implement successful public health measures that reach large numbers of people living in a small area. Although urbanization is inevitable, the inequalities typical of rapid urbanization are not. These are related to, and influenced by, the urban policies that are adopted. Understanding urban growth and the health and well-being of populations in urban environments will be essential in the search for more sustainable forms of development.

Beyond increases in numbers, improvements to social equity and to the quality of life of people in cities will become an increasing challenge. The poor will continue to make up a very large part of this urban growth, and this is likely to occur in the more irregular and unplanned settlements. Today, about 830 million people (24% of

A. Sánchez (✉)
International Development Research Centre, Ottawa, ON, Canada
e-mail: ecohealth@idrc.ca

the world's urban population) live in informal settlements and slums (UN-Habitat 2010), often on marginal lands such as flood plains or steep hillsides, where they are exposed to significantly greater health risks from degraded working and living environments.

Wide regional disparities persist across the world. Latest estimates indicate than in sub-Saharan Africa, more than 60% of the urban population live in slums, 35% in Southern Asia, 31% in South-Eastern Asia, 28% in Eastern Asia and 24% in Latin America and the Caribbean (UN-Habitat 2010). This situation contrasts with a 6% slum-living urban population in developed regions (WUF 2004). These are not passing trends. According to the latest UN State of the World Cities report 2010/2011, the world slum population may grow by as much as six million each year, to reach a total of 889 million by 2020 (UN-HABITAT 2010).

The case studies presented in this part illustrate common challenges faced by poor urban dwellers around the world. They also complement experiences of urban ecohealth research presented in other chapters. The vulnerability of poor urban communities, whether in Kathmandu or Yaoundé, is linked to the lack of security in land tenure, unplanned and unregulated land use, deficient health and education services, and an absence of political voice. These are places with high rates of infant and maternal morbidity and inadequate access to water and sanitation. They are where young and old shoulder multiple and relentless exposures to health hazards, including pollutants and pathogens, and face violence and unsafe working and living conditions. The town of Bebnine in northern Lebanon, although relatively better off in relation to the others, shares many of the same poor environmental conditions affecting its people, including poor or limited access to safe drinking water, deficient wastewater collection and treatment, poor waste disposal, curative rather than preventive health services, and weak local government.

In these poor urban settings, community-based organizations and other forms of civil society organizations exist but may be fragmented along ethnic or religious lines. This is often accompanied by a high level of unemployment and a sizable transient population of renters and squatters, all of which weaken social cohesion. Social imbalances are determined by a multitude of factors that stress the environment and heighten risks to human health. Not only are poor neighbourhoods vulnerable and degraded places, but they also house vulnerable people, trapping them in perverse feedback loops of poverty, marginalization, disease, and poor physical and mental development. In all case studies, people were busy surviving any way they could at the onset of the ecohealth research, in ever-deteriorating environments while local authorities watched impassive and impotent.

The case studies vary in the length of time that ecohealth research efforts have been on-going to bring about change. Not surprisingly, the depth of change achieved in each case varies significantly with the longevity of the change effort. But in all, change was facilitated by project interventions in spite of what appeared at first glance to be situations with little hope for success.

Herein lie two key contributions of ecohealth research in urban settings. In the more individualistic and depersonalized type of social environment in slums, ecohealth approaches link health and environmental concerns into a concrete entry point of common and immediate relevance to diverse local actors. People – whether land owners, renters, or squatters, and regardless of ethnic background or cast – care about and value their health and the health of those close to them. In many of the communities or neighbourhoods where the studies took place, it was the quality of the urban surroundings (wastes, air and water quality, equity, and security) that were of first concern to people. Connecting the dots between these social, environmental, and health issues helped transform people's engagement into a collective project on quality of life. The problem that originated the social mobilization was assimilated into the broader problematic of advancing towards a more sustainable, healthy, and equitable community. This is particularly evident in the Kathmandu case study.

Ill health and disease are sources of stress, grief, and expense. Making explicit environmental and social connections to health brings with it the possibility of going beyond curative approaches towards actions that achieve a different and better reality. The other key contribution is the process of change that can be engendered. An ecohealth approach offers a nuanced understanding (a diagnosis) of social and environmental determinants of health, and possible collective actions for change that are guided by scientific evidence.

In Yaoundé, Cameroon, the lack of services and planning creates living environments heavily contaminated by human waste. Growing at an annual rate of 5.6%, the development of slum neighbourhoods was the way newcomers settled in the city. These slums have precarious sanitation systems. Because they are prone to regular flooding, the lowlands in particular bear heavily polluted ground and surface waters. This, in combination with crowded and haphazard housing, low access to safe drinking water, and indiscriminate disposal of garbage, causes high rates of diarrhoea and intestinal parasitoses in children. The project produced a genuine change in the project neighbourhoods by influencing hygiene practices and environmental sanitation interventions through a participatory approach that mobilized the community and engaged local city authorities as well as civil society organizations and other international donors. It became clear that people's involvement in managing the development of their living environment, increased resident awareness, and education (especially that of mothers) were important factors in building healthier neighbourhoods.

The Kathmandu project is the longest running and marks a progression. In spite of political strife and severe economic challenges in the country, the case study presents a remarkable example of self-organized change in two city wards. The open-air butchering system that once so dominated the landscape along the banks of the Bishnumati River has evolved beyond recognition. Local governance and forms of community engagement and citizenship have also been radically transformed.

In contrast, the case of Bebnine in northern Lebanon is more recent and is in a more affluent setting, but this does not make changes any easier. The project took place at a time of high political conflict in and around the town and in the country as a whole. The 2006 war persisted throughout the study, and at times made it difficult for researchers to carry out the research in the field and provoked delays in project activities. However, despite the social–political instability and associated risks and delays, the project adapted its strategies and was able to achieve its main goals.

The research team set out to explore the links between people, water, and health, and focused particularly on water-related factors associated with diarrhoea at the household level. Implementing participatory action research in a town where local communities, local services providers, and local authorities were not used to working together, was, from the beginning of the project, a challenge for the team.

A progression in different forms of participation can be perceived in the various case studies. Passive participation occurs when the community is a source of information and the object of educational campaigns and sensitization efforts. This form of participation appears to dominate the first years of relationship-building between scientists and community actors. A second form of engagement then follows, which focuses on interactions and negotiations between different stakeholders that are key to bringing about some predetermined change. The older projects (Kathmandu and Yaoundé) go beyond these, to the level of building capacity for local organizations with the intent that people lead and pursue their own vision of change. Helping develop and negotiate that vision of change from a social–ecological systems perspective is one of the main contributions of the ecohealth approach.

The case studies offer a wide spectrum of possibilities and challenges in the organization of collective action in the domain of health protection and restoration of urban ecosystems. They illustrate the scope and reach of the ecohealth paradigm and the translation of the knowledge gained into action that does not lose its systems perspective. For this to occur, it is necessary to understand both the local "system of action" (the engagement of key community and city actors, and processes of consultation and planning), and the formal "framework of action" (institutional and organizational context). The dynamic interaction between these elements is central to the resolution of health–environment problems at the local level (e.g., the dilemma in Kathmandu between collective action and individual group action).

The integrative nature of an ecosystem approach to health places the inhabitants of a place or territory in physical interaction with the other components of their surroundings. It also allows researchers to transcend the limits of sector-based, monodimensional approaches that focus on pathologies and that trap them into a curative form of reasoning that is anchored in a health-system intervention.

References

Population Bulletin. (2007). World Population Highlights: Key Findings from PRB's 2007 World Population Data Sheet. Population Bulletin, 62(3), 10–11. Available at: http://www.prb.org/pdf07/62.3Highlights.pdf.

UN-HABITAT. (2010). State of the World Cities 2010/2011 — Cities for All: Bridging the Urban Divide. UN-HABITAT, Nairobi, Kenya. Available at: http://www.unhabitat.org/documents/SOWC10/R1.pdf.

WUF (World Urban Forum). (2004). Dialogue on the Urban Poor: Improving the Lives of Slum-Dwellers. World Urban Forum, UN-HABITAT, Nairobi, Kenya. Available at: http://www.unhabitat.org/downloads/docs/3076_44679_K0471801%20WUF2-6c.pdf.

Chapter 18
Rebuilding Urban Ecosystems for Better Community Health in Kathmandu

D.D. Joshi, Minu Sharma, and David Waltner-Toews

In 1991, the National Zoonoses and Food Hygiene Research Centre (NZFHRC) in Nepal and the University of Guelph, Canada, began to work together to understand and control the transmission of a tapeworm infection called echinococcosis or hydatidosis in Wards 19 and 20 of Kathmandu. Ecchinococcocus species are small tapeworms that depend on 2 animal hosts to complete their reproductive lifecycle. The mature tapeworm lives in the intestine of a dog or other canine. The worm eggs are shed in dog faeces, contaminating the environment. A herbivorous animal may ingest the eggs while grazing, thus becoming an intermediate host for the immature stage of the worm. This immature tapeworm forms large thick-walled cysts in the organs of the intermediate host, not usually causing much harm to the animal, depending on the location and number of cysts. When the intermediate host is eaten by a dog or canine, the immature tapeworms take hold in the carnivore's intestine and develop into mature worms, completing the lifecycle.

People who accidentally ingest tapeworm eggs can become very ill from the large cysts that may develop. In humans, the two main types of the disease are known as cystic echinococcosis (CE) and alveolar echinococcosis (AE). CE is found worldwide and its total burden is estimated to be about one million disability-adjusted life years (DALYs) per year with approximately 200,000 new diagnosed cases reported annually (WHO 2010). The burden of human AE was recently estimated at more than 600,000 DALYs per year with more than 18,000 new cases globally per year, of which 91% occur in China (Torgerson et al. 2010).

In the 1980s, NZFHRC found that many of the water buffalo slaughtered along the banks of the Bishnumati River were infected with hydatid cysts. In Wards 19 and

---

D.D. Joshi (✉) • M. Sharma
National Zoonoses and Food Hygiene Research Centre (NZFHRC), Chagal, Kathmandu, Nepal
e-mail: joshi.durgadatt@yahoo.com

D. Waltner-Toews
Ontario Veterinary College, University of Guelph, Guelph, ON, Canada

20, 50 butchers worked in more than two dozen open-air slaughter sites along the river banks, and supplied more than 60% of the meat sold in the city. About half of them disposed of the offal into open and unguarded garbage containers, while the rest tossed the waste to the side or threw it directly into the river. Thus the tapeworm was well established and disease transmission was easy between livestock and dogs: dogs ate the hydatid cysts in the slaughter waste, became infected, and contaminated the surroundings with tapeworm eggs. The conditions along the riverbank promoted human exposure to echinococcus: infected dogs roamed freely in and around the dirt-floor houses; children played in the dirt; women washed clothes and bathed in the polluted river; and men worked as casual labour in the butchering areas. These two wards were also home to a large number of meat shops (over 40 in both wards). Most did not have offal disposal containers and it was common practice for butchers to cut out cysts found in the meat and feed them to the dogs. Not surprisingly, large numbers of dogs could always be seen patiently waiting by the meat shops.

Building on this knowledge, NZFHRC and the University of Guelph led a series of epidemiological studies (1992–1998) to determine disease infection rates in animals and people, and to identify risk factors that could theoretically be manipulated to prevent the disease (Baronet et al. 1994). These risk factors concerned canine and human behaviour, including human–dog interactions in the communities, and open-air slaughtering along the banks of the Bishnumati River. A number of interventions for breaking the cycle of disease transmission were recommended and some were attempted from this epidemiological perspective but had little success. Butchers were not organized into formal associations and did not see a valid reason to change. Dogs were valued by the community, and people did not want to see them killed. Neither the ward offices nor city government had the resources to implement a proper waste-collection and management system. These failings forced a rethinking of the project team's approach from a strategy based on epidemiological surveillance and public-health education campaigns to one based on an urban-ecosystem health initiative that integrated animal and human health with local community development.

To prevent this tapeworm disease, different actors in the community needed to change the ways in which they interacted with one another and their environment. This required a participatory approach to learning and behaviour change. Engaging with the community also required negotiating needs and priorities among and between the different types of stakeholders. This, in turn, opened the scope of the project to address many pressing issues of community health through the same systems perspective.

To strengthen capacity in social action and mobilization, a local NGO, Social Action for Grassroots Unity and Networking (SAGUN), was invited to join the project. SAGUN worked as part of the team during these critical transition years (1998–2001). In October 1997, the urban-ecosystem health project phase was launched in a workshop with ward and city officials, community leaders, and an array of community members, including butchers, street sweepers, meat-shop owners, and restaurant owners. The project had two main objectives that were addressed using a participatory action research (PAR) methodology. The first was to engage the communities as co-researchers to develop a richer understanding of the interactions between the socio-cultural, political, economic, and environmental determinants of human health in these two wards. The second was to set in place a sustainable process to enhance the capacity of local individuals and groups to improve existing poor public health conditions, articulate demands for better livelihoods, and lobby local and city authorities for change. A comprehensive picture of the current and historical state of the local urban ecosystem in these wards emerged from this work, along with a number of potential points or areas for action (Neudoerffer et al. 2005, 2008; Waltner-Toews et al. 2005).

What followed was a rich and sustained effort (2002 to present) of collaborative learning and action between NZFHRC and Canadian researchers, ward authorities, and a wide array of community groups.

Engaging in Community Participation and Organization

The identification of stakeholders began by approaching the ward offices, the most local level of government. In 1991, a multi-party political system emerged in Nepal, which had operated under a single-party "panchayat" system with the monarchy retaining absolute ruling power for 30 years. This transition was not easy, and political and civil strife ensued for many years, culminating in 2008 in the abolition of the monarchy and the creation of the new Federal Democratic Republic of Nepal. In 1998, in a bid to respond to demands for increased representation, a municipal electoral system was established and local elections were held. The elections attracted young and enthusiastic ward leaders, who were open to local development initiatives. The newly elected ward officers recommended the participation of religious and community leaders in the project, who brought legitimacy to the aims of the research and provided initial guidance on problems and priorities in the wards. Four community researchers where then engaged (one male and one female from each ward) based on the advice of the community leaders. They received training in participatory urban appraisal (PUA) and gender and stakeholder analysis (GSA) and identified community volunteers for the research support team.

A number of local organizations existed in Wards 19 and 20 in 1998, but they were not legally registered and did not interact with each other or with local ward offices. No shared community vision of problems or possible solution existed. A process to reinforce local organizations and participation was initiated by inviting 26 representatives involved with meat selling, food safety, waste management, and health. They included representatives from the ward offices, butchers, meat vendors, restaurant and teashop owners, sweepers, health clinics, squatters, and members of various community clubs. The group developed a picture of relationships between the urban environment, community health, and people's livelihoods with respect to the main community problems they perceived by using rich pictures, system maps, and other tools. The roles of each type of actor in

Table 18.1 Local community groups that participated in the urban ecosystem-health project (2001–2010)

Community stakeholder groups
Street-Vendors Society, Wards 19 and 20, KMC
Small Teashop, Hotel and Restaurant Society, Wards 19 and 20, KMC
Maruhity Club, Ward 19, KMC
Nhu-Phucha Club, Ward 20, KMC
Nepal Masu Byabasayi Samittee, Wards 19 and 20, KMC
Nepal Khadgi Sewa Samittee, Wards 19 and 20, KMC
Ward 19 Office, KMC
Ward 20 Office, KMC
Community Urban Health Clinic (CUHC), Ward 19, KMC
Community Urban Health Clinic (CUHC), Ward 20, KMC
Sweepers Association, Wards 19 and 20, KMC
Squatters Association, Wards 19 and 20, KMC
School Teachers Association, Wards 19 and 20, KMC
Kantipur Youth Club, Ward 20, KMC
Young Star Sports Club, Ward 20, KMC
Kankeshwori Bhajan Mandali, Ward 19, KMC
Bal Shikshhan Kendra, Ward 20, KMC
Indra Binayak Club, Ward 19, KMC

contributing to improved community health and well-being were identified and discussed. Stakeholder groups were then established, largely according to their profession or experience.

A total of 18 stakeholder groups became project partners (Table 18.1). Stakeholder groups were encouraged to seek formal legal status so that each group would have the capacity to engage with different levels of government and with national and international development donors. This also encouraged group autonomy and lessened their dependence on the project, both in terms of the groups' goals and access to resources. Most groups obtained legal status by 2006, though longer and more sustained support was required for the most marginalized groups (sweepers, squatters, teashop owners, and street-vendor associations).

Different forms of coordination between researchers and the ward communities evolved over the course of 8 years. Overall direction on research and community development activities was first provided by the Ward Committees in Wards 19 and 20. These committees included members recommended by the ward offices and community leaders, and were chaired by the ward chairpersons. With the identification and engagement of stakeholder groups in the years that followed, the Ward Committees were closed in 2002 and were replaced by the Implementing Coordination Committee (ICC). The ICC includes one representative from each of the first 12 stakeholder groups that were officially organized and registered, a representative from each ward office, and a representative from the NZFHRC, and remains self-organizing and self-funded till today.

Community Action Plans and Local Capacity Building

Many members who joined the different stakeholder groups came from low-income groups or from occupational sectors without any history of formal organization or interactions with government. Each group developed its own workplan based on interests and needs, and worked with the project team to implement a lobbying strategy that included joint meetings with city officials to present their plans, explore government support, policy and regulatory implications, and follow-up actions. The project also provided small grants or seed support to the stakeholders' initiatives based on their workplans, and in many instances helped broker collaboration with different funders, provided assistance to prepare funding requests, and offered technical input into the design of activities. Different tools and methods were used to strengthen the capacities of different groups, depending on needs. These included the REFLECT approach for adult learning and social change (see ActionAid, www.actionaid.org), Social Analysis Systems techniques (see SAS, www.sas2.net), and gender analysis. These helped raise awareness and build local capacity and skills that were used by the different groups to address their own goals and visions (Joshi et al. 2003a, b). At the same time, a significant level of effort was needed to ensure that a broader and inclusive community perspective was not lost.

Weaving the Work of Different Stakeholder Groups into Community Development

The project was confronted with a dilemma early on with this multi-stakeholder group approach. By choosing to strengthen individual groups, opportunities were reduced to work together to address collective problems. From a systems perspective, when each group works independently, the potential emerges to act at cross-purposes, or at least for opportunities for synergies to be lost (Neudoerffer et al. 2005, 2008). Cultural and social legacy compounded this issue and needed to be addressed in a pragmatic manner. The traditional caste system was officially abolished in Nepal in 1963; however, many people continued to practice occupations traditionally associated with their caste. In the first community meeting, for example, it was discovered that sharing a table with one group that was considered "untouchable" meant being rejected at another group's table that saw itself as having higher status.

In hindsight, the decision to first strengthen group cohesion among people with similar problems and needs by fostering their group vision and articulation of their own interests and aspirations was perhaps key to the overall success of the project. It also meant that the project had to counterbalance a tendency to fragment by using a number of means, beginning with the establishment of the ICC. With the ICC, project coordination was transferred from a group of individuals (some with more others with less voice) who were appointed or recommended by community leaders, to a committee of group representatives. Dialogue between the different groups was encouraged, especially during capacity-building activities on community health, project review meetings, and lobbying efforts with different levels of government (Joshi et al. 2003c, 2006; Regmi 2009a, b).

Some Achievements Along the Way

The following may appear at first glance to be a hodgepodge of changes in day-to-day community life that were observed over the years. For reasons of space, only an overview of a small set of achievements can be presented; together they reveal an important shift in how people in these two wards relate to one another and to their environment.

It is fitting to begin with the breaking of the once intractable cycle of echinococcosis transmission because this was the project's point of entry more than a decade ago. It came about over time, not through a pre-planned course of events, but rather as an emerging outcome from a number of interconnected social, economic, and environmental changes (Fig. 18.1). These included: a new national policy (Animal Slaughterhouse and Meat Inspection Act); local regulations for butchers and meat sellers; a stronger organization of these groups and their increased capacity to value and adopt new rules; and community awareness and pressure. In combination, these changes facilitated the removal of livestock from the riverbanks (now a contractor keeps the livestock outside the city and supplies butchers according to need) and the transformation of the once heavily contaminated and degraded city environment.

The riverbanks in both wards have been transformed into community gardens with controlled access and are now maintained by local community groups under the supervision of ward offices. Caretakers plant and sell flowers and seedlings to finance their upkeep. Open-air slaughtering has essentially stopped in both wards, and mostly all over Kathmandu City. Ward offices now ban open-air slaughtering of goats, pigs, and water buffalo, and require more hygienic butchering methods and the proper disposal of wastes in identified locations. The butchers have also stopped throwing blood, bones, and skins in the streets and riverbanks. Instead, they now sell much of the waste, including hides, bones, and ligaments to contractors who recycle these products into glue, fertilizer and other by-products, thus reducing pollution and increasing incomes. A biogas plant was also built to treat organic slaughter waste. This plant was constructed with a combination of government funding (Kathmandu Metropolitan City, KMC), seed funding and technical guidance from the project, and co-funding from World Vision International. It is now run jointly by the butchers' and meat associations, and although it is not free from operating problems, it helps to convert organic waste into compost.

Table 18.2 presents the changes in hygiene and sanitation of slaughtering sites observed from 2000 to 2009. The most important changes were that livestock were

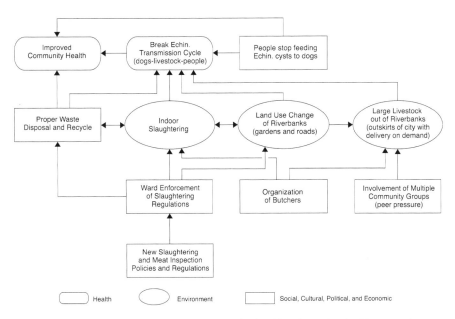

Fig. 18.1 Social and ecological outcome pathway for breaking the cycle of echinococcosis transmission in Wards 19 and 20, Kathmandu, Nepal

Table 18.2 Changes in hygiene and sanitation conditions at slaughtering sites in Wards 19 and 20 from 2000 to 2009

Sanitary indicators	2000 (25 sites) (%)	2009 (20 sites) (%)
Livestock to slaughter kept on riverbank	100	0
Open field slaughtering	43	0.5
Under roof slaughtering, but not enclosed	14	1
Enclosed slaughtering	43	90
Brick soiling floor	29	30
Cement concrete soiling floor	71	50
Butchers using paddy straw or wheat straw for burning skin and hair of the carcass	14	0
Butchers using clean water (from public piped system instead of river) for washing meat	71	90
Butchers disposing of offal and waste into river	43	0.0
Butchers disposing of offal into waste containers	57	75
Good hygienic and sanitary conditions	14	90

no longer kept along the river, and that slaughtering moved inside buildings dedicated to this purpose, with various types of flooring. The apparent decrease in the use of cement flooring reflects the high cost of pouring new concrete. While open concrete pads by the river were abandoned in favour of much more hygienic enclosed slaughtering facilities, not all of these had concrete floors. The enclosed slaughtering areas were equipped with drainage to capture and control wastes. Equally important was the improvement in meat processing hygiene and sanitation, including proper disposal of wastes and offal for rendering.

The first modern hygienic slaughterhouse in Kathmandu was built in Thankot (about 12 km away from the wards) with private and government funding and began operations in 2007. The plant is now running at only 40% capacity because many butchers preferred to go back to their wards. Nonetheless, as they return to their wards, they are organizing themselves into medium and small slaughter houses that are now indoors and have proper waste management and improved hygienic conditions.

Two stakeholder groups, the Nepal Butchers Association (Nepal Khadgi Sewa Samiti) and the Nepal Meat Association (Nepal Masu Byabasawa Samiti), stand out in their evolution. Although both organizations existed when the project began, they were not legal entities and functioned as semi-organized trade groups. Through participation in the project, both organizations were able to obtain their legal status. The butchers' association has grown to have a national presence in the last 8 years, and has members in 23 districts of the country. It has secured significant funding from government and international donors, and joined the meat association to create a cooperative to provide much needed credit to its members. It has a major role in promoting adherence to new national regulations for the sector.

Changes in garbage collection and disposal have also been significant and involved ward offices, the sweepers' association, households, and butchers. It was common in the late 1990s to see households throwing waste directly onto the street rather than into garbage bins. Moreover, most butchers were not aware of the contamination dangers posed by meat wastes, and their butchering techniques facilitated the transmission of animal diseases to humans. Over time, residents curbed the habit of throwing garbage from windows or doors into the streets, which was creating extra work for street sweepers. Mass awareness campaigns by the project led to better waste management at the household level, with people storing garbage in bags and depositing the bags in specific collection points, or holding them until the garbage collectors come. A 2005 waste-collection survey found that there were an average of four garbage collection centres in each "tole" (neighbourhood) of Wards 19 and 20. Among these, a total of 15 garbage collection centres were studied with the help of a questionnaire and observation guide. In all toles, most households reported that they deposited their garbage at the collection points on a regular basis, and in 40% (6) of the cases sweepers also collected household garbage. In all cases, KMC regularly transported the household waste from these points to a dump site. No garbage piles in either ward can be seen today along the riverbanks (Fig. 18.2).

Improved health services and health awareness were also facilitated in both wards by this community-centred development effort. Before the project, health services were provided by either private doctors or drugstores. For mass immunization

Fig. 18.2 Riverbanks in Wards 19 and 20 were transformed into gardens (**b**), replacing the contaminated riverbanks that served earlier as repositories of garbage (**a**)

programmes, people had to go to the public health department of KMC. Early in the project, a number of meetings were organized with ward offices, city government, and several donors to secure land and buildings to set up a health clinic in each ward. A clinic-building construction committee was put in place to prepare a proposal, oversee the design, and seek approval from both the ward office and the KMC physical planning division. These efforts were successful, and two clinics (one in each ward) were built and staffed. Both health clinics own their land and buildings, and have been running for a number of years with joint funding from KMC and the operating budgets of each ward. The clinics provide preventive and curative services and have outreach programmes on family planning, HIV-AIDS, maternal and child health, childhood immunizations, TB drug treatment, and deworming for children. They also provide basic medicines with controlled mark-ups, so they sell drugs at a lower cost than the private pharmacies.

Perhaps one of the most significant changes that emerged over the course of the years is the increase in social equity and the decrease in class-based discrimination in these wards. The community planning and dialogue facilitated by the ICC and joint group activities seem to have helped improve the voice and influence of the more disadvantaged groups, who are now represented through registered associations. An example of this is the changes in the squatter community. In the early 1990s, highly polluted environments with a strong stench from slaughter blood and offal decomposing under the sun were unwanted by others and therefore offered safe spaces for squatters. Butchering activities were a source of jobs, but these were squalid living and working conditions. Today, the housing quality in this part of the community is considerably improved as are their surroundings. The squatters have not yet been able to obtain titles to the land, but they are now recognized ward citizens (with voice and representation) and have access to basic services (education, water, sanitation, and health). Most are also members of the registered community organization.

What We Learned

Before the project, there were no local mechanisms to address community problems, and people were reluctant to organize themselves. Community input in official planning did not exist. Donors and international NGOs would make agreements with city government officials without involving the ward offices or other local stakeholders. Community development was haphazard and led by outside implementers. The project's journey over the last 20 years illustrates how much more can be achieved through an approach that engages with the impacted communities and responds to their social and environmental realities. Nepal's democratic transition, and new forms of local government representation provided a window of opportunity for community participation in building healthier local urban environments.

The political changes in Nepal in the last 10 years were characterized by political and economic instability, several parliamentary crises, and civil unrest that reached even into Wards 19 and 20. For example, both ward offices and health clinics were badly damaged by bombings in 2007, but were rebuilt with contributions from government, private, and civil society organizations, including many local groups, a heartening indication of the social capital that has been built there.

In spite of this political instability, and perhaps in part because of the positive aspects of the democratic movement, the project succeeded in bringing about important changes in the environment and community life in both project wards. Through the project, the researchers, community members, NGOs, and representatives of various levels of government came to recognize the need for shared responsibility in transforming a community. In a disorganized setting, everyone may feel responsible but no one will take responsibility. The project's experience in Kathmandu Wards 19 and 20 has shown that strengthening community organization can empower all parties to take on this responsibility for addressing community problems. Different stakeholder groups (including local government) can then work together to define their respective roles and take steps to address collective problems in wards 19 and 20, this identification of roles and division of responsibilities was facilitated by adopting an ecosystem approach to health and to urban ecosystem health early in the project. This work engaged different actors in developing a joint understanding of problems and negotiating a common vision for the future, which was based on a better understanding of social and environmental interactions, and their health implications. The project was not only able to deal successfully with the echinococcus problem, but went far beyond to permanently and positively transform and empower the ward communities, the city, and even some national meat-sector laws.

Acknowledgments We are grateful to a large number of researchers and institutions who worked with us over the years. Special thanks are due to our partners from SAGUN, all members from the 18 stakeholder groups from Wards 19 and 20 of Kathmandu, students and other colleagues from the University of Guelph, Canada, and all staff, consultants, advisors, and resource persons who worked directly with NZFHRC. IDRC support was provided through projects 003320, 101277, and 104659.

References

Baronet, D., Waltner-Toews, D., Craig, P.S., and Joshi, D.D. (1994). *Echinococcus granulosus* Infection in Dog Populations in Kathmandu, Nepal. Annals of Tropical Medicine and Hygiene, 88, 485–492.

Joshi, D.D., Sherpa, C.L., and Waltner-Toews, D. (2003a). Participatory Action Research on Urban Ecosystem Health in Kathmandu Inner City Neighbourhoods. National Zoonoses and Food Hygiene Research Centre (NZFHRC), Kathmandu, Nepal. Available at: http://idl-bnc.idrc.ca/dspace/handle/10625/32067.

Joshi, D.D., Sharma, M., Maharjan, M., Joshi, H., and Chand, P.B. (2003b). Urban Ecosystem Health Approaches to Local Initiatives in Kathmandu Ward No. 19 and 20, Nepal. National Zoonoses and Food Hygiene Research Centre (NZFHRC), Kathmandu, Nepal.

Joshi, D.D., Regmi, D.N., Thapa, D.B., and Gurung, C.K. (2003c). Sanitary Condition of Households in Kathmandu City. National Zoonoses and Food Hygiene Research Centre (NZFHRC), Kathmandu, Nepal.

Joshi, D.D., Gupta, R., Sharma, M., and Dahal, M. (2006). Three Years Progress Review of Urban Ecosystem Health Project, Phase II. National Zoonoses and Food Hygiene Research Centre (NZFHRC), Kathmandu, Nepal.

Neudoerffer, R.C., Waltner-Toews, D., Kay, J., Joshi, D.D., and Tamang, M.S. (2005). A Diagrammatic Approach to Understanding Complex Eco-Social Interactions in Kathmandu, Nepal. Ecology and Society, 10(2), 12. Available at: http://www.ecologyandsociety.org/vol10/iss2/art12/.

Neudoerffer, R.C., Waltner-Toews, D., and Kay, J. (2008). Chapter 15: Return to Kathmandu, A Post Hoc Application of AMESH. In: David Waltner-Toews, D., Kay, J.J., and Lister, N-M.E. (Editors), The Ecosystem Approach: Complexity, Uncertainty, and Managing for Sustainability. Columbia University Press, New York, NY, USA.

Regmi (2009a). Assistance in Synthesis and Documentation of Health, Environment and Development Outcomes in Urban Ecohealth Project Kathamdu. Consultancy Report. International Development Research Centre (IDRC), Ottawa, Canada. Available at: http://idl-bnc.idrc.ca/dspace/handle/10625/45392.

Regmi (2009b). Assistance in Synthesis and Documentation of Health, Environment and Development Outcomes in Urban Ecohealth Project, Kathmandu. Workshop Report. International Development Research Centre (IDRC), Ottawa, Canada. Available at: http://idl-bnc.idrc.ca/dspace/handle/10625/45391.

Torgerson, P.R., Keller, K., Magnotta, M. and Ragland, N. (2010). The Global Burden of Alveolar Echinococcosis. PLOS Neglected Tropical Diseases, 4(6), e722. Available at: http://www.plosntds.org/article/info:doi/10.1371/journal.pntd.0000722.

Waltner-Toews, D., Neudoerffer, C., Joshi, D.D., and Tamang, M.S. (2005). Agro-Urban Ecosystem Health Assessment in Kathmandu, Nepal: Epidemiology, Systems, Narratives. EcoHealth, 2(2), 155–164.

WHO (World Health Organization). (2010). Report on Neglected Tropical Diseases 2010: Working to Overcome the Global Impact of Neglected Tropical Diseases. World Health Organization, Geneva, Switzerland.

Chapter 19
Understanding Water, Understanding Health: The Case of Bebnine, Lebanon

Rima R. Habib*

In the town of Bebnine (population approximately 16,000) in northern Lebanon, a team of researchers from the American University of Beirut and local residents worked together on a community-based research project that used an ecosystem approach to better understand the links between people, water, and health (Forget and Lebel 2001; Lebel 2003). Initially, the study focused on the impact of polluted irrigation water and the use of agrochemicals on the quality of local produce. However, preliminary meetings with key community stakeholders unearthed community concerns about the apparently high rate of occurrence of gastrointestinal diseases among Bebnine residents. This more urgent issue led to a shift in project scope to align better with community priorities and address the relationship between drinking water quality and community health and well-being. The results of the work guided the development and setup of community-based short- and medium-term interventions to alleviate the effects of water-related health problems.

Although community participation proved exceptionally valuable in research design and implementation, for example in collecting and interpreting the data, most community stakeholders remained reluctant to carry forth study activities without direct support from the research team. It can be speculated that the research team was a catalyst to community activism in a region where collaboration between civil society and local government in community development has been historically poor. Further, the project took place at a time of high political conflict in and around the town and in the country as a whole. Political instability began in 2005 but peaked with the July–August 2006 war and continued through most of 2007. As can be expected, progress of the work was significantly affected and field activities were delayed during times of hostilities.

---

*On behalf of the Bebnine Research Team.

R.R. Habib (✉)
Department of Environmental Health, Faculty of Health Sciences,
American University of Beirut, Beirut, Lebanon
e-mail: rima.habib@aub.edu.lb

Overview

The project first sought to develop an understanding of how water is used by residents, determine the water quality of different water sources used by the community, and assess the impact of microbial contamination of water on community health. Diarrhoeal diseases were selected as the principal health indicator. The contribution of non-water sources to disease transmission was also investigated.

Bebnine is located in the province of Akkar in northern Lebanon, 95 km from the capital, Beirut. It is a large town with an area of about 4 km^2 of which 3.3 km^2 are agricultural lands including olive, citrus, and fruit orchards, greenhouses and organic farms. The town is inhabited by 3,300 families, with an average of 5–6 persons per household. A demographic survey carried out by the research team in 2005 revealed that nearly half of the residents were less than 30 years old and 3% were more than 65 years old. Illiteracy rates were relatively high (19% for males and 34% for female household heads), and almost all women (94%) between 18 and 65 years of age and 34% of men in the same age group were unemployed. Residents reported a number of health problems, especially chronic diseases such as diabetes, heart disease, and high blood pressure. Environmental problems included solid-waste disposal, water pollution, and sewage disposal. Despite the prevailing low socio-economic conditions compared with the rest of the country, there are seven primary health care centres, 11 private schools, and eight public schools in the town.

An Ecosystem Approach to Health

Based on many visits to the town and meetings with various community members, the research team realized the complexity of the physical, social, and environmental factors and their links at multiple levels to the water–health issues in Bebnine. These links included relationships between families and between neighbourhoods, the level of popularity and perceived performance of representatives on the municipal council, the number and location of health centres, and the seasonal changes that affect water availability. To make sense of this complexity, the project adopted an ecosystem approach to health to untangle the multifaceted factors that contribute to health problems, and to frame recommendations for community solutions.

Acknowledging the social, ecological, and public–health interactions involved in water–people–health links, a multidisciplinary research team from diverse backgrounds worked together to develop and execute the project. The team consisted of experts from health sciences, environmental engineering, agricultural sciences, and social policy from the American University of Beirut (AUB).

Community participation was a cornerstone of the study. Residents and community authorities contributed throughout the project by helping to select the research topic, participating in the mapping of water sources, helping to identify the sources of contamination, assessing the exposure of family members to water-borne pathogens, evaluating people's knowledge about the problems, and recommending culturally sensitive and economically feasible interventions. This participatory process was based on an ongoing dialogue with community representatives; their contributions consistently informed and improved research activities. An advisory committee consisting of local men and women was formed at the beginning of the project to provide local knowledge and guidance. Initially, seven community members formed the advisory committee, with two male members from the municipal council, two women from a local nongovernmental organization (NGO), a female school director, the former male mayor of Bebnine, and the male head of a former Bebnine water committee that had been dismantled. The committee closely oversaw the progress of the research through monthly meetings, where research activities were discussed and findings were shared.

During the course of the project, some members of the advisory committee were replaced by other members from the community who showed greater interest in the project. The new members were socially active women, well known in Bebnine and familiar with the local situation. The president of a local NGO (Women's Charity Association) became highly engaged in the project activities, establishing useful contacts with other women, establishing trust within the team, and informing the team about local perceptions and opinions.

A field coordinator for the project was present in Bebnine on a daily basis and provided the required regular updates to both members of the advisory committee and the research team. She performed and supervised most field activities. Having previously worked in the town on another project, she was well known to the community and familiar with the local situation, understood the relationships between the families, and was perceptive of local attitudes. Her daily responsibilities included inviting people to join meetings and conducting public presentations, training, discussions, and focus groups. In addition, she worked in the field, collected water samples, installed and maintained a water-disinfection unit, disseminated test results to individual households, recruited interviewers, and visited health centres and the municipality to gather and supply data and information. A group of 15 local women trained by the research team participated in field activities and data collection.

Research Methods and Community Interventions

During the project's lifetime, the team tried to ensure that all community groups were represented in the project. Bebnine is a conservative and male-dominated community with distinct roles and responsibilities for men and women, in both the public and private spheres. Project staff encouraged both men and women to participate in the focus group meetings, where both had the opportunity to discuss patterns of water use, behaviours, and perceptions of other environment-related problems. From the women, the team learned about levels of women's autonomy, freedom of

Table 19.1 Frequency of use (expressed as the number of users and percentage) of different water sources among 2,223 households in Bebnine (2005)

Water source	Drinking water		Service water	
	Number	Percentage	Number	Percentage
Main spring	443	20	462	21
Artesian well(s)	465	21	640	29
Bottled water	357	16	–	–
Other sources (small springs)	197	9	221	10
Main spring and artesian wells	139	3	259	12
Combination of sources	622	28	641	29

movement, and the scope of their responsibility in household decisions about water practices, child care, and control of expenses.

The research activities were completed in several phases, each building on the experiences of previous ones. In the first year (2005), a household survey was carried out of most households in Bebnine ($n=2223$, 94% of all households), and more than 20 sources of drinking water were monitored monthly for microbiological contamination. The survey aimed to understand the distribution and use of different sources of drinking and service water (used for cleaning, bathing, cooking, and laundry) according to neighbourhood and socio-economic status.

A more in-depth survey with a smaller sample of households was carried out the following year (2006) to explore water quality and diarrhoea relationships. A random sample of 462 out of 2,223 Bebnine households was drawn, representing 50% of all households that relied solely on "wells" or on the "main town spring" (Table 19.1) during the summer. The survey identified water-related risk factors for diarrhoea, such as the cleaning (or not) of water storage containers, consumption of bottled water, methods of treating household drinking water, and local perceptions of water quality. Included in this survey were questions about factors not related to water, such as cooking and eating habits, breastfeeding practices, food storage, perception of health, bathing and hand washing practices, and solid-waste and wastewater disposal (Abou Mourad 2004; Al-Ghazawi 2004; Wright et al. 2004; Al-Medhwahi et al. 2005; Fewtrell et al. 2005).

The project measured the number of diarrhoeal episodes among family members in a 4-week period, accompanying symptoms, duration of the episode, treatment methods, and related beliefs. Afterwards, focus groups were held for a portion of the participants (men and women) from the 2006 study, to acquire qualitative information on water-related household decisions, practices in response to diarrhoea episodes, people's understanding of causes and prevention of diarrhoea, water handling in the household, and attitudes about personal and domestic hygiene.

In 2007, the study was interrupted by a new conflict near Bebnine. When the study resumed, a cross-sectional survey was used as before to assess occurrence of diarrhoea. This survey also addressed physiological and mental health, and the social and economic impact of the conflict on residents in Bebnine. The sample for this 2007 survey was the same as the households that participated in the in-depth survey in 2006 (423 out of the 462 households participated in the study).

Results from the surveys revealed an important number of diarrhoea cases: 23% of households in 2005 (501/2,223 households) reported at least one case of diarrhoea in the past year; and 32% of households in 2006 (150/462 households) reported at least one case of diarrhoea in a 4-week period. A similar rate (31% or 133/423 households) was measured in 2007. It is possible that these reported rates underestimate the true rate of illness, because of respondents' inability to accurately recall diarrhoeal episodes.

In response to these results, a surveillance study was undertaken of diarrhoea cases reported to six public and private health centres in Bebnine. The surveillance study was initially planned for the summer of 2006 as recommended by the advisory committee and the local physicians, summer being a peak season for diarrhoea cases. This activity was postponed to October–November 2006 because of the war. The study measured 160 cases of diarrhoea reported to the centres over the 2-month period. Cases were distributed equally by gender and most of them were reported among children less than 5 years old. Most cases (85%) were reported from two health centres. This reflected a higher use of these health centres, rather than higher rate of illness among the population using these centres, presumably because of greater community trust in their services. Although no statistical associations with the sources of water could be established, a greater proportion of diarrhoea cases reported drinking from the main spring of Bebnine.

The findings of all previous activities inspired the development of a series of community actions, including health awareness and promotion activities, water-quality monitoring in some locations, and the installation of a water-disinfection unit that served one of the neighbourhoods and benefiting more than 140 households.

Water Sources in Bebnine

Bebnine residents obtain their drinking and service water from various sources, including springs inside and outside of town and private and public artesian wells. Some households purchase bottled water for drinking and others purchase water carried by tankers that transport water for domestic use from sources outside town. Table 19.1 shows sources of water as reported by female homemakers in the 2005 survey ($n = 2223$). About 20% of households rely on the main water spring of Bebnine as a source of water; whereas 21–29% of households rely on private or public artesian wells. The main water spring is located in the south of the town in the neighbourhood El-Ain/Baddou, with a higher elevation than other neighbourhoods. Initial water quality analyses indicated high contamination of several sources.

Residents had the option to join a new piped drinking water network installed in 2004 that carries (presumably very high quality, clean) water from the mountains of the Province of Akkar and is managed by the regional water authority. Residents were expected to pay a connection fee of about 400,000 LBP (US$ 267), which included the provision of a meter box, and a yearly water tariff of 250,000 LBP (US$ 167). However, the 2005 survey showed that nearly half of residents were unable or unwilling to pay for this service because of its high price. This finding prompted the mayor of Bebnine to lobby and convince the water authority to exempt the town from the connection fee. Unfortunately, for unknown reasons, the network is still not operational at the time of this publication (December 2010).

Water Quality

To understand the extent of the water problem and its root causes, the quality of drinking water at the source and in selected households was assessed. About 40 water samples were taken once per month over 3 years (2005–2008) from 20 locations around Bebnine. Water samples were taken from public wells (in schools), private wells, public reservoirs, the main water spring of Bebnine, and other secondary springs around the town. Sampling locations were selected based on suggestions from local community members and the advisory committee. The samples were analysed for faecal and total coliforms at the Environmental Engineering Research Center at AUB. The assessment of the microbial quality of selected water sources showed significant levels of faecal contamination, as high as 5,000 CFU/100 mL. All water sources were found to be contaminated at least once in a year (Table 19.2).

During October and November 2007, samples of drinking water were taken from 423 households and tested for microbiological contamination. Faecal coliforms were found in 83% of samples. Table 19.3 shows the distribution of faecal contamination of the samples. Although water is heavily contaminated in Bebnine, the study also identified non-water related factors that were strongly correlated with risk of diarrhoea – especially poor hygiene practices in the home and inadequate neighbourhood sanitation (El Azar et al. 2009).

Discussions with heads of households, and during the presentation of project results, revealed that the community was largely aware of the widespread microbiological contamination of water sources. It also attributed the relatively high prevalence of enteric diseases in the town to the poor quality of the water used. Water-quality results confirmed microbial contamination of Bebnine's water sources, although some neighbourhoods experienced higher levels of contamination than others. A relatively low percentage of the community took precautions regarding water quality: about 16% bought bottled water and 7% disinfected water before drinking ($n=2223$).

Table 19.2 Summary of results from samples taken from different water sources in Bebnine on a monthly basis between 2005 and 2008

Neighbourhood	Water source	Faecal coliform range[a] (CFU/100 mL)	Percentage of samples contaminated[b]
1	A local spring	32–4000	100
2	A public well	8–2260	100
3	A public well	0–5000	75
4	A public well	0–144	64
5	A public well	0–400	45
6	Water from the pipe coming from the main spring in Bebnine, before it enters the storage tank used for water collection	0–500	56
7	A public well	0–114	43
8	Water from the main storage tank that receives water from the main spring and a local well. The samples were taken *before* the installation of a disinfection unit in April 2006	2–400	100
8	Water from the main storage tank that receives water from the main spring and a local well. The samples were taken *after* the installation of a disinfection unit in April 2006	0–1	9
9	Water taken from the pipe coming from the main spring of Bebnine, before it enters the storage tank used for water collection	0–70	24

[a] Range of coliform-forming units (CFU) from the total number of samples taken from each source between 2005 and 2008. Samples were taken on a monthly basis. The US-EPA standard for maximum contaminant level for total coliforms, including faecal coliforms, is zero (US-EPA 2010; WHO 2006)
[b] Percentage of samples contaminated, out of the total number of samples taken from each source, over the 3-year period (2005–2008)

Table 19.3 Distribution of levels of faecal contamination in 423 samples of drinking water in Bebnine (2007)

Microbiological contamination	Number of households	Percentage of households
0 CFU/100 mL	70	17
0–10 CFU/100 mL	94	22
10–500 CFU/100 mL	148	35
>500 CFU/100 mL	111	26

CFU: Colony-Forming Units

A Community Intervention: Water-Disinfection Unit

On the basis of these findings, the research team and community members developed an intervention. The team installed a water-disinfection unit in a water-storage tank that served Neighbourhood 8. The community contributed by building a structure to house the unit. Neighbourhood 8 was chosen because of the presence of high contamination of all samples in this location (Table 19.2) and the large number of households that used this water source ($n = 140$). The municipality agreed to maintain the unit and ensure its proper operation. After the intervention, community members were confident of the quality of the treated water. Anecdotal evidence obtained by the researchers since then indicates that word of this improved clean water soon spread beyond Neighbourhood 8 and attracted people from other neighbourhoods and towns. Despite increased demand, there have not been any complaints of reduced availability of water. People appear to be willing to invest time and energy for water they trust to be safe.

Water–Diarrhoea Relationships

The 2006 survey sought to identify water-related and other factors associated with diarrhoea at the household level. Data analysis revealed that the absence of water disinfection in the household increased the risk of diarrhoea among children aged 6–14. Young children (less than 5 years old) who walked barefoot also had an increased risk of diarrhoea. Poor hygiene and sanitation were commonly noticed in the town where children as young as 3 years of age would spend their play time outdoors, barefoot, playing with dirty soil, and eating on the ground. According to testimonials from focus groups, women had become used to the occurrence of diarrhoea, perceiving diarrhoea as a fact of life rather than a health problem. Some believed that *having diarrhoea is God's will* and that *it is not affected by [personal] behaviour.*

Health Awareness and Promotion

Following the household surveys in 2005 and 2006, the data were collected and used to design targeted health messages and awareness campaigns. Several steps were taken to inform the local community about the progress of the study. One of these activities consisted of a series of lectures entitled Water and Diarrhoea in Bebnine: Prevention and Methods of Disinfection. These lectures presented the results of the research project and provided information about affordable and simple methods to prevent diarrhoea among infants and children. The lectures, attended by approximately 520 female homemakers and young female adults, began in spring 2007 and continued until spring 2008. They were also given in 11 schools across several neighbourhoods. In addition, the field coordinator visited all residents that participated in the 2007 survey to share the results of the water analyses and to explain simple water-disinfection methods at the household level.

Lessons Learned

Over the course of the project, the research team completed three surveys, monitored several local water sources each month, conducted awareness campaigns, and installed a water-disinfection unit. In addition, the team began to understand and raise awareness of the socio-ecological picture of vulnerability to faecal–oral disease transmission through water in the town. The partnership with the community was essential to foster their participation in the planning process, interpretation of study findings, and development and adoption of appropriate new strategies for dealing with the problems. It was essential to weave together a network of neighbourhood leaders, families, parents, students, teachers, professionals, and others to stimulate community participation in environmental and public-health issues of importance to the community itself.

The research found that major community decisions are made by the municipal council, which, although democratically elected, is not considered representative of the community, partly because of political alliances for or against the ruling families. Still, the project strove to work with the municipal council because of the access it provided to the community and to other resources. Although most community members accepted the involvement of the municipal council in the project, 4% of residents refused to participate, citing the involvement of the municipal council as the primary reason.

The advisory committee also proved useful throughout the study. It raised important community concerns and provided comments on the surveys and input on the location of the disinfection unit. The research team considered encouraging this committee to become an advocate of water-health issues. However, the committee quickly became inactive without support from the research team. Community members and local organizations do not have a history of self-organization for advocacy. For example, although a new water network has been developed, and connection fees were waived, the community did not lobby for managing its operation. Although it could be stated that the study was able to implement a sense of awareness and change for better health for a number of community members, further changes are still needed in the community. In addition, the government's capacity to ensure services (like drinking water) is not yet consistent, and needs to be further developed. This factor is one of the major stumbling blocks to the development and implementation of effective policy, and also to improving trust in democratic institutions in Lebanon.

The research activities addressed the complexity of the people–water links, which were elucidated through a mix of quantitative and qualitative research. Water access in Bebnine is further complicated by the distribution of families by neighbourhood. Family feuds hindered the progress in solving problems within the water network and led to tensions between some neighbourhoods and the municipal council (led by people from other families). This experience highlighted how community access can be both a blessing and a challenge: a blessing, when the community grants access, participates, and informs research decisions; but a challenge when such involvement is fraught with political tensions that impede positive change.

The complex relationships between water and health and the multifaceted causes of diarrhoea require multilevel interventions that focus on both the physical and social environments. Diarrhoea is endemic in Bebnine, especially among younger children. However, apart from water provision, hygiene and sanitation are also important issues. Diarrhoea in Bebnine will not be controlled if the efforts by the municipality and the community focus only on the quality and quantity of water and ignore other social and environmental factors.

Although water is a public good and a life-sustaining service, social, cultural, and economic barriers often stand in the way of its equitable provision. This is where this ecohealth project facilitated success. It demonstrated the many elements and perspectives that make group action for change a possibility.

Acknowledgments I thank my fellow research team members from the American University of Beirut, Iman Nuwayhid, Mutasem El-Fadel, Rami Zurayk, Dima Jamali, Mona Haidar, Hind Farah, Grace El-Azar, and Safa Hojeij, and the project's Advisory Committee members. The contributions of the community of Bebnine, the mayor and the elected municipal council, school directors, women's associations, and directors of health care centres were also very important. IDRC support was provided through project 101815.

References

Abou Mourad, T.A. (2004). Palestinian Refugee Conditions Associated with Intestinal Parasites and Diarrhea: Nuseirat Refugee Camp as a Case Study. Public Health, 118, 131–142.

Al-Ghazawi, Z. (2004). Ecosystem Approach to Human Health in Two Villages of the North Jordan Valley: Scoping the Problems. EcoHealth, 1(Suppl. 2), 97–108.

Al-Medhwahi, E., Briggs C., and Keane, S. (2005). Household Hygiene Improvement Survey in Yemen: Knowledge, Practices, and Coverage of Water Supply, Sanitation and Hygiene. US Agency for International Development (USAID), Washington, DC, USA.

El Azar, G., Habib, R.R., Mahfoud, Z., El-Fadel, M., Zurayk, R., Jurdi, M., and Nuwayhid, I. (2009). Effect of Women's Perceptions and Household Practices on Children's Waterborne Illness in a Low Income Community. EcoHealth, 6(2), 169–179.

Forget, G., and Lebel, J. (2001). An Ecosystem Approach to Human Health. International Journal of Occupational and Environmental Health, 17, S3–S35.

Fewtrell, L., Kaufmann, R.B., Kay, D., Enanoria, W., Haller, N., and Colford, J.M. (2005). Water, Sanitation, and Hygiene Interventions to Reduce Diarrhea in Less Developed Countries: A Systematic Review and Meta-Analysis. The Lancet Infectious Diseases, 5, 42–52.

Lebel, J. (2003). Health: An Ecosystem Approach. In Focus Series. International Development Research Centre (IDRC), Ottawa, Canada. Available at: http://www.idrc.ca/in_focus_health/.

US-EPA (United States Environmental Protection Agency). (2010). Drinking Water Contaminants. US-EPA, Washington, DC, USA. Available at: http://www.epa.gov/safewater/contaminants/index.html.

WHO (World Health Organization). (2006). Guidelines for Drinking-Water Quality. First Addendum to Third Edition. Volume 1, Recommendations. WHO, Geneva, Switzerland. Available at: http://www.who.int/water_sanitation_health/dwq/gdwq0506.pdf.

Wright, J., Gundry, S., and Conroy, R.M. (2004). Household Drinking Water in Developing Countries: A Systematic Review of Microbiological Contamination Between Source and Point-Of-Use. Tropical Medicine and International Health, 9, 106–117.

Chapter 20
Water, Wastes, and Children's Health in Low-Income Neighbourhoods of Yaoundé

Emmanuel Ngnikam, Benoît Mougoué, Roger Feumba, Isidore Noumba, Ghislain Tabue, and Jean Meli

The city of Yaoundé in Cameroon, like many cities in Africa, is experiencing rapid urban growth (5.6% per year). Quickly expanding slums and poorly serviced neighbourhoods now make up 40% of the current total land area of the city.

In 2001, a team of researchers from the Ecole Nationale Supérieure Polytechnique and the Université de Yaoundé I and II began to work in 12 underprivileged neighbourhoods of the city, located in the upstream part of the Mingoa River watershed. These were neighbourhoods with poor housing, uncertain land tenure, and a lack of urban infrastructure. For years, the population has been stable at about 21,000 inhabitants in 4,500 households – due in part to strong pressure from city government to prevent further urban growth and relocation of squatters. The vast majority of houses are rented (75%) and almost 70% of household heads work in the informal sector (e.g. temporary jobs or self-employment in restaurants and catering, shoe repairs, hair salons, and street markets).

When the project began, 90% of homes could only be accessed via poorly constructed and maintained footpaths that meandered between buildings, many of these of haphazard construction. Overall environmental conditions were very poor, and characterized by poor drainage, recurrent flooding in low-lying areas, and heavy contamination of both surface- and groundwater from traditional pit latrines and disposal of household wastewater. Domestic garbage was piled up in any open space

---

E. Ngnikam (✉) • R. Feumba • G. Tabue
Environment and Water Sciences Laboratory, Ecole Nationale Supérieure Polytechnique, Yaoundé, Cameroon
e-mail: emma_ngnikam@yahoo.fr

B. Mougoué
Department of Geography, Université de Yaoundé I, Yaoundé, Cameroon

I. Noumba
Faculté des Sciences Economiques, Université de Yaoundé II, Yaoundé, Cameroon

J. Meli
Faculty of Medicine and Biomedical Sciences, Université de Yaoundé I, Yaoundé, Cameroon

available, including vacant lots, or was thrown directly into ditches or "natural" streams formed by runoff, a black and fetid mixture of rainwater, raw sewage, and other household wastes. Inadequate access to safe drinking water forced most households to draw and store water from shallow, heavily polluted traditional wells or nearby springs. Not surprisingly, the prevalence rates of diarrhoea and intestinal parasitoses in children were very high. The majority of those involved in activities to improve neighbourhood environmental conditions were individuals driven to action by their family's need for water.

Exposing Critical Links among Water, Waste, and Children's Health

The initial impetus for the project came from concerns expressed by residents of the Melen 4 neighbourhood about the safety of children as they walked to and from school, and of women in their journey to the market. The main complaint was about the difficult and hazardous conditions people endured when moving about their crowded and polluted neighbourhood. In addition, people had often complained to researchers about diarrhoeal illnesses, but without necessarily recognizing or making a direct link with their unsanitary environment.

A multidisciplinary team with expertise from eight different academic fields (civil engineering, hydrogeology, geography, sociology, economics, paediatrics, epidemiology, and statistics) began to explore these challenges. Applying an ecohealth approach (Lebel 2003), they engaged various local representatives in project design, including neighbourhood and city authorities (Yaounde VI City council and Yaoundé City Hall (CUY)), community members (heads of neighbourhood and local development committees), and a national nongovernmental organization (NGO), ERA-Cameroon (Environment: Research and Action in Cameroon). The focus and vision of the project grew from an originally local concern in Melen 4 to a broader goal of exploring and understanding the existing relationships between the watershed's physical environment, household sanitation and hygiene, and the health of children less than 5 years old. Research and development activities evolved over two project phases implemented during a 6-year period (2003–2009).

The project's main premise was that improving environmental sanitation in these urban slums would reduce the prevalence of diarrhoea and intestinal parasitoses in children. This improvement in sanitation was not going to be easily achieved given the size of the challenge and the human, financial, and technical resources available. It would require new forms of engagement of local residents and government officials to identify and act on critical intervention points in this degraded urban setting. The project set out to understand the extent of the problem, find out how to change the situation for the better, and facilitate the needed changes. Concurrent research studies were conducted and their findings integrated to build an understanding of water, sanitation, hygiene, and health links. Children less than 5 years old were targeted because this age group is the most vulnerable to diarrhoeal diseases.

Mapping Traditional Water Sources

A geographic information system (GIS) for the project area was developed for spatial–temporal analyses of links between water and environmental sanitation. This included a first-ever survey of water sources for the whole Mingoa basin. A total of 35 traditional water sources such as wells or springs (i.e. sources other than the municipal piped distribution system) were selected and monitored for 1 year, and changes over time were tracked for water quality, water yields, and water levels. The water sources were selected based on their location (1 source per 1 ha grid area) and level of use (more than 50 users per day). Water samples were collected from all 35 sources during the peak of the dry season (December) and during the first rainy season (May).

Water-quality tests[1] followed standard methods (WHO 2008). Potential sources of pollution (e.g. latrines and garbage) around the sources were also mapped, and in five cases, tracer studies with kitchen salt were conducted to reveal contamination pathways. Direct observation studies of water-collection behaviour and number of users between 6 am and 6 pm for 7 consecutive days in each source were completed, once per source. Observations included the relative age and sex of persons fetching water (men, women, boys, or girls), length of queues, types of containers used for water collection, and sanitary conditions around water sources.

Monitoring Diarrhoea and Intestinal Parasitoses in Children

To better understand the pathways of disease transmission and assess the impact of interventions (including the level of community satisfaction, behavioural changes in selected households, and changes in child health), a first 2-year longitudinal study was carried out in 2003–2005. It followed 360 children who were less than 3 years old at the beginning and who lived in the Mingoa basin (from a population of 1970). The children were selected based on a stratified random sampling procedure with the following criteria: geographic location of family; household source of drinking water (traditional source versus connection to municipal system); availability of household sanitation; and informed consent from parents for children to participate in the study. Of the original sample, complete (or near-complete) data were obtained for 279 children. Several children were lost from the original sample when their parents moved out of the project area.

Monitoring was conducted monthly by a team of eight nurses and two medical students. Health indicators tracked included: nutritional anthropometric measurements (weight, height, head circumference, upper arm circumference,

---

[1] Tests for pH, total and dissolved solids, dissolved nitrate – nitrogen, phosphate, ammonia, sulphate and iron levels, faecal coliforms and streptococci.

and chest circumference); diarrhoeal episodes (in the 2 weeks preceding the visit); and prevalence of intestinal helminthiases (count of children with symptomatic intestinal parasitoses at the time of the visit, or in the 2 weeks preceding the visit). During these household visits, parasite infections were assessed symptomatically, and suspected infections were confirmed by tests at the local hospital and treated.

A follow-up 2-year study (July 2007 to June 2009) continued health data collection, using the 2007 census. A similar sample size of 360 children was selected through stratified random sampling from a total population of 1,497 children less than 3 years old in the project area. The average number of children monitored throughout the 2 years was 277, with the total number of children fluctuating considerably because of families moving (permanent or temporary relocations) and deaths (7 children). From July 2007 to October 2008, children leaving the cohort were replaced in the longitudinal study using the same selection criteria. No new children were recruited during the last 8 months of this second longitudinal study, with 133 children from the original sample being followed for the 2-year period.

Throughout the study, nurses conducted direct observations, collected water samples from sources and homes, and recorded diarrhoeal episodes and symptoms of intestinal parasitoses. Selection of the children was again based on their age (less than 3 years old), parent's informed consent, the geographic location of the household, type of household water access, and tenure of the dwelling (i.e. owned or rented). Stool samples from children were taken and analysed once in 2008 and again in 2009. Suspected cases were referred to hospitals during the first nurse visit. Sixty-one children were referred to the hospital for stool tests in this first visit, of which 79% were confirmed with a parasitic infection and treated.

Hygiene-promotion campaigns in schools and community meetings on water protection, sanitation, and hygiene were organized every 6 months. This also included focus group discussions on these same topics with parents of children who were participating in the longitudinal study. Community workshops in 2005 (at the end of the first phase) presented research results and engaged local actors in discussions on the safety of different community water sources and possible actions to improve sanitary conditions in the watershed. These successful workshops were continued annually for the rest of the project.

Water-Handling Practices and Household Hygiene in High-Risk Homes

The rapport developed with the families of the children allowed the team to conduct other in-depth investigations. The team explored the level of drinking water contamination in households that lacked direct connection to the municipal water system, but that still collected water from this public system (e.g. from neighbours, water kiosks, or nearby businesses) and stored the water in containers at home. Information collected included water-collection practices, types of containers used, place and modes of water storage in the home, and water-quality parameters (ammonia content, faecal coliforms, and streptococci) of the stored water. The team also explored the hygiene practices in households in which children had experienced more than three bouts of diarrhoea between January and June 2004. A total of 50 households were followed for a period of 6 months through two household surveys conducted before and after the team's hygiene-promotion campaigns.

The project explored the social dimensions of the sanitation problem by studying the attitudes, behaviours, and practices of neighbourhood residents and government workers and officials, both in terms of their contribution to existing sanitation problems and their solution. This involved identifying stakeholders in sanitation and health, their intervention logic, and the existing synergy among their actions; establishing, through group discussions and neighbourhood assemblies, how a participatory approach could contribute to local sanitation and health interventions; exploring, from direct observations and targeted surveys, the role of mothers and older children in the transmission of diarrhoeal diseases and helminth infections in the younger children; and suggesting ways to reduce the identified negative impacts.

Relationships established with different community groups, civil society organizations, and government institutions were instrumental in developing and carrying out a number of interventions to improve sanitation and reducing the risk of childhood diarrhoea. The project team observed how inequalities affected different groups in the community, particularly women. Gender and social equity were integrated into all activities, including the selection and siting of all infrastructure provided by the project. For example, the participation of the team of women nurses and their monthly visits to homes created a closer relationship with women than might have been achieved by male researchers interacting with male heads of households. This contact led to a better understanding of the health status of children, the mothers' concerns, and other health and hygiene problems in the households.

As the project grew in scope and reach, so did the size of the team. By 2006, the core transdisciplinary group comprised 32 members (10 academics, 8 community representatives from different neighbourhoods, 8 nurses, 1 secretary, and 5 members of civil society organizations). Two NGOs – ERA Cameroon and Engineers Without Borders, Catalonia (EWB) – helped strengthen local organization and community actions. The project helped establish local development committees in each neighbourhood. These committees engaged local residents in defining sanitation intervention priorities and their contributions in cash or in-kind materials and labour. Two local youth organizations, Tam Tam Mobile and GIC Le Vert, provided solid-waste collection and disposal services. Government representatives joined multi-stakeholder committees held every 3 months to provide strategic advice on interventions, and facilitated coordination with other development projects in the city.

What We Learned

The project documented the widespread use and impact of traditional deep-pit latrines on water quality and child health (LESEAU 2005). Tracer studies with table salt confirmed direct contamination of groundwater by the latrines. Less than 40% of the population had access to the municipal water system and relied on other, highly polluted sources. Water-quality monitoring of the 35 selected traditional water sources indicated gross levels of contamination, with an average faecal (thermo-tolerant) coliforms above 28,000 CFU/100 mL. The average coliform count for the dry season for all sources was 14,500 CFU/100 mL, whereas the average for all sources during the rainy season increased to 120,000 CFU/100 mL. These findings confirmed the high vulnerability of these water sources to faecal contamination and their unsuitability for human consumption. Current WHO guidelines stipulate that no faecal coliforms be detectable in any 100 mL sample of drinking water (WHO 2008).

In terms of health indicators, important relationships were established between diarrhoea in children and access to water and sanitation (LESEAU 2005). For example, a significant correlation was found between the use of water not obtained from the public system and a higher rate of diarrhoea in children (unpublished data, $n=279$, $p=0.002$). Sanitation method (type of latrines, septic tanks, and sumps) and the state of completion of the latrine also had a significant impact on the frequency of diarrhoea (unpublished data, $n=192$), with higher rates of diarrhoea occurring in households with poor sanitation infrastructure. Spatial and temporal variations were also observed. Households in neighbourhoods located on higher ground tended to report fewer bouts of diarrhoea than those in low areas. Neighbourhoods with more connections to the public water system also reported fewer bouts of diarrhoea in children. During the dry months (December to February) the incidence of diarrhoea increased in all neighbourhoods, whereas in March and April (beginning of the rainy season) it decreased. One factor that may contribute to these trends is the more frequent interruption in the public water supply during the dry months.

In addition, stool tests identified three types of parasitoses present in children in the project area: yeast (33–40% of stool samples analysed), amoebiases (5–8%), and other parasitoses (4–5%). The latter were primarily roundworm.

Project activities helped improve child health to the extent that a steady decline in bouts of diarrhoea was observed (Fig. 20.1). The prevalence of diarrhoea decreased from 44% at the beginning of the second phase in 2007 to 12% in May–June 2008, and 6% in June 2009. A similar trend was observed for symptoms of intestinal parasitoses in children (ranging between 45% in 2008 and 50% in 2009, down from 79% in 2007) (LESEAU 2009).

Due to the lack of a control group, it is not possible to attribute observed trends solely to improved hygiene awareness (health education and advice given to households during home visits by nurses). These trends remain indicative of positive contributions to better health. The infrastructure provided by the project (e.g. safer

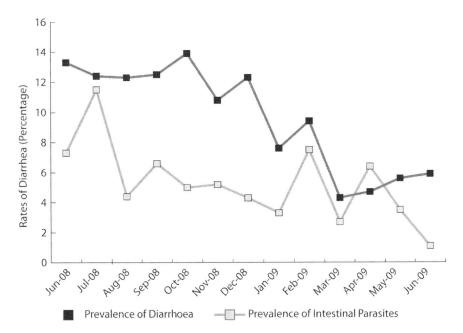

Fig. 20.1 Decrease in the rates of diarrhoea in a cohort of children under 5 years old (LESEAU 2009)

water sources, improved latrines, drains, and walkways) brought well-being to the people, reduced potential disease transmission paths, and increased local economic activities.

Mapping existing water-handling practices in high-risk households showed potential points of contamination throughout the cycle of drinking water use. In the initial survey (2002), only a small number of households were connected to the public water system in the Mingoa watershed (25%). Households with no connection to the public system obtained drinking water from their neighbours (59%), from nearby military barracks (5%), or from car wash sites (3%), whereas the rest did not indicate how they got their drinking water. Many, especially during times of water shortages in the public system, resorted to using water from contaminated wells and natural springs (LESEAU 2005).

Water samples collected from the public system, but at points away from the home (e.g. tap in yard, public tap, or from neighbours), were also tested to better understand the deterioration of water quality after collection from a safe water source. Although the sample size (78 water samples from 26 households with triplicate water tests per sample, for a total of 234 water tests) was insufficient to establish levels of significance in comparing collection and storage handling practices in different homes, important insights were obtained. For example, containers with no lid were found to be more contaminated with faecal coliforms (60% of 234 water tests done) than containers with lids (50%); more containers stored on the ground were contaminated (60%), compared with those kept above ground in shelves or on tables (50%); most containers stored in a refrigerator had water of good quality (85%); and the longer the storage period in the home, the lower the water quality (e.g. 68% of the containers used to store water for more than 3 days were found to be contaminated). Distance travelled from point of collection was also important. Samples of water collected in the user's own yard were more likely to be safe when compared with samples collected beyond a distance of 30 m from the home (LESEAU 2005). These results informed the project's hygiene promotion activities and the development of alternative water-storage containers.

Local Actions for Cleaner and Safer Communities

Gathering reliable information and developing meaningful engagement from a range of stakeholders led the project to design and implement activities that reduced environmental contamination and transmission of diarrhoeal diseases. Improvements in neighbourhood infrastructure (e.g. pedestrian walkways, gutters, improved latrines, and extension of the drinking water system) were targeted in five neighbourhoods based on need and potential for collaboration with other partners (e.g. possibility of complementing local and externally funded works by ERA-Cameroon (2002) and CUY). Health and hygiene promotion activities were carried out in all 12 neighbourhoods of the project.

Sanitation Interventions

Traditional pit latrines surrounding the most-used traditional water sources were replaced with improved latrine designs (double-vault urine-diverting composting latrines with removable latrine slabs) to help reduce contamination of groundwater. A total of 41 new latrines were built. The pits are interchangeable, which allows for easy emptying when they are full. They are also well ventilated and minimize odour and flies, making them attractive to users. The latrines cost two to three times more than the traditional (polluting) pit latrines. In spite of this, five households built the new type at their own cost, indicating a more than casual interest in this new technology. To assess the impact of the new latrine design on groundwater quality, three different traditional water sources were monitored near where new latrines had been installed by the project. In all three cases, water quality improved dramatically over 4 years. Faecal coliforms and faecal streptococci were reduced by two to three orders of magnitude, and ammonium content in the water decreased 90% from initial levels. This confirmed the old pit latrines to be the main source of groundwater contamination, and the effectiveness of these new latrines in reducing pollution.

Improving the Collection and Disposal of Community Wastes

Historically, poor hygiene practices and bad access roads and pathways led to widespread dumping of household sewage as well as liquid and solid wastes in the surroundings of homes. This accumulation of wastes contributed to the deterioration of the urban ecosystem and residents' health. To reduce the stagnation of wastewater in the neighbourhoods and minimize their contact with people, 4.5 km of paved walkways, 400 m of paved access roads, 800 m of gutters, and five bridges were constructed by ERA-Cameroon with financial support from the project and the Yaoundé City Council (CUY), which assumed more than two-thirds of the costs.

Youth associations led cleanup operations and regular garbage collection. This effort began in 2002, with the local youth organization Tam Tam Mobile, which received technical and financial support from the project to provide solid-waste collection in Melen 3 and 4 neighbourhoods. Household garbage was collected and transported for a fee by wheelbarrow to specified temporary disposal points chosen in agreement with the CUY and the contractor responsible for transporting it from the collection point to the municipal landfill (Ngnikam and Tanawa 2006).

This small business activity enabled the delivery of an essential service – so much so that it has continued for over 7 years without any external support and despite little involvement or coordination from municipal authorities. Households using this service also reported a decline in the number of cockroaches and mice in house surroundings and an overall improvement of the neighbourhood environment (Ngnikam et al. 2009). Drainage was also improved with the reduction of garbage in pathways and drains, which reduced the number and size of fetid pools of stagnant water and waste mixtures. This waste-collection service was scaled up in 2007 to the entire watershed, with another youth organization GIC Le Vert joining the effort, in response to a request for proposals co-funded by the project and UNDP. Improved garbage collection also reduced the risk of flooding during the rainy season in the lowlands by preventing damming and blockage of runoff streams and drainage canals caused by the accumulation of garbage.

Promotion of Health and Environmental Hygiene

Monthly educational sessions were conducted by project staff in households that participated in the children's cohort longitudinal study. Information campaigns were also organized in neighbourhoods and schools to raise awareness among households, students, and teachers about pathways of water contamination, the importance of sanitation to health, and hygiene steps to reduce transmission of diarrhoeal and intestinal parasitic infections. More than 2,000 community members and 1,800 students were reached with posters, role playing, and theatre plays that used the results of water tests conducted in households, wells, and springs in the project area as illustrative examples. Educational workshops were organized for women in each neighbourhood to communicate the potential hazards and recommended uses of the various water sources available in the community.

Interventions on Drinking Water

Given the high contamination levels found in traditional water sources used for drinking, both municipal and private water authorities were lobbied to increase the piped water supply. As a result, the number of households in the Mingoa watershed connected to the public water system by the national water authority (SNEC) went from 25% in 2002 to 32% in 2007. Connection rates varied: better-off neighbourhoods, like Elig Effa 2 and Melen 1, reached connection rates of 50 and 42%, respectively. Three other neighbourhoods benefited less: Elig Effa 5 (8% connected), Elig Effa 7 (11%), and Elig Effa 6 (18%).

To increase connection rates, new public water-distribution pipes were laid by EWB (860 m) and CUY (900 m). The residents then paid for each individual household connection (almost 200 new household connections in the 5-year period). In addition, 15 drinking-water kiosks were constructed with project funding, and they operate today in a self-sustaining fashion, supplying 7,500 people. These kiosks are small businesses that first obtain a connection to the water system and then install a set of faucets and staff the kiosk with a caretaker who sells the water. Part of the funds is used in maintaining the kiosk. Being a city connection, the water tends to be potable as long as the municipal treatment system is functioning properly. All together, these interventions contributed to the supply of safe drinking water to nearly 9,500 people, or 45% of the area's total population.

Still, nearly 60% of the households remained without connection to the public system. Given earlier results on the risk of contamination of water during collection and home storage, suitable water containers were designed and tested to meet the needs of households without connection.

Prototype water-storage containers were tested in 80 households selected by a stratified random sample, by neighbourhood. This pilot group received the containers free of charge, but the containers were also made available for affordable purchase from local manufacturers and retailers. The pilot study sought to assess household satisfaction with the new containers, to improve the functionality of the prototypes, to assess ease of use and maintenance, and to identify possible forms of misuse that could lead to the contamination of the stored water.

Of the 80 households selected, 55 consented to the monitoring of water quality in the water-storage containers. Physical, chemical, and bacteriological tests were conducted. At least one water sample per household in October 2008 and in June 2009 was collected for 49 households, whereas in six households one sample per day for 5 consecutive days was taken in each of the two testing months. Bacteriological tests revealed that in about a quarter of the new containers sampled, faecal coliforms were present exceeding WHO guidelines. Although most households complied with the usage guidelines, observations and interviews with household members revealed various potential routes of contamination. These informed the formulation of health-promoting messages for good water-use practices in the home, including avoiding poor-quality water to wash the containers; not using containers to draw water from contaminated sources; keeping containers out of the reach of children; using the spout instead of immersing cups or glasses to draw water; encouraging regular cleaning of containers; and ensuring that water is not stored beyond 3 days.

The results were communicated to the community and improved water storage containers were distributed to the remaining families participating in the children's study.

Conclusion: Tracking Outcomes to Energize and Guide Action

The project sought to link drinking water protection and environmental sanitation with community organization for better health and community development – a difficult challenge in a setting where poverty and absence of urban planning led to high unmet demands for basic services. To build an ecohealth perspective, the project moved beyond single-sector approaches and combined conventional water and sanitation interventions with health and environment research. This required flexibility from all involved but bore fruit. Multifactor and multisector systemic situation analysis showed how people's concerns, interests, and visions for better and healthier communities can lead to a new plan of action. All participants gained a better understanding of how their urban environment was being degraded and what needed to change to build a healthier community.

The project addressed a wide range of environmental health issues, including solid-waste collection, drainage of stagnant waters, improved pedestrian walkways, reduced groundwater contamination through safer disposal of human waste, improved household hygiene, and safe drinking water, all of which led to a reduction in disease transmission. Lower rates of infection in children, improved hygiene, cleaner environments, and increased representation and organization of residents contributed to improving the quality of life. Seemingly intractable problems were eventually addressed by the community with an energy and determination that were lacking when the project began. The project helped galvanize the communities and obtained unprecedented levels of engagement from city and government authorities responsible for providing basic services.

The ecosystem approach to health used by the project helped overcome the inertia of unplanned growth where everyone fends for themselves in an ever-deteriorating environment. It produced a genuine change in people's attitudes and hygiene and sanitation practices. Moreover, project partners bought into the idea of collaborating to improve the health and environmental conditions of the neighbourhoods and watershed. Community leaders, community members, and government officials each took a share of the responsibility for making and maintaining the improvements. The project illustrates that despite the diverse social strata in these neighbourhoods, residents and local authorities contributed to local development projects. But this required the strengthening of local organizations (e.g. neighbourhood development committees and youth organizations), the building of trust (e.g. through community assemblies and home visits by nurses), and coordination.

In the end, financial and in-kind contributions for community projects were made possible largely by the support provided by social organizations. Tracking the realtime impact of community investments in health and environment helped ensure their continued involvement. It also helped attract the support of other institutions, and allowed the project to expand its work. This was the case for the CUY, and also for Engineers Without Borders, Catalonia, which obtained funding from the European Union to support sanitation and infrastructure development in all 12 neighbourhoods and to replicate the project's work in other parts of the city. The project also received the 2010 Grand Prize Award from the Suez Environment's Water for All Foundation.[2] It is a tribute to the spirit of collaboration and dedication of a large group of people who joined in this work, including communities, municipal and city authorities, the university, and civil society organizations.

Acknowledgments We thank community members who participated in the research and contributed to the results presented in this paper. In particular, the authors would like to thank their late colleagues, Professor Henri Bosko Djeuda Tchapnga (Project Leader until January 2007), Professor Félix Tiecthe, Director of the "Mere et Enfants Chantal Biya" Foundation, and Mrs. Pemboura Djiasse, state-certified nurse. This paper is dedicated to the parents of the seven children who died during the project's second phase (2006–2009). IDRC support was provided through projects 100772–06 and 103605.

References

ERA-Cameroon (Environment: Research and Action in Cameroon). (2002). Mise en place de structure de précollecte et de traitement des déchets solides urbains dans une capitale tropicale: cas de Yaoundé au Cameroun. Pilot Action Report, ERA-Cameroon, Yaoundé, Cameroon.
Ngnikam, E., and Tanawa, E. (2006). Les villes d'Afrique face à leur déchets [How Cities Address the Problem of Solid Waste]. Edition de l'UTBM (Université de Technologie de Belfort-Montbéliard), Belfort Cedex, France.
Ngnikam, E., Tanawa, E., Mogoue, B., and Etoga, S.M. (2009). Pre-Collection of Domestic Waste in Slum Districts of Yaoundé. Communication, 8th International Conference on Urban Health, 19–23 October, Nairobi, Kenya.
Lebel, J. (2003). Health: An Ecosystem Approach. In Focus Series. International Development Research Centre (IDRC), Ottawa, Canada. Available at: http://www.idrc.ca/in_focus_health/.

---

[2] For more see http://www.concourseaupourtous.fr/en/les-laureats-2010/grand-prize-2010/grand-prize-2010/.

LESEAU (Laboratoire Environnement et Sciences de l'Eau, Ecole Nationale Supérieure Polytechnique de Yaoundé). (2005). Maîtrise de L'assainissement dans un Écosystème Urbain à Yaoundé et Impact sur la Santé des Enfants Âgés de Moins de Cinq Ans, Phase I. (Project 100772–06). Final Technical Report to IDRC. Available at: http://idl-bnc.idrc.ca/dspace/handle/10625/45404.

LESEAU (Laboratoire Environnement et Sciences de l'Eau, Ecole Nationale Supérieure Polytechnique de Yaoundé). (2009). Maîtrise de L'assainissement dans un Écosystème Urbain à Yaoundé et Impact sur la Santé des Enfants Âgés de Moins de Cinq Ans, Phase II. (Project 103605). Final Technical Report to IDRC. Available at: http://idl-bnc.idrc.ca/dspace/handle/10625/42032.

WHO (World Health Organization). (2008). Guidelines for Drinking Water Quality, Third Edition. World Health Organization. Geneva, Switzerland.

Part V
Building a New Field

Chapter 21
Better Together: Field-Building Networks at the Frontiers of Ecohealth Research

Margot W. Parkes, Dominique F. Charron, and Andrés Sánchez

Ecosystem approaches to health frame systemic relationships across scales and accentuate a range of interactions between various actors, from the local to the global. In their transdisciplinary and participatory implementation, ecosystem approaches to health reveal both the richness of bringing together multiple perspectives and the power of collaboration and partnership. It is not surprising that many ecohealth research practitioners place considerable value on networking as a means of expanding knowledge and enhancing capability to bring about change.

Taken together, the experiences illustrated by the case studies exemplify lessons learned from the application of the six principles outlined in Chap. 1. These principles inform the practice (or doing) of ecohealth research (transdisciplinarity, systems thinking, and multistakeholder participation) as well as inform ecohealth research goals (sustainability, equity, and the application of scientific knowledge to guide change). The case studies focus on what happened, and illustrate how these principles were applied and manifested by different research teams in a range of contexts. They neither dwell on the principles nor on how they were relevant or built on throughout the research process. This chapter explores how the six principles are closely linked with processes of working and learning together.

In addition to applying principles of ecosystem approaches to health in individual projects, many of the case studies presented in this book share other relationships. The importance of transdisciplinarity and participation is evident in collaborations between projects and researchers in networks and communities of

M.W. Parkes (✉)
Ecosystems and Society, Health Sciences Programs, University of Northern British Columbia, Prince George, BC, Canada
e-mail: parkesm@unbc.ca

D.F. Charron • A. Sánchez
International Development Research Centre, Ottawa, ON, Canada

practice (CoP). This chapter explores how networks are also a manifestation of ecohealth principles, and how they have contributed to the evolution of ecohealth as a field of endeavor newly recognized in science. The phenomena of collaborative association and partnership that are typical of ecohealth research practice are also situated in broader scientific trends and literature ranging from the theoretical basis of fields in sociology (Fligstein 2001) to field-building experiences in public health (Ottoson et al. 2009) and evaluation (King 2010).

Field-building is an intuitive process that seems very much informed by hindsight. There is no clear moment when a sub-speciality becomes a field in its own right. Rather, a new field or domain appears to achieve recognition when sufficient numbers of experts engaging in related activities affiliate themselves with this domain, and produce sufficient collective high-quality evidence of their particular contribution to be recognized as distinctly valuable by their peers. There is no recipe to build a field, and the idea that there may be criteria and other elements required to build a new field is relatively new. Because fields tend to evolve into being, there are very few examples of new fields having been deliberately built from the outset (although ecohealth may be one such example). In any case, having a group of peers engaged in promoting excellence and in advancing a field with a unique contribution appear to be minimum requirements. The establishment of a journal or society is indicative of a peer group (e.g., Green 2009; McBride et al. 2004). Other characteristics of a field include the use of common competencies and standards of practice (King 2010). The establishment of the International Association for Ecology & Health and the journal *EcoHealth* is among other favorable indications that ecohealth is emerging as a field – the result of converging lineages of scholarship and practice that include work that is the focus of this book as well as others' (Aguirre et al. 2002; Waltner-Toews 2004; Webb et al. 2010; Wilcox et al. 2004).

This chapter examines the process of field-building in ecohealth by drawing on lessons from individual case studies, and linking these with insights gained when experience is shared through networks and CoP in ecosystem approaches to health. The field-building contribution of the case studies and networks is their relevance to complementary developments in literature and scholarship. A matrix of key concepts is presented and discussed, drawing attention to the broader implications of ecohealth approaches in relation to knowledge integration, different facets of participatory processes, and the role of research as part of larger processes of collaborative learning and action.

The chapter presents some of the challenges and opportunities that arise in a transition from individual studies to CoP, action, and scholarly impact. These field-building lessons from across the spectrum of ecosystem approaches to health are seen as contributions to addressing the twenty-first century challenge that: *the escalating complexity of science and engineering is moving research toward a collaborative mode, with greater focus on intellectual integration* (US National Science Foundation 2001).

Networks and Communities of Practice in Ecohealth

The evolution of ecohealth as a field has been marked by collaboration and collective endeavor. Bringing different people and their contributions together in pursuit of a shared goal is a widely recognized strategy to harness capacity to address complex societal challenges (Brown 2007; McKnight and Kretzmann 1996; Pohl 2008). The case studies in this book illustrate examples of research and impact being enhanced by partnerships among researchers, community members, and other stakeholders.

The examples from Kathmandu, Nepal, and Ekwendeni, Malawi, demonstrate that strong and lasting partnerships can emerge from research-oriented collaborations and lead to substantially improved health and well-being for the communities under study. These partnerships arise from local, and sometimes informal, relationships between a research team (at least initially) and civil society groups, government, and other organizations with a stake in the issue. In other words, the processes of association and joint work are key characteristics of transdisciplinary, participatory multistakeholder research that characterize ecohealth research and enhance its outcomes (Mertens et al. 2005).

IDRC's nearly 10 years of support to network initiatives in ecohealth informs both the exploration of their contributions to field-building and reflection on the trade-offs that these expanded forms of knowledge production and use entail. Before the 2003 International Forum on Ecosystem Approaches to Human Health in Montreal, IDRC's Ecohealth program led an electronic consultation with more than 60 research and donor organizations around the world to address needs and expectations in furthering ecohealth research. The consultation was prompted by feelings of intellectual isolation apparently widely shared among researchers, and a need for joint learning and cooperation to overcome the many challenges of ecohealth research (De Plaen and Kilelu 2004). IDRC responded with targeted support to networking and capacity-building initiatives (Table 21.1). These investments were intended to foster the development of an ecohealth peer group including North–South and South–South knowledge exchange and partnerships. The consultation identified three core functions of any ecohealth network or community of practice in ecohealth (CoPEH) (Flynn-Dapaah 2003).

- *Create an ecohealth peer community.* Provide opportunities for researchers to learn and exchange ideas in ecohealth. Capture and share existing tacit knowledge, improve scientific rigor and relevance of research by the formation of peer groups, share experiences using ecohealth and like-minded approaches, and foster dialogues among research, policy, and practice.
- *Develop research capacities in ecohealth.* Further develop skills in ecohealth research and transdisciplinary methods and techniques, and help young researchers, project teams, and policymakers understand and use ecohealth approaches to achieve intended outcomes in their projects and programs.

Table 21.1 Ecohealth networking initiatives co-funded by IDRC (2004–2010)

Network name	Initiation date	Case study in this book	Network web page (active)
CoPEH-LAC (Community of Practice in Ecosystem Approaches to Human Health in Latin America and the Caribbean)	2004	Mercury in the Amazon; Manganese in Mexico; Gold mining Ecuador	http://www.una.ac.cr/copehlac/
SIMA (Systemwide Initiative on Malaria & Agriculture)	2004	Malaria in Tanzania and Uganda	
RENEWAL (Regional Network on HIV/AIDS, Rural Livelihoods and Food Security)	2004	N/A	
CoPEH-MENA (Community of Practice in Ecosystem Approaches to Human Health Middle-East and North Africa)	2005	Water quality Lebanon	
CoPES-AOC Africa (Communauté de practique écosanté en Afrique de l'Ouest et du Centre)	2006	Sanitation in Cameroon	http://www.copes-aoc.org/
IDRC-TDR (WHO Special Programme for Research and Training in Tropical Diseases) – *Eco-Bio-Social Research on Dengue Fever in Asia*	2006	Dengue in Asia	
IAEH (International Association for Ecology and Health)	2006	Example in this chapter	http://www.ecohealth.net/
APEIR (Asia Partnership on Emerging Infectious Diseases Research)	2007	Example in this chapter	http://www.apeiresearch.net/main.php
CoPEH-Canada (Community of Practice in Ecosystem Approaches to Human Health – Canada)	2007	Example in this chapter	http://www.copeh-canada.org/index_en.php
CDLAC (Communicable Diseases in Latin America and Caribbean)	2007	Chagas in Guatemala	
IDRC-TDR (WHO Special Programme for Research and Training in Tropical Diseases) – *Eco-Bio-Social Research on Dengue Fever and Chagas Disease in Latin America and the Caribbean*	2009	N/A	

- *Enhance the uptake of ecohealth research and its influence on policy and practice.* Foster opportunities for dialogue, dissemination of tools, and development of capabilities of both researchers and research end-users to enhance the uptake of findings by policy, community, and relevant professional practices (e.g., in public health and environmental management).

With these aims in mind, IDRC eventually supported four Communities of Practice in Ecosystems Approaches to Health (CoPEHs) in Canada, Latin America and the Caribbean, the Middle East and North Africa, and West and Central Africa, and a number of additional networking activities (for example, the Asian Partnership on Emerging Infectious Diseases Research [APEIR] and the International Association for Ecology & Health). In parallel, other networks developed independently from IDRC, but with a focus on ecosystem approaches to health, generating cross-fertilization and scholarly debate among different networks. Examples include work initiated by the Network for Ecosystem Sustainability and Health (Waltner-Toews and Kay 2005), and the development of the project Sustainably Managing Environmental Health Risks in Ecuador (Parkes et al. 2009).

A community of practice is an alternative model to traditional academic networking (De Plaen and Kilelu 2004). As initially described by Lave and Wenger (1991), a community of practice promotes a shared *domain* or common interest (in this case, ecosystem approaches to health), a sense of *community* among participants or members, and a shared purpose of building a *practice*. Practice, in this sense, refers to the agreed-upon ways of formalizing and implementing collectively developed knowledge and solutions that further the community's mission (Wenger et al. 2002). CoP generally involve: *groups of people who share a concern or a passion for something they do and learn how to do it better as they interact regularly* (Wenger et al. 2002, p. 4).

As in previous chapters, the diversity and unique experiences of networks and CoPEHs are illustrated with different examples around the world. They share what Bunch et al. (2008) describe as a "family of origin" in ecosystem approaches to health and illustrate the sense of momentum and worth created by researchers coming together to learn in ways that they had not previously attempted or imagined.

Most networks or CoPEHs developed around two main purposes: doing better research on a specific theme linking health and environment, and teaching others how to do ecohealth research. The CoPEH in Latin America and Caribbean (CoPEH-LAC, Example 21.1) initially became organized around both research on environmental toxics and ecohealth (as well as relevant specific disciplinary) training. CoPEH-LAC introduced and adapted ecosystem approaches into existing research projects; provided numerous short courses (e.g., on an ecosystem approach to health, and on methods for detecting sub-clinical neuro-behavioral effects from chronic exposures to heavy metals); and is now developing transdisciplinary graduate and undergraduate curricula throughout the LAC region.

As a long-standing network, CoPEH-LAC represents some of the key benefits of this model of network, notably its capacity to evolve strategically while strengthening interactions within the network. CoPEH-LAC formally examined changes over time among collaborative relationships between members (e.g., co-publishing; co-organizing a conference, a course, or similar events; or working together on a project). By year 2 of the CoPEH-LAC project, there was a remarkable increase in the size of the membership and a shift in the pattern of collaborative relationships, which moved from being centered on the Canadian node to a more "horizontal" pattern of collaboration between all regions of Latin America (Fig. 21.1). Research on the development and progression of the types of relationships between members

> **Example 21.1** Communities of Practice in Ecosystem Approaches to Health in Latin America and the Caribbean (CoPEH-LAC)
>
> In August 2004 in Santiago, Chile, 13 people debated how to respond to a Canadian call for proposals to establish a Canada–Latin America collaboration for a CoPEH-LAC. Five years later, CoPEH-LAC has centers (nodes) in Mexico, Central America and the Caribbean, the Andean region, the Southern Cone, Brazil, and Canada. This cohesive community of 150 people from academia, NGO, and governmental organizations share the goals of incorporating ecohealth concepts of interdisciplinary methodology, gender and social equity, and community participation into research and public policies. Members are involved in regional and inter-regional workshops and collaborations on research and training, curriculum development, participation in government-organized research or programs, public health debates, outreach to community groups, and active participation in regional, national, and international events. Success has depended on CoPEH-LAC's decentralized structure, which provides for autonomous nodal planning and implementation that is grounded in the social, political, and environmental realities of each region. Each node has evolved to build on regional strengths and capacities. The network has grown in size and complexity, with horizontal communication growing between nodes, creating a resource network for ecohealth research in the region.

(e.g., from information exchange to a diversity of collaborative endeavors) was an explicit CoPEH-LAC objective, and these research findings have informed the evolution and work of the CoP throughout. Inspired by CoPEH-LAC, CoPEH-Canada is engaging in a similar self-evaluative research.

External catalysts often provide the motivation and means for network formation. Sometimes they are convened to address a policy issue (international research on avian influenza for APEIR, for example) or a particular opportunity – CoPEH-LAC was born out of a joint call for proposals by IDRC and the Canadian Institutes of Health Research (CIHR), prompted by IDRC's global consultation described earlier. The strength of participating research teams and the leverage of their resources were also important in the success of this CoPEH. Several case studies in this book informed the development of, benefited from, and were influenced by CoPEH LAC. Three case studies in the environmental pollution section were led by leaders of this community of practice: gold mining in Ecuador; manganese mining pollution in Mexico; and mercury in the Amazon. Indeed, the neuro-behavioral assessment tools first used to diagnose subtle impairments among fishing communities of the Amazon, and the methods used to link these to mercury exposure, were also applied in Mexico

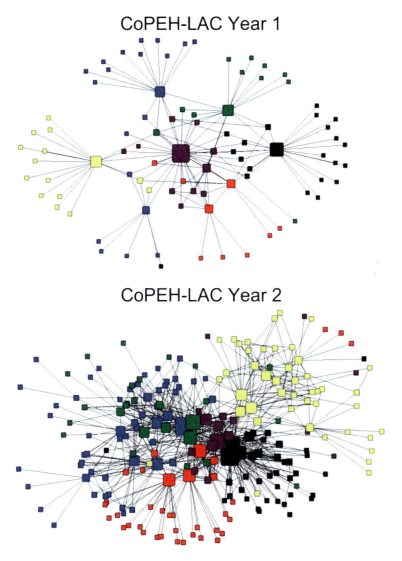

Fig. 21.1 Changes over time in collaborative relationships between members of CoPEH-LAC. The larger the symbol, the greater the number of collaborative relationships. The different shades of the *symbols* represent different sub-regions in the network

and in Ecuador, among other places. Supported by CoPEH-LAC, researchers in one country traveled to train researchers in other countries. At the same time, further refinements were made to guiding concepts and the application of ecosystem approaches to health. This learning was then integrated and applied in other research projects and also in direct collaboration with policy partners who were active participants across all regions of CoPEH-LAC.

> **Example 21.2** Canadian Community of Practice in Ecosystem Approaches to Health (CoPEH-Canada)
>
> CoPEH-Canada began in 2007 with ten investigators in the Universities of Guelph, Quebec in Montreal (UQAM), and British Columbia. In the first 3 years, the three "founding" universities expanded to include the University of Moncton and University of North British Columbia. Previously, these academics constituted a loose, unevenly distributed and poorly linked network of people working on ecohealth research. Many had never met, but were linked through common interests and experiences in ecosystem approaches to health. Some were involved with international ecohealth networks and CoP. CoPEH-Canada's initial raison d'être and focus was the collective design and delivery of an intensive course on ecosystems approaches to health. In the initial phase of CoPEH-Canada (2008–2010), the design and delivery of the graduate-level course was hosted by one of the three founding universities each summer, and provided key opportunities to foster links at the local level. This was achieved through the inclusion of instructors and experts from regional, academic, government, private, and community organizations, and helped to attract outstanding students and working professionals from across the country. The skills and experiences of the members were therefore complemented by locally engaged guests who shaped each iteration of the course and have also become part of the emerging community. Although the short course is the most obvious "product" of CoPEH-Canada, it is far more than a project output to this community. Rather, the annual course has become a feeding ground for creating a community of practice where practitioners, policymakers, and scholars (both students and faculty) share practices and ideas and confront common challenges. The short courses provide a rare opportunity for all participants (teaching team and students) to experience transdisciplinary, systems-oriented and participatory approaches to teaching and learning, with a view to sustainable and equitable solutions to local and global ecohealth challenges in Canada and beyond. CoPEH-Canada embraces its diverse cultural legacy through a commitment to bilingualism and emphasis on aboriginal perspectives.

Relationships and forms of collaboration continue to expand and increase in scale, progressing to increased interactions between ecohealth teams, and more recently between networks. Cases in point are the relationships between the LAC and Canadian CoPEH (Example 21.2) that share key members and CoPEH-LAC's participation in a nascent consortium of regional organizations aiming to strengthen scientific ecohealth leadership in Latin America and the Caribbean (LAC) on prevention and control of vector-borne diseases.

CoPEH-LAC and CoPEH-Canada both have a strong identity and a mission promoted by a core cadre of researchers. They are typical CoPEHs in this regard. They have adopted similar "nodal" structures that favor horizontal interactions between community members within regional nodes, while being committed to creating opportunities for interactions between these regions.

Nodal structures have proved useful to enabling progressive development of CoPEH's but are also seen in other kind of networks, such as the three nodes of activity focused around the provincial universities involved in the project in Ecuador on Sustainably Managing Environmental Health Risks (Example 21.3). APEIR was built around several related multi-country research teams (Example 21.4), and also developed a national nodal structure in most of the member countries to enhance national coordination and exchange of ideas between teams. In each of these examples, what began as nodal interactions among researchers involved in distinct research projects grew and expanded, to create increased awareness of innovations, facilitate new and expanded research collaboration, and strengthen ecohealth research.

In networks focused on training, such as CoPEH-Canada and the Sustainably Managing Environmental Health Risks initiative in Ecuador (Examples 21.2 and 21.3), collaborative relationships were oriented toward the collective design and delivery of ecohealth training programs. These led to new research collaborations. Cross-fertilization has also begun to occur between these initiatives. A workshop for the launch of a new phase of a master's program at the Universidad de Cuenca in Ecuador brought together program alumni and members of CoPEH-LAC and CoPEH-Canada with a common agenda to build the field of ecohealth at the regional, national, and international levels. This event also included raising awareness of related opportunities for networking, including growing opportunities for involvement with the International Association for Ecology and Health (IAEH).

The ability to tackle challenging and complex environmental health problems from an integrated and transdisciplinary perspective is neither intuitive nor easy – hence the importance of education in building the field of ecohealth. Students represent the future of any field in science. Their engagement with concepts, challenging questions, and eventual leadership in applying and further developing the field are all essential to its continued relevance and effectiveness. The Student Section of the IAEH has provided a hub for cross-fertilization among different networks and CoPEHs, and a common emphasis on the value of capacity building, mentoring, and exchange among different "generations" of ecohealth researchers and practitioners.

The 2008 International EcoHealth Forum in Mérida, Mexico, co-convened by IDRC and IAEH, brought students from around the world to mix with the 600 researchers participating in the Forum. This event provided a new realm of networking opportunities, including fruitful interaction between mentors and students. This new generation of researchers and practitioners is playing a leadership role and is supported by peers faced with similar challenges.

Following the Mérida Forum, the international reach of the IAEH's active and diverse student membership expanded from 5 to 19 countries. Interest in maintaining fertile interdisciplinary exchange in the field of ecohealth has led CoPEH-Canada

> **Example 21.3** Training a New Generation of Ecohealth Researchers in Ecuador
>
> The Sustainably Managing Environmental Health Risks project in Ecuador was launched in 2005. It has trained a new generation of researchers across different regions of Ecuador. They are involved in community-based, policy-relevant research that is informed by an ecosystem approach to health. Through a network involving four Ecuadorian universities, ten centers at the University of British Columbia, and institutions from Cuba and Mexico, this 6-year initiative was established to build human resources and institutional capabilities to reduce health impacts in areas such as pesticide poisoning, heavy metal contamination, solid waste pollution, sanitation, air pollution, and vector-borne disease. By 2009, a nationally accredited Master's in Health with an Ecosystem Focus, based at three provincial universities in Cuenca, Machala, and Guaranda produced ten graduates per university, enhanced the capacities and qualifications of an equal number of instructors and community members, and produced research that engaged more than 1,200 participants in 15 different communities. A second cohort was launched at the University of Cuenca in 2009, with graduates from the first cohort participating in the international faculty team. The network was strengthened by the introduction of an innovative PhD program (Collective Health, Environment, and Society) at the Universidad Andina Simón Bolívar in Quito with 19 doctoral students from the Andean region, including graduates of the master's program.

alumni to establish a virtual-poster forum for peer-to-peer exchange to help refine their graduate research with an ecohealth orientation. A variety of educational, training, and mentoring strategies are being adopted and implemented in the name of next-generation field-building in ecohealth. Some graduate programs are informed by ecosystems approaches to health (Example 21.3; Parkes et al. 2009). Ecohealth training, using intensive hybrid approaches such as field schools, summer schools, and professional development workshops, is also multiplying and providing graduate or professional development credits where possible.

Training and capacity development was also an important dimension of networks organized around a specific theme or entry point such as health impacts of toxic substances in the environment (CoPEH-LAC) or avian influenza (H5N1) in Asia (APEIR). With APEIR, IDRC's intent to foster regional collaboration among agencies responsible for research coincided with, and complemented a need for, policymakers and researchers to build a regional research agenda to guide disease prevention.

> **Example 21.4** Building Strong Social Capital Against Emerging Infectious Diseases
>
> The Asian Partnership on Emerging Infectious Diseases Research (APEIR) includes researchers from a range of disciplines (e.g. public health, veterinary medicine, sociology, political sciences, and economics) and official leaders from several sectors (e.g. health, livestock, and wildlife protection). The partnership involves six countries (Cambodia, China, Lao PDR, Indonesia, Thailand, and Vietnam) and was conceived in 2006 in response to the spread of Avian Influenza (H5N1). By 2009, it had expanded to examine a range of emerging infectious diseases (EIDs). Its vision is to be the leading knowledge and research network in Asia for EIDs based on ecohealth concepts by 2013 (www.apeiresearch.net). Directed by a Steering Committee of senior officials and researchers from each country, APEIR has expanded to 30 partner institutions engaged in multi-country applied research. APEIR is focused on bridging the research–policy interface and addressing the determinants and consequences of EIDs. An emphasis on socio-economic, environment, and health links makes lessons from avian influenza research applicable to new and future emerging diseases. Understanding the drivers of disease emergence, prevention, detection, and control are all crucial to address EIDs. APEIR aims to facilitate the flow of information and knowledge and foster multi-sector collaboration throughout the region.

Like many of the networks in Table 21.1, both APEIR and CoPEH LAC developed internal relationships and common ground over a short time. Trust, respect, and common understanding of shared goals have grown, and in both these cases, supported an expansion over time to a wider thematic focus. These changes reflect the emerging capacity and priorities of their members. This evolution may also increase the potential influence and impact of the missions of such networks.

Principles, Processes, and Capacities for Ecohealth

The field of ecohealth is partly defined in relation to other fields, and by its contribution to a wider body of scholarship. Current conceptualizations of integration, participation, and collaboration that address capacity to deal with complex health, environment, and equity concerns are of particular relevance.

Integration of perspectives, knowledge, and methods; full participation of stakeholders; and a wide range of collaborations are instrumental in the production and use of knowledge that is grounded in social–ecological systems thinking (Bunch et al. 2011; Parkes et al. 2010; Waltner-Toews 2009; Williams and Hummelbrunner 2010). As discussed in Chap. 1, framing research in a (complex) systems perspective demands a focus on interrelationships between people and their environment; engagement with a diversity of views in understanding and facilitating change; scrutiny of boundaries set around systems and subsystems; understanding change as a dynamic process; and an awareness of interactions and links across scales (social, geographic, and temporal). The resulting innovative research and action crosses disciplinary boundaries and integrates different forms of knowledge; engages multiple actors with different interests, needs, and potential contributions; and approaches change as a collaborative, adaptive, and learning-oriented endeavor.

The concepts of integration, participation, and collaboration help situate aspects of the practice of ecohealth in a broader scholarly debate (see Parkes and Panelli 2001; Brown et al. 2005; Brown 2008; Pohl and Hirsch Hadorn 2008; Parkes et al. 2010). As shown in Table 21.2, there are many points of intersection between the three concepts and ecohealth principles, processes, and capacities. It also connects the field-building contributions of networks with these concepts. Knowledge integration, participation, and collaboration are linked to related concepts in CoP (Lave and Wenger 1991; Wenger et al. 2002) and expanded forms of scholarship (Boyer 1997; Woollard 2006). Table 21.2 highlights convergences and makes explicit links to broader literature that can inform what it means to be "better together" when applying ecohealth approaches.

This focus on integration, participation, and collaboration is intended to link the "how to" of the ecohealth principles described in Chap. 1 with other, similar concepts, and demonstrate their applicability across different approaches. The principle of *transdisciplinarity* depends on knowledge integration, with new knowledge arising through the synthesis of diverse participant knowledge (disciplinary, experiential, and tacit). Where, how and to what extent stakeholders become involved influence these processes. The principle of *multistakeholder participation* addresses the processes of building relationships and negotiating explicit roles and responsibilities for action. The principles of *knowledge-to-action, equity*, and *sustainability* pertain to an iterative process of change based on collaborative learning and action toward common goals.

The benefits of networking in ecohealth can therefore be seen to relate to integration (that which is being "combined to form a whole"), participation (different dynamics of taking part and sharing), and collaboration (working together). These interrelated concepts feature in scholarly discussion of interdisciplinarity and transdisciplinarity (Hirsch Hadorn et al. 2008; Jantsch 1972; Klein et al. 2001; Somerville and Rapport 2000; Wilcox and Kueffer 2008), collaborative innovation (Gross Stein et al. 2001), communication models and knowledge communities (Campos 2003), integration and implementation sciences (Bammer 2005). Similar emphasis can be found in a review of 10-years of research proposals to NSF for integration across and beyond disciplinary boundaries to encourage innovation and development in research and technology (US-NSF 2001), (STEPS Centre 2010).

Table 21.2 A framework for analyzing the added value of networks and communities of practice with the concepts of integration, participation, and collaboration, and their relationships to other constructs from the literature

	Integration	Participation	Collaboration	References
Process	New knowledge generation through the synthesis of participants knowledge and understanding	Type, place, and mode of participation … an explicit process of defining and negotiating roles and responsibilities	Collaborative learning and actions are an iterative process rather than an endpoint, determined by the approach to integration and participation	Adapted from Parkes and Panelli (2001)
Some principles of ecohealth	Transdisciplinarity requires awareness of diverse knowledge relevant to the systemic understanding of health and environment issues – goes beyond academic disciplines, and leads to new types of participation and processes for knowledge generation	Multistakeholder participation requires value and respect for different perspectives, roles, and responsibilities – involves sharing within and between communities, researchers, and decision-making groups and also among them	Achieving knowledge-to-action requires learning and working together, a process which also demands attention to equity and respect for diversity, with explicit focus on gender disparities and variations among social groups involved. Equity between generations supports ecosystem sustainability	See Chap. 1
Communities of practice	Domain – group identity defined by a shared domain of interest. Membership implies commitment to that domain and a shared competence	Community – the social fabric of learning, and relationships that enable them to learn from each other	Practice – a shared repertoire of resources (experiences, stories, tools, ways of addressing recurring problems) – in short a shared practice	Lave and Wenger (1991) and Wenger et al. (2002)
Expanded definitions of scholarship	Scholarship of integration consists of making connections across disciplines and, through this synthesis, advancing what we know	Scholarship of engagement connects (other) dimensions of scholarship to the understanding and solving of pressing social, civic, and ethical problems	Scholarship of application asks how knowledge can be practically applied in a dynamic process whereby new understandings emerge from the act of applying knowledge through an ongoing cycle of theory-to-practice-to-theory	Boyer (1997) and Woollard (2006)
Dictionary definition	To combine with another to form a whole	To be involved, take part	To work jointly on an activity or project	Oxford Dictionary (2010)

In the public health and environment fields, these same concepts are reflected in: increased demand for integrated, community-based, participatory, and collaborative approaches to research and practice (Barten et al. 2007; Israel et al. 1998; Koné et al. 2000; O'Fallon and Dearry 2002; Sauvé and Godmaire 2004); calls for multistakeholder processes (Hemmati 2002); and growing attention to knowledge translation and exchange (Lavis 2006; Roux et al. 2006). The explicit contribution of nonacademic voices to integration, participation, and collaboration has focused attention on designing processes of collective action that span researchers, communities, policy, and practice (Brown 2007, 2008; Brown et al. 2005) and recognition of the complex terrain of crossing different knowledge cultures (McDonell 2000; Melin 2000; Ziman 1994). These developments in contemporary literature provide an important backdrop to any reflection on the field-building contributions of ecohealth, for example Lebel's presentation of the three pillars of an ecosystem approach to health (Lebel 2003), Wilcox and Kueffer's (2008) treatment of transdisciplinarity and its application, and the concepts and thinking advanced in this book.

The experiences of ecohealth networks and CoPEHs, as well as those in the earlier case studies, challenge the idea that integration (or participation or collaboration) is an end unto itself. Rather, in the context of ecohealth, collaboration is a process that results from – and facilitates – integration across disciplines and ways of knowing, and is founded on the participation of different stakeholders. Returning to the examples of capacity building and training within the CoPEHs and networks, researchers focused on participation go beyond convening a multistakeholder gathering, and pay explicit attention to which participants, and what kinds of knowledge are being included or excluded. New cadres of ecohealth researchers are challenged to see collaboration as more than a period of interaction, and instead as a basis for joint learning among the different knowledge bases of participants. The capacity to combine integration, participation, and collaboration goes beyond "traditional" definitions of academic scholarship (discovery and teaching) and places increasing value on the scholarship of integration, engagement, and application.

From Concepts to Practice: Ecohealth Networks

Capacity building for transdisciplinarity and integration of knowledge was a shared orientation and priority across each of the four network examples given in this chapter. The example networks illustrate that the need to integrate knowledge among previously dispersed groups can be a prime motivation for bringing different people together. In addition to the integration of knowledge that enhances the experiences and brings benefits into specific projects, each ecohealth network points to the added value of group processes that maintain integration through collaborative learning, interaction, and exchange beyond an isolated research project.

Each network made explicit and implicit decisions about participation, including for whom, and with whom to develop research, training, or policy. This leads to decisions about who is included in the participatory process (type of participant), where participants take part, share, and exchange (place of participation), and how different participants will be involved, including structure, roles, and responsibilities (mode of participation) (Table 21.3).

The example networks demonstrate a wide range of *types of participants*, who span disciplinary, sectoral, and national boundaries, in direct reflection of the complexity of health issues grounded in both social and ecological systems. Beyond listing different groups, some authors have found it helpful to distinguish participants according to their different knowledge cultures. Brown (2007, 2008) distinguishes individual, community, specialized, organizational, and holistic knowledge; whereas, Pohl and Hirsch Hadon (2008) refer to the importance of both abstract–theoretical, and case-specific–practical knowledge as features of transdisciplinary participatory processes. Boelen's "partnership pentagram" (involving policymakers, administrators, communities, academic institutions, and professionals) also helps to focus on the types of participant who may influence capacity for integration, engagement, and application of knowledge (Boelen 2000; Woollard 2006).

For each of the examples in Table 21.3, *place of participation* is a key consideration. Both the Ecuador training network and CoPEH-Canada designed graduate-level education and training initiatives to engage with national issues and priorities. To achieve this, the initial phase of both projects was based around three universities in different provinces. Hosting the course in these different locations encouraged engagement with the specific issues, people, and relationships arising in each place. Although resource intensive, such a combined national and regional approach has created a stronger collective capacity than might have been achieved in a single location.

Responding to the challenge of field-building across even larger geographic areas, APEIR and CoPEH-LAC both highlight the importance of building nonhierarchical, inclusive, and trust-based relationships between participating individuals and organizations. There are benefits of face-to-face interaction (however brief) to develop and nurture relationships in all ecohealth networks. Many of these groups (notably the IAEH Student Section) are using electronic means such as social networking, interactive websites, virtual classrooms, and blogs to continue and expand their relationships. Although online communications were a strong focus for early CoP (Johnson 2001; Sherer et al. 2003), the CoPEHs and related training courses have also highlighted the value of planning activities that bring people together to engage with the specifics of place for teaching and research planning.

The different network structures reflect different *modes of participation* described in Table 21.3. The functions and roles match the needs and priorities of each group. These considerations are apparent in the nodal structures of CoPEH-LAC and CoPEH-Canada. In addition to the perceived benefits of the nodal structure outlined above, CoPEH-Canada alumni have developed links between nodes, roles, and working groups through collaboration. Student approaches to adapting traditional

Table 21.3 Key characteristics of example networks relative to founding theme and to aspects of participation

	Example 1: Community of practice in ecosystem approaches to health in Latin America and the Caribbean	Example 2: Canadian community of practice in ecosystem approaches to health	Example 3: Sustainably managing environmental health risks in Ecuador	Example 4: Asian partnership on emerging infectious diseases research
Founding theme, problem field or domain	Link ecohealth research to reduce exposure to toxic substances in the environment	Collective design and delivery of a pan-Canadian ecohealth short course	Capacity for community-based environmental health research in Ecuador	Regional research-to-policy collaboration on avian influenza in Southeast Asia
Types of participant	Academic research, policymakers, nongovernment and government organizations	Researchers, policymakers, practitioners, educators, and students (alumni)	Researchers, practitioners, educators, students in master's program in health with an ecosystem focus	Researchers from multiple disciplines, policymakers, and politicians
Place of participation	Fourteen countries with nodes in Mexico, Central America – Caribbean, Andean Region, Southern Cone, Brazil, and Quebec (Canada)	Three nodes in Western Canada, Ontario, Quebec-Atlantic-Acadian. Course rotates among core university hosts	Three hubs in Ecuador provincial universities and one in Quito	Cambodia, China, Lao People's Democratic Republic, Indonesia, Thailand, and Vietnam
Modes of participating (includes structure of the group)	Nested nodal structure, with local network developed around each core member – many teleconferences and research planning meetings within or between nodes	A short course for graduate students delivered by core faculty team, other ecohealth researchers, alumni, and working groups involved with nodal, national, or course activities – field work bursary program	MSc and PhD training program delivered by faculty team jointly across four universities, with alumni becoming faculty	Five different multicountry research projects involving 22 participating institutions – country and regional meetings with steering committee and country focal persons

governance structures for their own needs offer valuable glimpses to future developments.

The example networks illustrate ecohealth practices and applications that range from research and education to policy, and blur distinctions among them. For example, through its steering committee, APEIR helps link researchers to decision-making processes. CoPEH-LAC has engaged diverse policy, nongovernmental, and research organizations and individual actors through research projects, and training and dissemination workshops, to consolidate and share experiences about ecohealth. But documenting policy influence (and all the intermediate steps on the way to policy change) is a challenge for CoPEH-LAC, indeed for most networks striving for broader uptake of research results (Carden 2009). The combined challenges of revisiting priorities for the future and evaluating progress over time are an integral part of becoming "better together."

Trade-Offs of Working Together

Involvement with ecohealth networks and CoPEHs accrues benefits and costs at several levels – that of the individual members, the network as a whole, and the research outcomes. Because the costs of interaction are not negligible for members or the network, and researchers have generally overcommitted schedules and resources, it should follow that the benefits of association with ecohealth networks must be worth the effort and costs.

The nature and amount of investment required vary across the networks and CoPEHs. In general, active research networks incur costs in time and other resources such as coordination, travel, meetings, self-promotion, and dissemination. They may also demand different social skills of their members, especially in CoP where collaboration is a fundamental *raison d'être* of the network, and require a range of tacit and learned skills such as flexibility, ability to negotiate and compromise, ability and capacity to convene, and communication skills. The magnitude and distribution of costs depend on the priorities and activities of each network. Some groups place greater emphasis on formal in-person meetings to achieve their goals; others may focus on training programs or on a mix of presence-based and virtual interactions between members. Some networks and CoPEHs have formal coordination structures with inherent, but variable, operating costs.

Benefits to individual members include: overcoming feelings of intellectual or practical isolation; furthering professional relationships; learning (e.g., easier and faster access to state-of-the-art thinking by peers, ideas on how to deal with challenges, avoiding reinventing the wheel, and access to new tools and methods); exposure to experiences in different contexts (e.g., geographic and thematic); broadening professional opportunities (e.g., different types of collaboration, pursuit of joint funding, co-publishing, and joint supervision of students); motivation and peer support; and increased voice and influence. It follows that these benefits are primary sources of motivation for individuals to become engaged in formal CoPEHs and networks.

The examples presented earlier also point to benefits that accrue at the network level. These include enhanced creativity and innovation (asking better research questions and defining relevant research agendas) and development of the field. Networks can also: enhance recognition of project results and uptake of results on a wider scale than could be achieved by individual projects; debate for the advancement of the field and furthering of theory and concepts; offer greater visibility and recognition of the field; and develop a greater pool of trainers and opportunities for graduate study and new researchers.

There is clearly a need for more formal assessments of the value and impact of the networks and CoPEHs. CoPEH-LAC's first cycle evaluation (Willard and Finkelman 2009) commended the learning- and trust-based aspects of their activities and the high level of collaboration, but found the need for more strategic approaches to joint action, particularly for greater policy influence and diversification of the funding base.

How are these costs and benefits, as well as added-value assessed? CoPEH-LAC and CoPEH-Canada have adopted internal evaluative processes informed by social network analysis to guide their evolution, along with longitudinal qualitative analysis of learning and collaboration. Similarly, APEIR has performed a self-reflection exercise with all of its members to achieve a qualitative and quantitative review of the impacts of the partnerships, and to form the foundation for a strategic thinking process to guide its future development. Although precedents for long-term analysis of collaborative processes are rare, they would contribute considerably to validating the benefits (as well as cost and effort required) of such groups. Gross Stein et al. (2001) propose a series of questions to assess added-value:

- Would we know less if this collaboration had not been created?
- Would we know differently if collaborators had not had the opportunity to work together?
- Would we have known what we know more slowly or less widely if the knowledge had not been disseminated by the research?

Given the action-research aspect of ecohealth, two further questions could be added to those posed by Gross Stein et al.:

- Would we do differently or not as well in the absence of collaboration?
- Would our actions have been fewer, slower or less widely applied if we had not had the opportunity to work together?

The combined insights from ecohealth networks suggest an affirmative answer to these questions. Each of the groups examined in this chapter resulted in opportunities to know more, differently, more quickly, and more widely than if they had not existed. In addition each contributed to field building at a higher level through knowledge integration, participation, and collaboration. Their practice of ecohealth research was also changed for the better because of opportunities and synergies provided in doing things together and learning from the process. Ecohealth networks can be seen to facilitate a move from "doing (the same) things better" to "doing better things" (Kravitz 2005).

On the whole, the examples in this chapter demonstrate three benefits of becoming and doing 'better together' by:

- *Strengthening peer groups through a shared learning commons* – the field becomes stronger when diverse ecohealth initiatives combine to create a foundation for debate, consolidation, and critique that leads to a deeper understanding of the principles, approaches, and tools of ecosystem approaches to health.
- *Learning beyond individuals or isolated projects* – new levels of impact become possible when collective experiences create opportunities for learning and exchange that would not otherwise have been possible, while leaving individuals feeling challenged, valued, enabled, motivated, and better equipped to engage with new types of challenges.
- *Fostering innovation and systematization* – creativity and imagination are essential if ecohealth is to thrive in the face of traditional academic and decision-making structures that emphasize individual, disciplinary expertise and orientations, despite demanding greater integration across sectors. Investments in collaborative ecohealth networks and communities provide a web of pathways, relationships, knowledge, trust, and courage to innovate in ways that might not otherwise have been imagined.

These benefits highlight an ongoing creative tension for those engaged in ecohealth research, education, and practice. Those who participate usually do so in an individual capacity, and usually not as representatives of their organizations. Although participants can share experiences, they cannot commit their organizations to a course of action. This is borne out by experience, wherein the different networking initiatives have achieved greater levels of knowledge sharing and capacity building than might be expected from single institution-led research projects. But, they have also had to reassess their functioning to better achieve enhanced uptake of research findings and proposed changes in new contexts, or at different scales. Ecohealth networks and CoPEHs offer an enabling, supportive, and practical means of developing new concepts, new knowledge, and new skills. They also allow their members to work within their own organizations to facilitate positive changes for more relevant and impactful development research and to build capacity to respond to future challenges.

Conclusion

Working together across sectors and disciplines to tackle complex interactions between health, environment, and equity is now widely recognized to be beneficial in development research. Yet, such research has tended to be single discipline or sector driven. More than ever, new means of facilitating knowledge exchange, co-learning, and collaboration between individuals and groups need to be found. In this sense, the proactive approach to supporting the networking initiatives presented in this chapter represents progress toward expanding policy and practice horizons.

The benefits of ecohealth research networks and CoPEHs are substantial, and worth the sometimes considerable investments and inconvenience to their membership. Each example provided glimpses into this culture of interaction and learning. Networks, and CoPEHs in particular, enhance the application of at least four ecohealth principles – those emphasizing the "how to" of ecohealth research: transdisciplinarity; participation; social equity; and research-to-action. The concepts of integration, participation, and collaboration were used to explore why members engaged in these communities, and what added value is achieved though this working together.

Shared benefits arising from ecohealth networks were described in relation to a shared learning commons, the potential for learning, knowledge exchange and application beyond the scale of the project, and opportunities for innovation, and institutionalization. The case studies and examples throughout this book also represent contributions to the development of ecohealth research as a new field. Without an active peer group seeking to expand and refine their understanding, methods, body of knowledge, and skill-set, any field of endeavor can be expected to stagnate. Ecohealth networks and CoPEHs are expanding and evolving and appear to be key to the further growth of the field. With this growth comes new challenges of fostering integration, participation, and collaboration within an ever-larger group, while developing relationships that provide the support, courage, and resources to explore new frontiers and increase innovation in ecohealth research for development.

Acknowledgments This paper draws on an internal APEIR report prepared by Chun Lai. Johanne Saint-Charles provided comments that improved the final text. We greatly appreciate contributions from: the Community of Practice in Ecosystem Approaches to Health in Latin America and the Caribbean (CoPEH-LAC) (IDRC projects 101818 and 105151); the Canadian Community of Practice in Ecosystem Approaches to Health (IDRC project 104277); the Sustainably Managing Environ Mental Health Risks in Ecuador Project; the APEIR steering committee (IDRC projects 103924, 104320-04, 106037, 106321); and the Students' Section of the International Association for Ecology and Health.

References

Aguirre, A. A., Ostfeld, R. S., Tabor, G. M., House, C., and Pearl, M. C. (Editors). (2002). Conservation Medicine: Ecological Health in Practice. New York, Oxford University Press.

Bammer, G. (2005). Integration and Implementation Sciences: Building a New Specialization. Ecology and Society, 10(2), 6. Available at: www.ecologyandsociety.org/vol10/iss2/art6.

Barten, F., Mitlin, D., Mulholland, C., Hardoy, A., and Stern, R. (2007). Integrated Approaches to Address the Social Determinants of Health for Reducing Health Inequity. Journal of Urban Health, 84, 164–173.

Boelen, C. (2000). Towards Unity for Health. Challenges and Opportunities for Partnership in Health Development. World Health Organisation, Geneva, Switzerland.

Boyer, E.L. (1997). Scholarship Reconsidered: Priorities of the Professoriate (2nd Edition). Carnegie Foundation for the Advancement for Teaching, Stanford, CA, USA.

Brown, V. (2007). Collective Decision-Making Bridging Public Health, Sustainability Governance, and Environmental Management. In: Soskolne, C., Westra, L., Kotzé, L.J., Mackey, B., Rees,

W.E., and Westra, R. (Editors). Sustaining Life on Earth: Environmental and Human Health through Global Governance. Lexington Books, Lanham, MD, USA.

Brown, V. (2008). Leonardo's Vision: A Guide to Collective Thinking and Action. Sense Publishers, Rotterdam, Netherlands.

Brown, V., Grootjans, J., Ritchie, J., Townsend, M., and Verrinder, G. (2005). Sustainability and Health — Supporting Global Ecological Integrity in Public Health. Allen and Unwin, St. Leonards, NSW, Australia.

Bunch, M., McCarthy, D., and Waltner-Toews, D. (2008). A Family of Origin for an Ecosystem Approach to Managing for Sustainability. In: Waltner-Toews, D., Kay, J.J., and Lister, N.M.E. (Editors). The Ecosystem Approach: Complexity, Uncertainty, and Managing for Sustainability. Columbia University Press, New York, NY, USA.

Bunch, M.J., Morrison, K., Parkes, M. and Venema, H. (2011). Promoting Health and Well-Being in Watersheds by Managing for Social-Ecological Resilience: The Potential of Integrating Ecohealth and Water Resources Management Approaches. Ecology and Society 16(1): 6. Available at www.ecologyandsociety.org/vol16/iss1/art6/.

Carden, F. (2009). Knowledge to Policy: Making the Most of Development Research. SAGE Publications, New Delhi and Thousand Oaks, CA. USA. Available at: http://www.idrc.ca/en/ev-135779-201-1-DO_TOPIC.html.

Campos, M. (2003). The Progressive Construction of Communication: Toward a Model of Cognitive Networked Communication and Knowledge Communities. Canadian Journal of Communication, 28, 1–13.

De Plaen, R., and Kilelu, C. (2004). From Multiple Voices to a Common Language: Ecosystem Approaches to Human Health as an Emerging Paradigm. EcoHealth, 1(Suppl. 2), S8–S15.

Fligstein, N. (2001). Social Skill and the Theory of Fields. Sociological Theory, 19, 105–125.

Flynn-Dapaah, K. (2003). Ecosystem Approaches to Human Health Global Community of Practice: Report on the Design Phase Consultations. Consultancy Report Submitted to IDRC. International Development Research Centre, Ottawa, Canada. Available at: http://idl-bnc.idrc.ca/dspace/handle/10625/45150.

Green, L.W. (2009). The Field-Building Role of a Journal About Participatory Medicine and Health, and the Evidence Needed. Journal of Participatory Medicine, 1(1), e11.

Gross Stein, J., Stren, R., Fitzgibbon, J., and MacLean, M. (Editors). (2001). Networks of Knowledge: Collaborative Innovation in International Learning. University of Toronto Press, Toronto, Canada.

Hemmati, M. (2002). Multi-Stakeholder Processes for Governance and Sustainability. Beyond Deadlock and Conflict. Earthscan, London, UK.

Hirsch Hadorn, G., Hoffmann-Riem, H., Biber-Klemm, S., Grossenbacher-Mansuy, W., Joye, D., Pohl, C., Wiesmann, U., and Zemp, E. (Editors). (2008). Handbook of Transdisciplinary Research. Springer, New York, NY, USA.

Israel, B., Schulz, A., Parker, E., and Becker, A. (1998). Review of Community-Based Research: Assessing Partnership Approaches to Improve Public Health. Annual Review of Public Health, 19, 173–202.

Jantsch, E. (1972). Towards Interdisciplinarity and Transdisciplinarity in Education and Innovation. In: Interdisciplinarity: Problems of Teaching and Research in Universities. Organisation for Economic Co-operation and Development (OECD), Paris, France.

Johnson, C.M. (2001). A Survey of Current Research on Online Communities of Practice. The Internet and Higher Education, 4(1), 45–60.

King, J.A. (2010). Response to Evaluation Field Building in South Asia: Reflections, Anecdotes, and Questions. American Journal of Evaluation, 31(2), 232–237. Available at: http://aje.sagepub.com/content/31/2/232.

Klein, J., Grossenbacher-Mansuy, W., Häberli, R., Bill, A., Scholz, R., and Welti, M. (2001). Transdisciplinarity: Joint Problem Solving Among Science, Technology and Society. Birkhäuser Verlag, Basel, Switzerland.

Koné, A., Siullivan, M., Senturia, K.D., Chrisman, N.J., Sandra, C.J., and Krieger, J.W. (2000). Improving Collaboration Between Researchers and Communities. Public Health Reports, 115, 243–248.

Kravitz, R.L. (2005). Doing Things Better vs Doing Better Things. Annals of Family Medicine, 3, 483–485.
Lave, J., and Wenger, E. (1991). Situated Learning: Legitimate Peripheral Participation. Cambridge University Press, New York, NY, USA.
Lavis, J.N. (2006). Research, Public Policymaking, and Knowledge-Translation Processes: Canadian Efforts to Build Bridges. Journal of Continuing Education in the Health Professions, 26, 37–45.
Lebel, J. (2003). Health: An Ecosystem Approach. In Focus Series. International Development Research Centre, Ottawa, Canada. Available at: http://www.idrc.ca/in_focus_health/.
McBride, A.M., Sherraden, M., Benítez, C., and Johnson, E. (2004). Civic Service Worldwide: Defining a Field, Building a Knowledge Base. Nonprofit and Voluntary Sector Quarterly, 33(Suppl. 4), 8S–21S.
McDonell, G. (2000). Disciplines as Cultures: Towards Reflection and Understanding. In: Somerville, M.A., and Rapport, D. (Editors). Transdisciplinarity: Recreating Integrated Knowledge. EOLSS Publishers, Oxford, UK.
McKnight, J.L., and Kretzmann, J.P. (1996). Mapping Community Capacity. Institute for Policy Research, Northwestern University, Evanston, IL, USA.
Melin, G. (2000). Pragmatism and Self-Organization: Research Collaboration on the Individual Level. Research Policy, 29, 31–40.
Mertens, F., Saint-Charles, J., Mergler, D., Passos, C., and Lucotte, M. (2005). Network Approach for Analyzing and Promoting Equity in Participatory Ecohealth Research. EcoHealth, 2(2), 113–126.
O'Fallon, L.R., and Dearry, A. (2002). Community-Based Participatory Research as a Tool to Advance Environmental Health Sciences. Environmental Health Perspectives, 110, 155–159.
Ottoson, J.M., Green, L.W., Beery, W.L., Senter, S.K., Cahill, C.L., Pearson, D.C., Greenwald, H.P., Hamre, R., and Leviton, L. (2009). Policy-Contribution Assessment and Field-Building Analysis of the Robert Wood Johnson Foundation's Active Living Research Program. American Journal of Preventive Medicine, 36(Suppl. 2), S34–S43.
Oxford Dictionary. (2010). Oxford English Dictionary, Oxford University Press, Oxford, UK.
Parkes, M., and Panelli, R. (2001). Integrating Catchment Ecosystems and Community Health: The Value of Participatory Action Research. Ecosystem Health 7(2), 85–106.
Parkes, M.W., Morrison K.E., Bunch, M.J., Hallström, L.K., Neudoerffer, R.C., Venema, H.D., Waltner-Toews, D. (2010). Towards Integrated Governance for Water, Health and Social-Ecological Systems: The Watershed Governance Prism. Global Environmental Change, 20, 693–704.
Parkes, M.W., Spiegel, J., Breilh, J., Cabarcas, F., Huish, R., and Yassi, A. (2009). Promoting the Health of Marginalized Populations in Ecuador through International Collaboration and Educational Innovations. Bulletin of the World Health Organization, 87(4), 312–319.
Pohl, C. (2008). From Science to Policy Through Transdisciplinary Research. Environmental Science and Policy, 11, 46–53.
Pohl, C., and Hirsch Hadorn, G. (2008). Methodological Challenges of Transdisciplinary Research. Natures Sciences Sociétés, 16, 111–121.
Roux, D.J., Rogers, K.H., Biggs, H.C., Ashton, P.J., and Sergeant, A. (2006). Bridging the Science–Management Divide: Moving from Unidirectional Knowledge Transfer to Knowledge Interfacing and Sharing. Ecology and Society, 11(1), Article 4. Available at: http://www.ecologyandsociety.org/vol11/iss1/art4/.
Sauvé, L., and Godmaire, H. (2004). Environmental Health Education: A Participatory Holistic Approach. EcoHealth, 1(4), 35–46.
Sherer, P.D., Shea, T.P., and Kirstensen, E. (2003). Online Communities of Practice: A Catalyst for Faculty Development. Innovative Higher Education, 27(3), 183–194.
Somerville, M.A., and Rapport, D. (Editors). (2000). Transdisciplinarity: Recreating Integrated Knowledge. EOLSS Publishers, Oxford, UK.

STEPS Centre (Social, Technological and Environmental Pathways to Sustainability Centre). (2010). Innovation, Sustainability, Development: A New Manifesto. STEPS Centre, Brighton, UK. Available at: http://anewmanifesto.org/wp-content/uploads/steps-manifesto_small-file.pdf.

US-NSF (United States National Science Foundation). (2001). 2001–2006 Strategic Plan. National Science Foundation, Arlington, VA, USA.

Waltner-Toews, D. (2009). Food, global Environmental Change and Health: Ecohealth to the Rescue? McGill Medical Journal, 12(1), 85–89.

Waltner-Toews, D. (2004). Ecosystem Sustainability and Health: A Practical Approach. Cambridge University Press, Cambridge.

Waltner-Toews, D., and Kay, J. (2005). The Evolution of an Ecosystem Approach: The Diamond Schematic and an Adaptive Methodology for Ecosystem Sustainability and Health. Ecology and Society, 10(1), Article 38. Available at: http://www.ecologyandsociety.org/vol10/iss1/art38/.

Webb, J., Mergler, D., Parkes, M.W., Saint-Charles, J., Spiegel, J., Waltner-Toews, D., Yassi, A., and Woollard, R. (2010). Tools for Thoughtful Action: The Role of Ecosystem Approaches to Health in Enhancing Public Health. Canadian Journal of Public Health 101 (6): 439-441. Available at: http://www.copeh-canada.org/documents/Volume_101-6_439-41.pdf.

Wenger, E., McDermott, R., and Snyder, W.M. (2002). Cultivating Communities of Practice: A Guide to Managing Knowledge. Harvard Business School Press, Boston, MA, USA.

Wilcox, B., Aguirre, A.A., Daszak, P., Horwitz, P., Martens, P., Parkes, M., Patz, P., Waltner-Toews, D. (2004). EcoHealth: A Transdisciplinary Imperative for a Sustainable Future. EcoHealth 1(1), 3–5.

Wilcox, B., and Kueffer, C. (2008). Transdisciplinarity in EcoHealth: Status and Future Prospects. EcoHealth, 5, 1–3.

Willard, T., and Finkelman, J. (2009). Findings Brief: External Review of a Community of Practice Development Project on Ecohealth in Latin America and the Caribbean. Consultancy Report Submitted to IDRC. International Development Research Centre, Ottawa, Canada. Available at: http://idl-bnc.idrc.ca/dspace/handle/10625/45349.

Williams, B., and Hummelbrunner, R. (2010). Systems Concepts in Action: A Practitioner's Toolkit. Stanford University Press, Stanford, California, USA.

Woollard, R.F. (2006). Caring for a Common Future: Medical Schools' Social Accountability. Medical Education, 40, 301–313.

Ziman, J. (1994). Prometheus Bound. Science in a Dynamic Steady State. Cambridge University Press, Cambridge, UK.

Chapter 22
Ecohealth Research in Practice

Dominique F. Charron

This book presents innovative research from the field of ecosystem approaches to health that addresses health problems arising at the nexus of economic development and deteriorated ecosystems or changing environments. In addition to seeking new knowledge and understanding, the research has sought to support change processes that will enable people to achieve better health as well as ecologically, socially, and economically sustainable development.

In Brazil, Malawi, Indonesia, and Lebanon, human health and well-being improved because of this research. Because of the process used (an ecosystem approach to health), these improvements are more likely to last. In Guatemala, Ecuador, Tanzania, and Kathmandu, laws and policies have been changed by this research, bringing improved health and well-being, and better environmental management to a much larger number of people that those who were touched by the projects. Along the way, a new field of research, education, and practice has emerged: ecohealth. From this, other similar improvements in health and ecosystems around the world can be achieved, now and in the future, as new researchers take on the challenges of the day.

This final chapter presents some lessons learned across different themes, places and phases of ecohealth research. Characteristics of a general practice of ecohealth research are proposed. The chapter ends with a discussion of ecohealth as a field of research and practice and identifies some opportunities for the continued development of the field.

D.F. Charron (✉)
International Development Research Centre, Ottawa, ON, Canada
e-mail: dcharron@idrc.ca

Findings

The case studies describe the richness and variation in applications of ecosystem approaches to health. Each is quite different; yet, a few strong and recurring themes can be discerned. They are innovation; community empowerment and voice; policy influence and systematization; social and gender equity; and improved health outcomes.

Innovation

As an applied process of inquiry predicated on the principle of knowledge-to-action, ecohealth research aims to achieve evidenced-based change in the health and well-being of people, at least partly through improved environmental conditions or better interactions with ecosystems. Ecohealth research innovates by effecting change through the application of new methods, ideas, or technologies. The principles underpinning ecohealth research are not new; it is in how these new ideas are arrived at that ecohealth makes a unique contribution. Innovation can be found in both the practice and outcomes of ecohealth research. The interventions described as part of almost all the case studies represent locally appropriate and ecologically sustainable innovations that make a difference to people's lives.

Innovation can simply mean developing and proving the worth of a new technology used for public good, and this occurred in at least two cases included in this book. Guatemalan homes were improved using a formula for long-lasting wall plaster developed using locally available and nontoxic materials, thus preventing infestations of the bug that transmits Chagas disease; and in Yaoundé, Cameroon, an inexpensive and hygienic water-storage container was developed and distributed to reduce childhood water-borne diseases. These important discoveries have been recognized widely, but in and of themselves do not represent the chief achievements of these projects. The process of their development, their application through an ecohealth process in the community development context, and the engagement of policymakers in their uptake for wider use were what really made the impact.

In several other cases, existing technology was applied in an innovative way through an ecohealth research process – for example, improved legume crops were used to enhance soil fertility and nutrition in an ecologically sound manner in Malawi; low-input farming practices improved yields and health of Ecuadorian farming communities in the Ecosalud project; and low-cost dust filters were installed in stone quarrying and crushing units in India. Other types of innovations were also developed and applied – a certification program for flower farms with acceptable occupational health practices has reduced exposure of flower workers to pesticides in Ecuador; in Yemen, an indigenous seed bank was created; and in Kathmandu, community-led ideas resulted in the design, construction, and operation of small, local and hygienic abattoirs.

Ecohealth research conducted in this way has positive spin-offs. Although the focus of the research is usually a health problem linked to a problem or change in an ecosystem, the process of situating this problem in a systems context with the participation of stakeholders can sometimes lead to community development activities with an entrepreneurial flavor. The case studies are full of examples: an alternative farming supply store in Ecuador's Ecosalud project; recycling services in Yaoundé; and catering with traditional foods in Lebanon.

The field of ecohealth has advanced because of the experiences presented in the case studies, as well as a growing body of similar work. Through networks and communities of practice and the dedication and leadership of individual researchers, there is more capacity to apply an ecosystem approach to health in developing regions. The capacity to execute research, and to apply research findings to new ways of doing things, is necessary for any society's success and further development. Research and its application are important parts of innovation systems. By its nature, ecohealth research contributes to the development of innovation systems: it fosters the testing and application of locally generated ideas and it engages with relevant decision-making processes.

Community Empowerment and Voice

Enhanced community organization and empowerment were frequently observed in case studies reporting on more mature research. In several cases, improved community organization was key to achieving positive changes in people's health and well-being – for example, in Kathmandu, great levels of community change and growth, organization of previously marginalized groups with no access to services, and formalization of different sectors (sweepers, slaughterers, and butchers) were achieved. A similar, albeit more diffuse, type of transformation was evident in the case study from Yaoundé, where a combination of improved responsiveness from various government agencies, input from self-organizing entrepreneurs, and better organization and leadership within the community resulted in improvements in the Melem neighborhood. In both of these cases, a collective process emerged that transcended individual health problems and the poor state of the ecosystems in the slums. This grass-roots action tackled broader issues of governance and relationships among groups in communities.

Ecohealth research similarly helped farming communities achieve positive changes. For example, in Malawi and Ecuador, farmers working together in field schools helped each other develop skills and improve yields while building social capital in their communities. In Malawi, gender relations within households were also improved, leading to greater involvement of women in farming decisions and to substantial improvements in child health. In Ecuador, smallholders realized that they could achieve better marketing opportunities, and develop their skills, by collaborating to manage and sell their organic produce. Greater social cohesion and representation, better equity among groups, and the ability to draw political response seen in the previous urban settings, were also obtained in these and other rural settings.

In other cases, the workplace was the setting for increased social cohesion and mobilization that led to improved health outcomes for workers and a range of improvements and opportunities for communities. The floriculture case study from Ecuador illustrates how ecohealth research led to economic gains and health benefits for the communities through reduced pesticide exposure and new export markets for flowers certified to have been grown in farms that adequately protect the health of workers. The health and well-being of workers in stone quarrying and crushing industries were also an entry point for community development in Bundelkhand, India.

The process of collaborative participation and the engagement of community and other stakeholders, which are typical of ecohealth research (see Chaps. 1 and 21), were essential to achieving empowering and transformative results. The participatory multistakeholder process, together with the evidence generated by the research, motivated the communities that were involved. A range of outcomes stem from this: communities seeking and obtaining action by government to assist them in making a change; more options and opportunities for previously voiceless marginalized communities (or groups within communities); and entrepreneurial opportunities that continue the process of change and create more opportunities for others.

Policy Influence and Systematization

Although the primary focus of most of the case studies was a problem located in a particular community and ecosystem, it is clear that for most studies, sustainable development outcomes were sought beyond the affected community. This is a peculiarity of ecohealth research – and of development research more broadly (Carden 2009). Applications of an ecosystem approach to health – being participatory in design and execution – tend to be highly localized, but the results are intended for a broader audience. Some of the case studies illustrate how results from participatory research can be applied to broader contexts, despite the specificity of the original research location. Others, like the dengue research in Asia, used a multi-center design repeating similar assessments and interventions in several locations. What makes ecohealth research scalable is that insights acquired from considering the systemic links between various elements of a problem are applicable to other contexts. Knowledge gained from a local, community-based knowledge-to-action research process can provide proof-of-concept for applicability to a wider scale. Knowledge gained from different studies in a variety of contexts contributes a weight of evidence that informs the uptake and systematization of research findings.

The case studies illustrate how ecohealth research can influence relevant policy processes by providing evidence for informed decision making. However, this link between research and policy is not easily achieved. In all settings – rich or poor – the movement of research results into policy depends on much more than just the production and effective communication of credible evidence. Evidence from research is arguably a minor factor in the complicated and dynamic process of governmental policymaking. In developing regions, many factors add extra challenges, including precarious democratic institutions in some places; high turnover of staff in both government and research institutions; lack of demand for research and lack of formal institutions (e.g., think tanks and the media) to disseminate results and advocate for policy change; and lack of confidence in domestic research combined with the sometimes overbearing influence of international financial institutions and donors in domestic policymaking (Carden 2009). However, developing countries also present unique opportunities for research to be taken up by policymakers, including more frequent radical policy changes, the greater interest and capacity some governments have to seek locally relevant evidence for decision making, receptiveness to local political movements or pressure, and the pervasiveness of information technologies (Carden 2009).

Influencing policy with research is not an unattainable goal; it is achievable using ecosystem approaches to health. Even among the limited number of experiences presented in this book, nearly half of the case studies[1] reported direct influence on policymaking. The experience of the case studies indicates that policy influence takes time, as reflected in the maturity of the research that achieved these kinds of results. The case of manganese pollution in Mexico's principal mining district documents the more than 10 years it has taken to make progress in municipal environmental and community engagement, mining company policies, and federal policymaking. Stricter national air-quality legislation is still pending. Policy change is also greatly aided by opportunity. The remarkable changes achieved in the policies governing butchering in Nepal were achieved during a period of increased democratization of the government. The uptake by national and international public-health agencies of housing improvements for Chagas control developed in rural Guatemala was enhanced by coincidental attention to the eradication of this disease by those same agencies. But it is also fair to say that without the manner of the research, and the evidence it generated, these beneficial policy outcomes would not have occurred.

Policy influence and change are not the only way for a wider audience to benefit from research results. As reported in the Malawi case study that developed field schools for farmers and other techniques to introduce improved legumes to the local farming practice, innovations can be socialized by word-of-mouth and other processes, without any policy change. Indeed, this contagiousness of positive experiences, supported by evidence, can help achieve policy change. In other cases, policy and policymaking processes may be part of the constellation of drivers that

---

[1] Ecuador – municipal agricultural and public policies and national pesticide legislation; Mexico – district and national air-quality policies; Tanzania – national agricultural policy; Guatemala – national and regional disease-control policies; Cuba – national disease-control policy; Nepal – provincial and national slaughtering and butchering policies; and Cameroon – municipal sanitation and infrastructure policies.

contribute to the problem, rather than the solution, or simply be inadequate to address the problem, as indicated in experiences around water-supply system management in Bebnine, Lebanon, and Yaoundé, Cameroon. Even when this is not the case, ecohealth researchers and their stakeholders often need to develop their own research-to-action processes to achieve positive changes in the conditions of people and the ecosystems they live in and depend on. For example, in the Amazon, ecohealth researchers over time have helped transform people's diets and their management of the land to both reduce dietary exposure to neurotoxic mercury and lessen the leaching of mercury into the river system. Both the remoteness of the location of this work, and the complexity of the different levels of government and policies (rural development and settlement of the Amazon, mining and transport, environmental conservation, fisheries, agriculture, and international affairs) present substantial challenges to effective policy influence.

Social and Gender Equity

Although policy influence and other large-scale social changes help research results reach and benefit many people, ecohealth research is particularly effective at making a lasting difference in the lives of the local communities involved in the research, and documenting these changes. Several case studies in this book present research that led to improved health and well-being of disadvantaged women, children, and social groups. Gender and other social differences were considered in most case studies, but a subset focused on women's health and livelihoods.[2] In Malawi, researchers (themselves mostly women) spent considerable time with village women to understand beliefs and practices around child-feeding practices and maternal health issues. Without this focused participatory process, it seems unlikely that the gains in child health attributable to this project would have been achieved. In addition, this research may one day help Malawian decision makers understand how to convert improved yields (a major agricultural priority) into better nutritional outcomes.

The project in the Brazilian Amazon, which aimed to reduce problems from mercury poisoning, illustrates how some projects harnessed the leadership of community women to better understand health problems. In this study, women's enthusiastic and consistent participation in a year-long diet study helped show that consumption of fruit might reduce some of the dangers of consuming mercury-laden fish. Other studies also addressed gender dimensions. The Chagas research in Guatemala worked with community women to develop the new plaster formula that would repel the disease vector because it is women who are traditionally responsible for keeping the house clean, including plastering the walls. In Bebnine, Lebanon,

---

[2]Malawi and Lebanon (dietary diversity); Brazilian Amazon; Uganda–Tanzania (malaria); and Ecuador (floriculture).

and in Yaoundé, Cameroon, the researchers chose to focus on educational programs for women to achieve changes in the behaviors of all household members to help reduce water-borne diseases.

Some case studies describe research that achieved major steps in reducing social inequities that exacerbate environmental health problems. The most dramatic of these is the project in Kathmandu, where previously marginalized groups (e.g., butchers and street sweepers) gained political voice, and with it opportunities for their own future development, and eventually for transforming the entire meat production sector in Nepal. Relationships between groups were also improved in case studies in agricultural communities in Malawi, Ecuador (both the floriculture and small-holder projects), and Yemen.

Improved Health Outcomes

Ultimately, improved health is necessarily the focus of research that implements an ecosystem approach to health. Most case studies present concrete evidence of better health outcomes resulting from actions and interventions based on ecohealth research.[3] Some infectious disease case studies focused on measuring reductions in disease vectors as a proxy for measuring disease outcomes in humans. Although it is important to track improvements in health outcomes to verify the positive impacts of changes based on ecohealth research, such data do not always reflect equally important gains in well-being. The case studies present research that led to reductions in diarrhoea, parisitoses, vector-borne diseases, and neurobehavioral and cognitive impairments. They also describe outcomes like improved nutrition and better maternal health indicators. However, health statistics alone do not describe changes in people's behavior – new ways of going about daily activities or livelihoods that materially improve other aspects of health and well-being (psychological, social, and physical health indicators) and prevent future poor health outcomes.

Ecohealth research achieves these kinds of positive outcomes through a new understanding of health and its dependence on ecosystems, or at the very least its relationship to degraded environments. This new understanding is visible in affected households and communities involved in the projects, and sometimes among other communities, stakeholders, and decision makers. Researchers are now grappling with the development of tools and methods to better understand and measure the full range of outcomes of ecohealth research through multi-scalar

---

[3] Malawi – improved anthropometric scores for children, improved birth weight; and improved maternal health indicators; Ecuador (Ecosalud) – reduced acute pesticide poisonings; Ecuador (floriculture and gold mining), Mexico (manganese), and Amazon (mercury) – reduced neurobehavioral problems from pesticide or heavy-metal poisoning; Nepal – reduced echinococcosis; and Lebanon (Bebnine) and Cameroon (Yaoundé) – reduced diarrhoea.

systems approaches. Researchers and donors as IDRC are also striving to understand the longer term impacts of ecohealth research to better understand the sustainability of improvements and change, and to assess how ecohealth research contributes to resilience and the adaptability of communities to new challenges that arise long after a project has ended.

Ecohealth Research in Practice

The practice of ecohealth is defined by its consideration of the social and economic activities of people together with the ecosystems they inhabit and use. This usually implies directing the process of inquiry substantially upstream from the immediate causes of ill-health – to be able to discern and understand the root causes of a problem and consider their systemic interactions. The practice of ecohealth research grapples with complexity. It strives not only to make sense of what is happening, but also to identify previously hidden levers or mechanisms that can be used to achieve some kind of material change for the better.

The case studies are indicative of the general application of ecosystem approaches to health from the mid-1990s until the present, with most of the research having been initiated before 2007. They illustrate the extent to which ecohealth researchers strive to avoid solely expert-driven approaches in favor of iterative, collaborative, and participatory multistakeholder engagement throughout the research process.

Ecohealth research is also characterized by a certain tension between (and merging of) research and application. The case studies are typical applications of ecosystem approaches to health and resemble other types of applied, implementation, and action research. From the experiences presented in this book, several elements emerge that together may inform researchers in their practice of ecohealth research.

- Ecohealth is research informed by a series of principles that aims to achieve sustainable change in human health and environmental conditions. The research generates knowledge and understanding that are used to inform local actions. It also strives to achieve changes beyond the local scale – such as change in policies and practices regionally, nationally, and internationally. For this, it is necessary to understand the local "system of action" (the engagement of key community and local government and private sector actors, and processes of discussion and planning), and the formal "framework of action" (institutional and organizational contexts at the subnational, national, and international levels).
- Knowledge generation, in ecohealth research, comes from a participatory, multistakeholder, integrative process that aims to achieve a systems understanding of the drivers (in various dimensions and at multiple scales) of a particular environmental health-related problem. In addition to scientifically

proven information, the knowledge gained may include new perspectives on established knowledge.
- Implementing ecosystem approaches to health is an iterative "learning-by-doing" process. This research process, illustrated throughout the case studies, and schematically represented in Chap. 1, is somewhat path dependent. Although ecohealth research can start from any phase, most of the experiences in the book reflect a similar start at the participatory design phase. Although the process moves back and forth between phases of design, data collection, analysis, inference, and action, on different scales and at different speeds, it does tend to move through these phases more or less in sequence, tending to achieve impacts at larger scales over time.
- Ecohealth research relies on and is enhanced by partnerships and coalitions between a wide range of actors (community stakeholders, decision makers at various scales, other researchers, and various organizations and institutions). These partnerships lead to networks and collaborations that enhance uptake and sharing of results, and often outlast projects.
- An emphasis on livelihoods is necessary to achieve sustainable improvements in both health and the management of ecosystems for better health.

Research conducted with an ecosystem approach to health can be effective in achieving ecologically sound, locally appropriate, pro-livelihood and pro-equity strategies for improving health problems that are rooted in degradation or poor management of the environment or in environmental change. Although as a matter of general principle, environmental sustainability underpins most of the research in this book, it is surprising that ecosystem-level outcomes are not emphasized in the case studies. Considerations of ecosystems (their quality, their functioning, and their implications for health and livelihoods) are somewhat explicit in the case studies, particularly in the design or conceptual stages, in descriptions of the context, and sometimes in the design of various interventions. For example, in the Amazon study of mercury mobilization pathways, the gold-mining pollution study in Ecuador, and the Mexican research on manganese pollution, research included the tracing of contaminants in the ecosystem. In the Asian dengue fever multicountry study, vector ecology of mosquitoes is described in detail. Vector ecology is also explored in a peridomestic context in Cuba (dengue fever) and Guatemala (Chagas disease) and in an agroecosystems context in Uganda and Tanzania (malaria).

However, wider ecological changes or dynamics, and their implications for vectors, animal hosts, or people are not explicitly addressed in most of the case studies. For example, the riparian zone of the Bishnumati River in Kathmandu was restored, but the project did not directly assess any wider ecological benefits of this change. Many of the studies exploring agriculture and health address some aspects of agroecosystems and their relationship to health. But in most case studies, few measurements are made of the condition and functioning of the ecosystem, except where these create exposures or risks to human health.

An aim of ecosystem approaches to human health is to foster better human health through healthier, more sustainable ecosystems, and not only through better environmental management. The Amazon mercury case study is among the most evocative and comprehensive projects ever supported by IDRC in this regard. It explored the cycle of mercury in the ecosystem, pinpointed deforestation as the key driver, and explored implications for water and food quality and the health of people. Due to space limitations in this book, some of this work was only cursorily mentioned. But, in general, the case studies are typical of a blind spot in many applications of an ecosystem approach to human health. There are explanations for this gap. An assessment of the condition of the ecosystem and its nonhuman inhabitants is often not part of the design, which is focused on assessing a health problem. The research may be narrowly focused on an "ecosystem" such as a poultry raising system, or an urban slum, or an arid farming area. These ecosystems are highly managed, and often severely degraded, and not necessarily the most intellectually appealing or appropriate settings for ecological research. Sometimes, the data may have been collected, but expertise, tools, or methodologies may be missing to properly use them in combined analyses with social, economic, and health data. In any case, it seems important for the users of ecosystems approaches to health to consider what more can be learned from ecology and how better to address ecosystems outcomes in ecohealth research. These considerations will generate stronger evidence of how ecosystem conditions and functions influence human health and well-being now, and into the future.

Although livelihoods and various income-generating strategies were often conduits for exposure to environmental health hazards and cornerstones of health-improving strategies, the case studies in this book reflect little in the way of formal economic considerations or analyses. Some of the research mentions cost–benefit analyses of various interventions (wall plastering for Chagas control, strategies for dengue-vector control in Asia). In other cases, such as the Kathmandu urban ecohealth project, economic analyses are implicit in the description of how various activities and improvements were chosen and implemented. But given that livelihoods (and by extension, household and community economies) are key to successfully and sustainably implement interventions designed through an ecohealth research process, there seems to be a clear need (and opportunity) to strengthen economic analyses and arguments for ecohealth. Armed with such analyses, researchers might find more traction with decision makers as an aid to systematizing innovative strategies for improved health.

The principle of equity is generally implicit in the design and conduct of ecohealth research. However, the depth of gender and social analysis in most cases remains superficial. To make substantial contributions to improving the health and well-being of women and other disadvantaged groups in society, a more thorough integration of gender and social difference and equity considerations seems in order, along with having these reflected more convincingly in research results. There is a clear need for tools in this area because existing literature is scant (see Chap. 1 for some examples). Similarly, although the research described in this book conformed to standards of ethical practice in the various countries, many ethical dimensions of ecohealth research are not assessed by ethics review boards. As a research approach informed by principles deeply rooted in ethical considerations (a better, healthier, fairer world), the practice of ecohealth would benefit from more discussion and tools to anticipate and manage ethical tensions.

Lessons Learned

This collection of case studies provides a rare opportunity to reflect on both some lessons learned and some challenges that are common to most applications of an ecosystem approach to health in developing countries. In addition to confirming the value of greater integration of ecological, economic, and gender and social analyses in ecohealth research, there are other lessons to be drawn.

Ecohealth research, as an iterative and integrative multistakeholder process, takes time, collaboration, and skill to do well. From the case studies, it may be apparent that many research teams struggled to develop a fully integrated and systemic research framework. As a result, some applications of ecosystem approaches to health began as parallel studies of different aspects of a problem, and later faced challenges linking these different parts of studies or making inferences between them. In some case studies, insights were attained and conclusions presented based on an intuitive construction of the experiences of the project, usually including (but not limited to) scientifically defensible evidence. This is not as heretical as it may first appear. Intuitive construction of knowledge and other explorations of knowledge acquisition and clinical decision making are regularly debated in medical fields (e.g., Borrell-Carrió et al. 2004; Pottier and Planchon 2011). Because of its value-oriented, participatory, and transdisciplinary (integrative) dimensions, ecohealth research has much in common with clinical reasoning. This parallel between ecohealth research and medical reasoning has been drawn by others (Waltner-Toews 2004; Waltner-Toews et al. 2004).

Nonetheless, a substantial challenge faces the field as it develops more sophisticated tools and methods to achieve better integration of different data and stronger analyses of systems. It is essential that conclusions drawn from ecohealth research be more consistently based on adequate levels of rigorously integrated analyses. More study is needed of the multistakeholder decision making occurring in ecohealth research – the equivalent to the debate over clinical reasoning in medicine. As the field matures, its peer community will naturally insist ever more convincing evidence to support published results. Many ecohealth research teams could make more effective use of their data by better planning integrated data collection and databases at the design phase. More explicit conceptual and analytical links could already be made between the knowledge generation and the intervention parts of ecohealth research. Integration of disciplines could receive more emphasis, within a participatory and on-going process, rather than waiting until after data are collected to merge them under a single objective. That the results of some case studies have yet to be published is indicative of this problem.

At the same time there is a tension, evident in the experiences presented in this book, between doing rigorous science for peer-review and conducting practical community-development interventions. The case studies present technical results and discuss how they were used to achieve community-development goals. But they also present many other development-oriented activities that do not appear in technical publications. For the community, the research question may not be a top priority and peer-reviewed publication will appear irrelevant to them. As part of the social contract inherent in this type of research, indeed, as a pillar of ethical research practice, the project and its resources need to also support the community's interests and priorities in improving their well-being. Ecohealth research may appear somewhat open-ended because its direction evolves over time. For example, in Guatemala, the Chagas disease project provided vaccination against common diseases of domestic poultry, an activity requested by the community but of no direct relevance to understanding or controlling Chagas disease, although examination of the chickens for evidence of Chagas (they had none, as dead-end hosts for the vector) and for evidence of vectors was relevant. Similarly, the leaders of the project in India's stone quarrying and crushing communities provided health assessments and advice unrelated to occupational health problems. There are many other examples in development research of this phenomenon of responding to and assisting communities in achieving improved conditions and well-being. Community participation builds expectations and requires a broader approach than that provided by conventional health or natural science research. Project teams can be either caught up in ethical dilemmas or consumed by efforts to make a difference in the lives of people sometimes living in unacceptably dire conditions. Many ecohealth research projects are successful exactly because of these efforts to make a lasting difference. However, the research process can (indeed, must) generate strong and credible evidence to be able to move more quickly into action.

Ecohealth research is a data-rich, iterative, dynamic, and adaptive process. There is much to be learned from experiences, the tests, successes, and failures of this process. Although technical findings generally find their way into peer-reviewed science publications, many additional results of ecohealth research (insights from participatory design, integrated approaches to knowledge generation, developing interventions, policy influence, social capital, and other, more indirect outcomes) are not fully captured as a matter of routine. Insights from developmental evaluation (Patton 2010), theory of change[4] (Weiss 1995; Connell and Kubisch 1995), outcome mapping (Earl et al. 2001), and from evaluation scholarship and knowledge translation fields more generally, can help the field of ecohealth become more systematic about capturing research outcomes.

No amount of diffusion and communication, no matter how innovative, will guarantee uptake of results beyond the immediate scope of the project. Systematization is a challenge, whether scaling up through levels of government or scaling out to other geographic or thematic contexts. Ecohealth research can contribute to meeting

---

[4]Many reference materials and resources are available at: www.theoryofchange.org.

this challenge. The field will benefit from greater levels of knowledge sharing around successes in systematizing results, and by acquiring a better understanding of the circumstances that favor uptake and multiplication of the impact of research results.

Building the Field

The experiences presented in this book comprise only a small part of the experiences in applying an ecosystem approach to health, and of the wider field of ecohealth. Yet in this modest set of case studies, common practices are still apparent, and taken together, they provide an initial insight into defining a practice. A defined practice shared by a peer group, as well as common principles, language, and goals, help to define a field. Debates around theory and methodologies are evidently occurring in ecohealth, but greater discussion in peer-reviewed forums would assist the field in moving forward. These forums need not be the traditional scientific format, although an established international journal like *EcoHealth*, while restricted to English, continues to provide a reliable venue for some of these discussions. Ecohealth practitioners and theorists have yet to fully embrace new information and communications technologies, which could provide more dynamic and inclusive platforms for furthering the debate to define the field.

A vibrant and active global community of interdisciplinary researchers and other stakeholders, including informed decision makers, is key to the future of ecohealth as a field of endeavor. This community is necessary for the sharing, peer reviews to conduct, and challenging of ideas, tools, and methods in ecohealth research. Members of this community already train new students, research assistants, and other academics. Through communities of practice and other ecohealth networks, as well as student groups, the field is being further developed. A growing number of people are learning about and using ecosystem approaches to health, including academics, representatives of communities, and both nongovernment and government organizations. A young International Association for Ecology and Health (www.ecohealth.net) provides still another platform for international debate and involvement.

Ecohealth is situated in a growing area of integrative and transdisciplinary practice, some of which was discussed in Chap. 21. The current interest around ideas like One Health and global health among multilateral organizations and governments, and the call for intersectoral approaches to address health vulnerabilities from climate change are indicative of this wider trend. In the private sector, integrative thinking is recognized as a necessary contribution to innovation and decision making (Martin 2007). In international development, innovation is also linked to integration, multistakeholder participation, and action for more equitable and sustainable outcomes (STEPS Centre 2010).

The field of ecohealth research, education, and practice makes welcome and valuable contributions to science and to development research in particular. This field, like other endeavors that include academic interests and specialized practice, involves building a body of knowledge and its application, to be sustained by skilled and trained practitioners. Based on the strengths and challenges, here are some key areas for further development of the field of ecohealth.

Stronger Evidence and a Wider Knowledge Base

Ecohealth research produces a credible body of results but could still benefit from stronger evidence. This seems an obvious priority to be taken on by the growing global peer community. Stronger evidence and a wider (more accessible) knowledge base can be achieved through attention to scientific rigor, integration between disciplines and types of knowledge, replication and weight-of-evidence, and further development of tools and methodologies to achieve more rigorous integration and analysis. Because the field is predicated on transdisciplinarity and inclusivity of perspectives, greater integration of ecology in the consideration of ecosystems can easily be achieved. Economic expertise can be sought from the fields of health and environmental economics. Experts in complex systems modeling (among others) can be recruited to contribute to improving tools for the design of ecohealth studies and integrated data analysis.

Communicating, Exchanging, and Disseminating Knowledge

Any field benefits from discussion and other means for collecting, analyzing, and sharing knowledge, information, and models of practice. Earlier parts of this book explored the considerable additional benefits that can be accrued from collaboration and exchange. Ecohealth communities of practice and other networks are well positioned to facilitate further dialogue. New media and technologies can enhance the capture, presentation, and diffusion of a broad range of results, and promote greater knowledge exchange. But beyond more effective dissemination of ecohealth research activities and results, ecohealth as a field has an opportunity to link more effectively to innovation systems – the national and international systems that link research, policymaking, and application.

Developing a Culture of Monitoring and Evaluation

The participatory, transdisciplinary, and iterative nature of ecohealth research sometimes leads to complicated research processes in the field. It is usually a challenge for members of research teams to keep abreast of findings, developments, interventions, and stakeholder positions. Cultivating evaluative thinking in ecohealth research can help establish order and sense to a sometimes overwhelming process. Rather than evaluation of results per se, evaluative thinking can assist research teams to understand what the research is finding (and doing), what is changing, and why. It can help structure the process of inquiry to establish: clear moments for taking stock of what new knowledge has been generated; what it might mean for stakeholders, decision makers, other researchers; and the direction of the research process itself. The field of monitoring and evaluation provides insights, methodologies, and tools for doing a better job of capturing results and outcomes of processes such as ecohealth research. Both outcome mapping and developmental evaluation, cited earlier, have shown some promise in recent applications in ecohealth research.

Developing Capacity and Strengthening the Peer Community

Training of graduate students and other academics is a fundamental aspect of any field of research, and ecohealth is exemplary in this regard. The Communities of Practice in Ecohealth have trained hundreds of graduate students and other professionals, some of whom have gone on to obtain positions as university faculty or ministerial science advisors. Members of the new generation of researchers are increasingly recognized for their transdisciplinary skills and are now working in international organizations and donor agencies. Greater opportunities are needed to engage stakeholders and policy actors in ecohealth training programs and to further enrich student training. These same stakeholders also learn and grow from exposure to the ideas of ecohealth research, and may become resources to help put ecohealth into practice in new settings and on new problems.

Ecohealth as a field lends itself to collaboration. Therefore, mechanisms and activities that connect scientists, stakeholders, and policy actors and practitioners for the exchange of ideas, knowledge, and information, and build individual and group learning opportunities and collaborative relationships, are key to its future development. Communities of Practice and other networks can foster greater North–South and South–South capacity-building partnerships. Through these types of networks, resources for the further development of the field can also be found, including leadership and human and financial resources.

Systematizing Results and Moving Upstream

Many ecohealth research projects make contributions beyond the local project-level scale. There are challenges in scaling up or out, but evidence from ecohealth research is increasingly of a maturity and sophistication appropriate to this challenge. However, there still appears to be a lack of published tools, techniques, and platforms for achieving up-scaling of results. The more distant the desirable impact is from the scale of the project, the greater the challenge to make lasting change because the drivers of that change are also more remote. For some changes to occur, ecohealth research may need to draw on economics, public administration, and political economy as much as it does on public health, ecology, and environmental sciences. In addition to collaboration and engagement with individuals involved in policy and decision making, more formal engagement with policy processes is needed. The field would be strengthened with contributions from relevant expertise and experience in political science.

An entire sub-area of the field is ripe for more research to understand how both health and environmental sustainability can benefit from co-responsibility among government, private sector, and civil society stakeholders, and how this shared responsibility for problem solving can move from the immediate scale of the project to wider applications. Leaders in ecohealth research and practice, communities of practice, and other networks can help provide the push, and share and sustain the effort often required to achieve wider impacts from ecohealth research. Engagement of these leaders and networks with decision making, from the local to international levels, can also help shape these decision making to be more receptive to research inputs and eventually develop a policy culture that seeks out evidence from research.

Over the years, concepts and ideas articulated around the need to link human health to environmental conditions, and to connect research to both local communities and decision-making processes, evolved into common research practice. Through this practice, researchers established the field of ecohcalth. IDRC has supported contributions from developing country researchers and others to this field. But more than this, IDRC has helped imbue the field with principles of integrated systems thinking, transdisciplinarity, participation, sustainability, equity, and knowledge-to-action. As a result, research using ecosystem approaches to health has made contributions to improving the health and well-being of people around the world.

As part of this growing field, ecohealth practitioners are well positioned to move research forward to meet more challenges of environment and health – both old and new. The experiences in this book show ecohealth research to be innovative and adaptable in its practice and in its outcomes. But the principles and elements of an ecosystem approach to health are not in themselves new. They draw on knowledge and understanding from many other domains. The understanding that human health depends on healthy ecosystems is a very old idea, represented in the beliefs of aboriginal peoples around the world, in the writings of Hippocrates from 2,500 years ago, and in the ancient writings and beliefs of many other cultures. The innovation lies in how these known elements, principles, and approaches are brought together.

It is clear from the experiences in this book that practitioners of ecohealth face many technical and other challenges as they strive to generate stronger evidence and to have greater or more long-lasting impact with research on tough, multi-faceted, value-laden and seemingly intractable problems. It is also apparent that the field has strong and deepening roots on every continent, and that connections and collaboration among practitioners are multiplying. Ecohealth research is in tune with some of the latest trends in innovation and international development thinking, and is in a position to contribute to pressing national and international environmental health agendas.

This is good news for a world in which unacceptable numbers of people live in poverty in degraded ecosystems, suffer an undue burden of ill-health, and have too few options for change. Ecohealth can assist in finding better and more sustainable ways to tackle health and well-being in the face of major environmental challenges from global climate change, to biodiversity loss, to the management of environmental disasters of increasingly greater impact. Much work remains to achieve healthier, more equitable and productive lives and livelihoods for people in healthy and sustainable ecosystems, now and for future generations. The field of ecohealth research appears ready for the challenge.

References

Borrell-Carrió, F., Suchman, A.L., and Epstein, R.M. (2004). The Biopsychosocial Model 25 Years Later: Principles, Practice, and Scientific Inquiry. Annals of Family Medicine, 2(6), 576–582.

Carden, F. (2009). Knowledge to Policy: Making the Most of Development Research. SAGE Publications, New Delhi, and Thousand Oaks, USA. Available at: http://www.idrc.ca/en/ev-135779-201-1-DO_TOPIC.html.

Connell, J. P. and Kubisch A. (1995). Applying a Theory of Change Approach to the Evaluation of Comprehensive Community Initiative. In: Connell, J.P., Kubisch, A.C., Schorr L.B., and Weiss C.H. (Editors). New Approaches to Evaluating Community Initiatives: Volume 1, Concepts, Methods, and Contexts. The Aspen Institute, Washington, DC.

Earl, S., Carden, F., and Smutylo, T. (2001). Outcome Mapping: Building Learning and Reflection into Development Programs. International Development Research Centre, Ottawa, Canada. Available at: http://www.idrc.ca/en/ev-9330-201-1-DO_TOPIC.html.

Martin, R.L. (2007). The Opposable Mind. How Successful Leaders Win through Integrative Thinking. Harvard Business School Press, Boston, USA.

Patton, M.Q. (2010). Developmental Evaluation. Applying Complexity Concepts to Enhance Innovation and Use. Guilford Press, New York, NY.

Pottier, P., and Planchon, B. (2011). [Description of the Mental Processes Occurring During Clinical Reasoning.] *In French*. La Revue de médicine interne, 32(6), 383–390.

STEPS Centre (Social, Technological and Environmental Pathways to Sustainability Centre). (2010). Innovation, Sustainability, Development: A New Manifesto. STEPS Centre, Brighton, UK. Available at: http://anewmanifesto.org/wp-content/uploads/steps-manifesto_small-file.pdf.

Waltner-Toews, D. (2004). Ecosystem Sustainability and Health: A Practical Approach. Cambridge University Press, Cambridge, UK.

Waltner-Toews, D., Kay, J., Tamsyn, P.M., and Neudoerffer, C. (2004). Adaptive Methodology for Ecosystem Sustainability and Health (AMESH): An Introduction. In: Midgley, G., and Ochoa-Arias, A.E. (Editors). Community Operational Research: Systems Thinking for Community Development. Kluwer and Plenum Press, New York, NY, USA.

Weiss, C. (1995). Nothing as Practical as Good Theory: Exploring Theory-based Evaluation for Comprehensive Community Initiatives for Children and Families. In: Connell, J.P., Kubisch, A.C., Schorr L.B., and Weiss C.H. (Editors). New Approaches to Evaluating Community Initiatives: Volume 1, Concepts, Methods, and Contexts. The Aspen Institute, Washington, DC.

Index

A
Abeyewickreme, W., 176
Accountability, 94
Acetylcholinesterase (AChE), 61, 64, 65
Acquired immunodeficiency syndrome (AIDS), 43, 44, 199, 234
Action-research, 2, 6, 8, 48, 49, 248
Adaptive methodology for ecosystem sustainability and health (AMESH), 19, 23
Aedes aegypti, other aedes, 163, 173
Agricultural intensification, 3, 33, 34
Agricultural livelihood, 34, 70
Agricultural transformations, 24, 34, 59
Agriculture
    extension, 47
    rainfed, 73, 75, 102
    slash and burn, 114–116
    small-holder, 37, 59
Agrochemicals, 33, 34, 54, 203
Agro-ecosystem, 34, 47–57, 74, 75, 140–144, 147, 148, 150
Agroforestry, 115
AIDS. *See* Acquired immunodeficiency syndrome
Air, air quality, 3, 5, 13, 88, 91, 92, 101, 103, 104, 106, 187, 189, 191, 192, 196, 240, 259
Alcohol, 37
Al-Hakimi, A., 69–79
Allen, T.H.F., 8
Alma Ata Declaration, 12
Amalgamating agents, 120
Amazon, 12, 20, 83–85, 109–116, 236, 260, 261, 263, 264

American University of Beirut (AUB), Lebanon, 69, 74, 203, 204, 208
AMESH. *See* Adaptive methodology for ecosystem sustainability and health
Andes, 47, 60
Animal slaughtering, 196
Animals, livestock, 141, 145
Anopheles arabiensis, 145
Anopheles gambiae, 145, 148
Anthropometric measurements, 38, 40, 218
Antimicrobial resistance, 34, 133
APEIR. *See* Asian Partnership on Emerging Infectious Diseases Research
Aquaculture, 6, 146–148
Arid land, 35, 77
Arsenic, 120, 125
Arunachalam, N., 173, 176
Asian Partnership on Emerging Infectious Diseases Research (APEIR), 234–236, 239–241, 245–248
Assessing chemical hazards, 84
Autlán, 87–89, 93–95
Avian influenza, 136, 236, 240, 241, 246
Ayala, V., 153–161
Azolla filiculoides, 145

B
Bacillus thuriengiensis, 148
Balakrishnan, K., 99–107
Barriga, R., 119–128
Batal, M., 69–79
Bazzani, R., 133–136
Beans, 38, 77
Bebnine. *See* Lebanon

Behaviour change, 192
Berti, P., 37–44
Betancourt, Ó., 119–128
Betancourt, S., 119–128
Bioaccumulation, 113
Biodiversity, 33, 74, 77, 78, 95, 271
Biomarkers of mercury exposure, 112
Bishnumati River. *See* Nepal
Boelen, C., 245
Boischio, A., 83–85
Bonatsos, C., 37–44
Bopp, J., 11
Bopp, M., 11
Boyer, E.L., 243
Branches, F., 110, 113
Brazil (or Amazon)
    Belem, 109
    Brasìlia legal, 110, 111, 114
    Cametá, 112
    Itaituba, 110
    Madeira, 109
    Santarém, 110, 111
    São Luis do Tapajos, 109, 110, 114
    Tapajos River, 109
Brazilian Agriculture and Livestock Research Agency, 116
Breast milk, 42
Breilh, J., 59–67
Brown, V., 245
Burundi, 139
Bustamante, D.M., 153–161
Butterfly model of health, 19

C

Cambodia, 241, 246
Cameroon, 14, 189, 215, 216, 219, 222, 223, 234, 256, 259–261
Canada, 1, 7, 56, 62, 67, 191, 200, 234, 236, 238, 239, 245, 246, 248
Canadian Institutes for Health Research (CIHR), 236
Capacity building, 51–52, 149, 166, 182, 190, 195, 196, 233, 239, 244, 249, 269
Cardiovascular diseases, 71, 133
Caruso project, 109, 110, 112–115
Case-control study, 17
Cash crops, 34, 70
Cassava, 38
Castro, X., 153–161
Catering, 74, 75, 215, 257
Cattle, cows, 141, 145, 148
Cattle ranching, 141

Cereal
    barley, 77, 78
    maize, 37–39, 41, 42
    sorghum, 77, 78
    wheat, 69, 77, 78
Cestodes. *See* Echinococcosis; Enteric parasites; Tap worms
CGIAR. *See* Consultative Group on International Agricultural Research
Chagas disease, 10, 14, 115, 134, 135, 153–161, 175, 234, 256, 263, 266
Charron, D.F., 1–24, 231–250, 255–271
Child health
    birthweight, 261
    child mortality, 70, 141
    child nutrition, malnutrition, 16, 37, 38, 40, 43, 51, 70, 77
    developmental delays, 91
    malaria, 2, 16, 20, 139, 141, 147
China, 191, 241, 245
Chione, E., 37–44
Chitaya, A., 37–44
Chronic diseases, 74, 76, 133, 204
Ciguatera, 13
CIHR. *See* Canadian Institutes for Health Research
CIP. *See* International Potato Centre
Cities, 8, 73, 110, 159, 187, 188, 215
Climate, climate change, climate extremes, 3–5, 9, 10, 16, 34, 44, 62, 70, 73, 134, 140, 174, 267, 271
Cohort study, 225
Cole, D.C., 47–57
Collaboration, 13, 55, 56, 74, 77, 111–112, 156, 222, 231, 233, 268, 269
    community, 77, 111–112
Communication, 12, 19, 26, 53, 56, 63, 89, 92, 94, 107, 113, 165, 236, 242, 245, 247, 259, 266
Community action plan, 195
Community development, 135, 157, 160, 177, 192, 194–196, 200, 203, 225, 256–258, 266
Community empowerment, 2, 122, 127–128, 256–258
Community of practice
    benefits, 102, 248
    community, 235, 238, 243
    COPEH-Canada, 234, 238
    core functions, 233
    costs, 102, 247, 248
    definition, 235
    domain, 235, 243
    evaluation, 181

Index

evolution, 112, 233
Latin America and the Caribbean
    (COPEH-LAC), 234–240, 245, 247
Middle-East and North Africa
    (COPEH- MENA), 234
practice, 235, 238, 243
structure, 236
training, 235, 240
West Africa (COPES-AOC), 234
Community of practice Canada, 238
COPEH-Canada, 238
Community organizations, 49, 60, 61, 63,
    65, 122, 178, 199, 200, 225, 238, 257
Complexity, complex systems, 6, 8, 10, 19,
    74, 78, 94, 204, 214, 232, 242, 245,
    260, 262, 268
Compost, 181, 196, 222
Conflict, 12, 13, 15, 42, 61, 70, 94, 189, 203,
    206, 207
Conflict over land, 70
Conservation
    biology, 9
    medicine, 9, 24
Consultative Group on International
    Agricultural Research (CGIAR), 140
Cooperatives, 75
Crop residues, 37, 39
Cuba, 13, 135, 163–165, 167, 169,
    240, 259, 263
    cotorro, 135
Cueva, E., 119–128
Cultural sensitivity, 63–65
Curriculum, 236
Cyanide, 120, 121, 126–128

D

Dairy farming, 141
Dakishoni, L., 37–44
DALYs. See Disability adjusted Life years
DDT, 18
Deforestation, 3, 112, 114, 115, 134, 264
    for agricultural purposes, 112
Dengue, 20, 133–136, 163–170, 173–183,
    234, 258, 263, 264
    epidemics, 164, 165
    integrated dengue surveillance system,
      164–168
Dengue hemorrhagic fever (DHF), 134, 163,
    164, 178, 179, 181
Determinants of health, 4–6, 15, 189
DHF. See Dengue hemorrhagic fever
Diarrhoeal disease, 204, 217, 219, 222
Díaz, C., 163
Dietary diversity, 35, 41, 43, 69–79, 260

Disability adjusted Life years (DALYs), 191
Disease control, prevention, 133–136, 142,
    147, 157, 159, 174–178, 206, 238, 259
Disease ecology, 5, 143
Disease emergence, 136, 241
Disease transmission, distribution, 133, 158,
    173, 192, 204, 212, 217, 221, 226
Dose response, Not found
DPSEEA framework, 19
Drama
    communication tool, 39
    puppet shows, 50
Drinking water, 3, 16, 20, 121, 124, 127, 128,
    188, 189, 206, 208, 209, 211, 216,
    218–226
Drug use, 61
Dust, 13, 83–85, 92, 96, 100, 103–107, 256
Dyslipidemia, 76

E

Earth summit, 1
Echinococcosis, 191, 196, 261
Echinococcus, 200
Eco-bio-social framework, 174, 175, 178, 183
Ecohealth
    field-building, 232, 233, 240, 244
    field definition, 5–6, 244
    journal, 7, 232
    principles, 2, 7, 9–19, 21, 23, 24, 62, 175,
      231, 232, 241–244, 250, 262, 265, 270
Ecohealth network, 24, 233, 244–250, 267
    core functions, 233
Ecohealth research, 1–6, 8–14, 16–24, 34, 85,
    94–95, 99–108, 188, 231–250,
    255–271
    frameworks, 94–95
    intervention phase, 22, 177
    practice, 232
    process, 11, 20, 23, 256, 264
École nationale supérieure polytechnique,
    Cameroon, 215
Ecology, 5–8, 10, 89, 115, 133–135, 143,
    165, 175, 177, 235, 239, 263, 264,
    267, 268, 270
Ecology of vectors, 133, 263
Economic burden, 3
Economic growth, 59, 83
Ecosystem
    change, 3
    health, 7, 20, 24, 62, 74, 78, 96, 100,
      192, 200
    intervention, 135, 148, 156, 159,
      180, 181
    services, 4, 6

Ecuador, 3, 4, 13, 20, 34, 35, 47–57, 59–67, 83–85, 119, 234–237, 239, 240, 245, 246, 255–259, 263
  Cananvalle, 62–65
  Cangahua, 60
  Carchi province, 47, 48, 50, 51, 54
  Cayambe, 60, 61, 63
  Chimborazo, 48, 50, 52–54
  floriculture, 59, 61, 66, 258, 260, 261
  gold mining, 20, 84, 85, 119–128, 234, 236, 261, 263
  Gramadal, 120–124, 126, 127
  horticulture, 47, 59
  Las Vegas, 120–124, 126, 127
  Municipal Environmental Management Units, 127
  Portovelo, 119, 120, 122–128
  Puyango Viejo, 120, 124, 127
  San Isidro, 62, 65
  Tungurahua, 48, 54
  Zaruma, 119, 120, 122–128
Ecuador graduate programs, 240
  Sustainably Managing Environmental Health Risks, 235, 239, 240
Education, 1, 5, 6, 9, 15–17, 42, 43, 47, 50, 54, 55, 65, 84, 89, 106, 123, 126, 143, 147–149, 163, 165, 167, 179, 188–190, 192, 199, 221, 224, 239, 240, 245, 247, 249, 255, 261, 267
EID. *See* Emerging infectious diseases
Ekwendeni. *See* Malawi
Electrification, 127
El-Fattal, L., 33–35
Emerging infectious diseases (EID), 136, 234, 235, 241, 246
Emigration, 159
Enteric parasites, 208
  worms, 191
Entomological surveillance, 165, 178
Environmental change, 3, 4, 9, 24, 34, 60, 133, 173, 187, 196, 263
Environmental contamination, 222
Environmental health, 7, 94, 95, 100, 102–103, 119–128, 165, 166, 226, 235, 239, 240, 261, 262, 264, 271
Environmental Management Roundtable of the Mining District of Molango, 93–96
Epidemiology, epidemiological, 7, 17, 48, 50, 66, 84, 88, 89, 96, 101, 114, 140, 143, 149, 164–166, 192, 216
Espino, F., 176
Ethics, ethical, 8, 14, 15, 17, 22, 23, 38, 243, 264–266
Ethnography, ethnographic, 76

Etoga, S.M., 215
Export markets, 59, 66, 258
Exposure assessment, 84
Extreme weather, 34, 187

F
FAO. *See* Food and Agriculture Organization of the United Nations
farinha (manioc flour), 111, 115
Farmer associations, 43, 44
Farmer field schools, 47, 51
Farmer Research Team (FRT), 38, 39, 43, 44
Fertilizer, 33, 34, 37, 39, 44, 47, 70, 142
Field-building, field-building education, 231–250
Fish, 13, 84, 109–116, 124–126, 128, 141, 146, 147, 181, 236, 260
Fish consumption, 84, 109–116, 124, 126
Fish-farming, aquaculture, 6, 141, 146–148
Floodplain lakes, 111, 112
Floriculture, flowers, 59, 61, 66, 196, 258, 260, 261
Flower Label Program (FLP), 62
Food and Agriculture Organization of the United Nations (FAO), 53
Food diversity, 35, 41, 74
Food insecurity, security, 17, 35, 37–44, 70, 72, 73, 78, 104, 135, 142
Food safety, 193
Forest, forestry, agroforestry, 99, 102, 110, 112, 113, 115
Forget, G., 1, 7
Framework, 2, 4, 6, 7, 9, 12, 13, 18–21, 23, 52, 88, 89, 94–95, 135, 149, 174–178, 182, 183, 190, 243, 262, 265
FRT. *See* Farmer Research Team
Fruit consumption, 114
Fruit production for export, 70

G
Gender
  differences, 15, 134, 260
  equity, 15–17, 62, 219, 236, 256, 260–261, 264
  gender and stakeholder analysis (GSA), 22, 179, 193
  GSA, 22, 179, 193
  inequity, 16, 17, 22
  relations, 40, 42, 43, 257
  roles (*see* Grandmothers)
Genetic resources, 33, 74, 76
Global health, 4, 5, 24, 53, 267

Index 277

Globalization, 3, 133
Gold, gold mining, 20, 84, 85, 119–128, 234, 236, 261, 263
Governance, 10, 15, 53, 54, 66, 78, 85, 93–94, 189, 247, 257
  definition, 93
Goy, J., 13
Grandmothers, 40, 42. *See also* Gender roles, Women
Granobles basin. *See* Ecuador
Grazing, over-grazing, 77
Greenhouse gas, 4
Greenhouses, 60, 61, 65, 157, 204
Green Revolution, 33
Gross Stein. J., 248
Groundnut, 37, 38
Guatemala, 10, 14, 22, 36, 134, 135, 153–161, 234, 255, 256, 259, 260, 263, 266
Guimarães, J.R.D., 109–116, 119–128

H
haciendas, 59
Harvest, harvesting, over-harvesting, 37–39, 42, 43, 65, 77, 141, 157
Havana. *See* Cuba
Health promotion, 5, 7, 50, 165
Health systems, 56, 61, 133, 166, 175, 190
Healthy kitchen network, 74
Heavy rainfall, 115
Helminths. *See* Worms
Hidalgo. *See* Mexico
Highlands, 35, 60, 63, 73, 76, 77
Hirsch Hadon, G., 245
HIV. *See* Human immuno deficiency virus
Hoekstra, T.W., 8
Horticulture, 47, 59
Household
  income, 16, 42, 73, 147
  resources, 42, 141
  as unit of analysis, 157
Housing, 134–136, 141, 146, 148, 154, 155, 160, 178, 199, 215, 259
  housing improvements, 135, 156, 259
Human ecology, 7, 177
Human immuno deficiency virus (HIV), 133, 199, 234
Hydatidosis. *See Echinococcosis*
Hygiene practices
  household hygiene, 219–220, 226
  hygiene and sanitation, 196, 198, 199, 212, 226
  hygiene in health care setting, 198–199
  meat hygiene, 198

I
IAEH. *See* International Association for Ecology and Health
IDSS. *See* Integrated dengue surveillance system
India, 3, 83, 85, 99–107, 176, 256, 258, 266
Indigenous
  indigenous people, groups, 48, 59, 65, 67, 178
  mestizo, 59, 65
Indigenous knowledge, 74, 75, 78
Indigenous plants, 75
Indonesia, 173–183, 241, 246, 255
Infectious disease, 7, 24, 84, 133, 134, 136, 187, 234, 235, 241, 261
Innovation, 2, 10–12, 17, 18, 39, 54, 73, 107, 135, 239, 242, 244, 248–250, 256–257, 259, 267, 268, 270, 271
Insecticides
  DDT, 18
  highly toxic insecticides, 34, 56
  insecticidal spraying, 133, 134, 153–156
Integrated dengue surveillance system (IDSS), 164–168
Integrated pest management (IPM), 50, 51, 54, 55
Integrated vector management, 174
Integration, 4, 8, 11, 20, 21, 95, 100, 149, 164, 166, 167, 169, 232, 241–245, 248–250, 264, 265, 267, 268
Intelligence quotient (IQ), 91
Intensive agricultural production, 47, 61
Intercropping, 37, 75, 77
Intergovernmental Panel on Climate Change (IPCC), 4
International Association for Ecology and Health (IAEH), 232, 234, 235, 239, 267
International Association for Ecology and health Student Section, 239, 245
International Code of Conduct for Cut Flower Production, 66
International EcoHealth Forum (Mérida 2008), 239
International Forum on Ecosystem Approaches to Human Health (Montreal 2003), 233
International Joint Commission, 7
International Potato Centre (CIP), 34
Intervention phase of ecohealth research, 22
IPCC. *See* Intergovernmental Panel on Climate Change
IPM. *See* Integrated pest management
IQ. *See* Intelligence Quotient

Iron, 42, 71
Irrigated rice farming, 147
Irrigation, 6, 61, 62, 70, 73, 143, 147, 149, 203

J
Joshi, D.D., 191
Jutiapa. *See* Guatemala

K
Kathmandu. *See* Nepal
Kay, J.J., 8
Kenya, 139
Kerr, R.B., 37–44
Kittayapong, P., 176
Knowledge, attitudes and practices (KAP) survey, 154, 156–159, 175, 179
Knowledge development phase, 22
Knowledge-to-action, 85, 242, 243, 256, 258, 270
Knowledge translation, 17, 18, 23, 48, 144, 149, 244, 266
Kraals, 140, 145
Kroeger, A., 176
Kueffer, C., 244
Kuhn, T., 8

L
Lakshmi, K.V., 99–107
Lao PDR, 241
Lave, J., 243
Lavis, J.N., 18
Lead
  contamination (cookware), 124–128, 225
  poisoning, 13
Lebanon
  Arsaal, 69, 70, 74
  Batloun, 74, 75
  Bebnine, 23, 203–212, 260
  Bebnine-water, 205, 207, 208
  Bekaa Valley, Not found
  Kuakh, 74, 77
Lebel, J., 1, 7
Legal frameworks, 18
Legitimacy, 193
Legume, 16, 18, 35, 37–44, 47, 77, 256, 259
Leishmaniaisis, 136
Lentils, 77
Livelihoods, 3, 5, 14, 15, 17, 34, 35, 51, 54, 59, 70, 78, 95, 101, 104, 111, 119, 134, 141, 142, 181, 193, 234, 260, 261, 263, 264, 271

Livestock intensification, 34
Local food systems, 35, 73–76

M
Maize, 37–39, 41, 42
Malaria, 2, 16, 20, 133–135, 139–150, 234, 260, 263
  epidemics, 141, 143
  treatment-seeking, 141, 147
Malawi, 12–14, 16, 18, 35, 37–44, 233, 255–257, 259–261
  Ekwendeni, 37, 38, 41, 43, 44, 233
Malnutrition, 16, 37, 38, 51, 70, 77
Manganese (Mn), 13, 18, 83–85, 87–96, 120, 124–126, 128, 234, 236, 259, 263
  air, 13, 88, 92, 259
  in fish, 125
  iron antagonism, 91
  soil, 18, 85, 88, 120
  water, 88, 124–126, 128, 234
Maona, E., 37–44
Map- ecosystem map, 2, 17, 20, 24, 88, 122
Marketing, 49, 56, 75, 257
Maya bees, 157
Mboera, L.E.G., 139–150
MDG. *See* Millennium development goal
MEA. *See* Millennium Ecosystem Assessment
Meat processing, 14
Media (journalism, reporting, as communication tool), 12, 19, 50, 52, 56, 75, 77, 89, 92, 94, 107, 113, 121, 135, 149, 178, 236, 242, 247, 257, 259, 266, 268
Media and policy influence, 18, 78, 247, 248, 256, 258–260, 266
Medical costs, 43
Mercury, 20, 83–85, 109–116, 120, 121, 124–128, 234, 236, 260, 261, 264
  in fish, 84, 109
  methylation process, 127
  poisoining, toxicity, 20
Mergler, D., 19, 109–116
Mertens, F., 12
Methyl mercury, 84
Mexico, 13, 18, 83, 84, 87–96, 234, 236, 239, 240, 246, 259
  Agua Blanca, 91
  Chiconcoac-Tolago, 89
  Naopa, 93
Micronutrients, micronutrient deficiency (ies), 38, 71
Midwife, midwives, 111, 112, 114

MIGA. *See* Environmental Management Roundtable of the Mining District of Molango
Migrant workers, 179
Millennium development goal (MDG), 2, 33
Millennium Ecosystem Assessment (MEA), 4, 5, 19
Mingoa River basin. *See* Cameroon
Mining, 13, 18, 20, 83–85, 87–96, 109, 112, 115, 119–128, 234, 236, 259, 260, 263
　employment, 83, 93, 99
　gold, 20, 84, 85, 109, 110, 112, 115, 119–128, 234, 236, 263
　informal, 83, 85, 119, 120
　manganese, 13, 18
　national legislation, 120, 259
　open-face ore extraction, 87
　silver, 119, 120
Ministries or Departments of Health, Environment, Agriculture, 61, 149, 161
Miranda, M.R., 127
Mithi, M., 37–44
Mogoue, B., 215
Moguel, B., 153–161
Monitoring and evaluation, 22, 181, 268–269
Monroy, C., 153–161
Morrison, K., 13
Mosquito breeding sites, 139, 142, 146, 147, 166, 167, 170
Motor and visual functions, 113
Msachi, R., 37–44
Mugisha, S., 139–150
Multi-criteria evaluation, 12
Myanmar, 173, 176, 178

N

National Institute of Hygiene, Epidemiology, and Microbiology, Cuba, 164
National Zoonoses and Food Hygiene Research Centre (NZFHRC), Nepal, 191–194
Natural resource development, 83
Natural resource extraction, 83, 85
Negotiation, 12, 13, 56, 93, 95, 157, 190
Neighbourhood, 14, 165, 167–168, 177–180, 188, 189, 198, 204, 206–211, 215–226
Nepal, 14, 191, 193–195, 197, 198, 200, 233, 260, 261
Network for Ecosystem Sustainability and Health (NESH), 235
Networks, 18, 24, 66, 113, 231–250, 257, 263, 267–270
　analysis, 248
　nodal structures, 239, 245

Neuro-behavioural assessment, 64
Neurological examinations, 113
Neurological functioning, 48
Neurotoxicity, 47
Ngnikam, E., 215
Nkhonya, Z., 37–44
Nutrition, 12, 14, 16, 35, 37–44, 49, 51, 70–78, 84, 104, 106, 126, 218, 256, 260, 261

O

Obesity, 71, 72, 76, 77, 133
Occupational pesticide exposure, 63
Occupation, occupational health, workplace health, 34, 35, 99, 100, 104–107, 119, 256, 266
Okello-Onen, J., 139–150
One Health, 24, 267
Orosz, Z., 83–85
Orozco, F.A., 47–57
Ottawa Charter, 5
Outcome mapping, 12, 17, 18, 266, 269

P

Pachanya, S., 37–44
Panamerican Health Organization (PAHO), 134
Pandemic influenza, 133
Pandemics, 3, 133
Parkes, M.W., 7, 19, 231–250
Parliament of Canada, Not found
Participation
　ecohealth networks, 24, 244–250, 267
　ecohealth principle, 19–21, 24, 232, 242, 250
Participatory action research, 2, 13, 165–167, 190, 192
Participatory design phase, 21–22, 263
　ecohealth research process, 11, 20, 23, 256, 262, 264
Participatory rural appraisal (PRA), 13, 75
Participatory urban appraisal (PUA), 193
Particulate matter (PM), suspended particles, suspendended particulate matter, 88, 103, 112, 124
Partnerships, 180, 233, 248, 263, 269
Pastoralists, pastoralism, 141, 145, 150
Pedro Kourì Tropical Medicine Institute, Cuba, 164
Pelat, F., 69–79
Peru, 13, 34, 119, 125

Pesticide
    carbamate, 61, 63–65
    chlorinates, 61, 64
    highly hazardous, 48, 50, 51, 54, 55, 61
    organophosphate, 61, 63, 64
    resistance, 34
Pesticide poisoning, acute, 34, 50, 52, 55, 240
Pests, 33, 44
Petzold, M., 176
Philippines, 176
Phthalate, 63, 65
Pigeon pea, 37
Pineda, S.S., 153–161
Pohl, C., 245
Policy change, 2, 23, 56, 60, 247, 259
Policy influence, 18, 78, 247, 248, 256, 258–260, 266
Policy makers, 19, 48, 50, 59, 78, 85, 96, 143, 144, 148, 149, 177, 233, 238, 240, 245, 246, 256, 259
Policy making, 116, 259, 268
Political economy (in ecohealth research), 62, 270
Political instability, 189, 200, 203
Political voice, 83, 188, 261
Pollution, 3, 5, 13, 18, 20, 24, 34, 64, 83–85, 88, 96, 101, 104, 112, 119–121
    definition, 84
    Perception, 84
Population
    growth, 2, 133
    health approach, 5
Potatoes, 47, 56, 63, 64, 77
Poverty alleviation, 78, 143
Power
    decision-making, 8, 49, 83, 123
    empowerment, 2, 13, 122, 127–128, 211, 256–258
PRA. *See* Participatory rural appraisal
Prevention, 133, 134, 136, 139, 142, 147, 153–161, 174, 175, 177–182, 206, 210, 238, 240, 241
Principles of ecohealth, 23, 24, 62, 232, 243
Prism research framework, 6, 7, 19, 20, 94–95, 175, 176, 178, 182, 265
Private sector, 18, 85, 262, 267, 270
PUA.*See* Participatory urban appraisal
Puya ngo River. *See* Ecuador

Q
Qat, 70
Quarry, 70, 85, 99–107, 256, 258, 266
Quiñonez, J., 153–161

R
Ramazzini, 119
Ranjan, R., 99–107
Regier, H., 8
Respiratory health, 85, 106
Rights
    farmer's rights, 53
    health rights, 66
    human rights, 5
Riojas-Rodríguez, H., 87–96
Risk-management, 13, 89, 92, 93, 96
Risk-perception, 91
Rivers, 115, 120, 127
Rodas, A., 153–161
Rodríguez-Dozal, S., 87–96
Rwanda, 139

S
Salinization (of groundwater, soils), 70
Sana'a. *See* Yeman
Sánchez, A., 33–35, 231–250
Sanitation, 3, 14, 102, 133, 166, 175, 187–189, 196–199, 208, 210, 212, 216–220, 222–226, 240
SAS. *See* Social analysis systems
Savannah dry, 143
Savannah, wet, 143
Scale, 3, 9, 10, 23, 48, 49, 55, 77, 85, 100, 119–128, 161, 175, 179, 187, 238, 248, 250, 258, 260, 262, 269, 270
Security, 5, 17, 37–44, 72–74, 78, 100, 104, 135, 142, 188, 189, 234
Sediments, river bottom, 125, 126
Seeds
    banks, 43, 76, 256
    storage, 40
Selenium, 114
Self-evaluation, 50, 53
Shumba, L., 37–44
SIMA. *See* Systemwide Initiative on Malaria and Agriculture
Slums, 14, 188, 189, 215, 216, 257
Small-scale mining, 119, 120
Social analysis systems (SAS), 12, 195
Social and gender analysis, 17
Social capital, 51, 53, 200, 241, 257, 266
Social class, 15
Social Determinants of Health CSDH, 4, 6, 15
Social-ecological systems, 136
Social innovation, 54
Social medicine, 7
Social unrest, 61
Socio-economic status, 71, 102, 206
Soils, Soil fertility, 12, 13, 18, 35, 37–44, 56,

60, 70, 109, 112, 115, 120
Sommerfeld, J., 176
Soya beans, Not found
Sri Lanka, 176
Stakeholder analysis, 13, 22, 88, 176, 179, 193
Stone crushing, 85, 99–107
Stunting, 16, 71
Subsidy, subsidies, 44, 78
Sudsawad, P., 18
Sulfur dioxide, 92
Sustainable agriculture, 49–51, 109–116
Sustainable development, 1, 4, 7–9, 14, 15, 78, 255, 258
Synthetic pyrethroids, 145
Systematization phase (in ecohealth research), 23
Systemwide Initiative on Malaria and Agriculture (SIMA), 142, 143, 146, 234

T
Tailings (mining), 120, 121, 127
Tana, S., 176
Tanawa, E., 215
Tanzania, 134, 139–150, 234, 255, 263
    Mvomero district, 140, 142, 144
Tapeworms. *See* Echinococcosis; Enteric parasites
Target framework, 19
Teasdale-Corti Foundation, 115
Thailand, 4, 176, 241, 246
Toxic substance(s), 3, 83, 85, 122, 126, 240, 246
Traditional cropping pattern, 75, 76
Traditional crops, 47
Traditional medicine, 111
Train-the-trainer program, training of trainers (TOT) 12, 56
Transparency, 93
Triatoma dimidiate, 136, 153
Triatomines, 153
Tropical Diseases Research and training, WHO Special Program (TDR), 134, 234
Trypanosoma cruzi, 134, 153
Tuberculosis, 2
Tumbes River, Peru. *See* Puyango River
Tyagi, B.K., 176

U
Uganda, 134, 139–150, 234, 263
    Mutara village, 148

United Nations, 1, 2, 9
    Agenda 21, 1
University
    Federal University of Pará, Brazil, 109
    Universidad de Cuenca, 239
    Universidade de Brasília, 111
    Universidade de São Paulo–Ribeirão Preto, 111
    Universidade Federal do Rio de Janeiro, 111
    Université de Montréal, 111
    Université de Yaoundé I and II, Cameroon, 215
    Université du Quebec à Montréal, 109
    University of British Columbia, 240
    University of Northern British Columbia, 238
    University of Guelph, 191, 192
Urbanization, 3, 24, 134, 163, 164, 175, 187

V
Vaccines, vaccination, 102, 133, 160, 163, 266
Vector, 79, 133–136, 139, 141, 142, 145, 148, 153, 154, 156, 158–160, 163–170, 173–180, 182, 183, 187, 238, 240, 260, 261, 263, 264, 266
Vector-borne diseases, 85, 133–136, 142, 154, 163, 165, 182, 238, 261
Vector ecology, 135, 175, 177, 263
Vector eradication campaigns, vector control, 133
Vehicular traffic, 92
Vietnam, 241, 246
Violence, 15, 123, 188

W
Wai, K.T., 176
Waltner-Toews, D., 1, 8, 13, 191
Waste, 34, 93, 96, 128, 166, 177, 179, 181, 188, 189, 191–193, 196–198, 204, 206, 216–217, 220, 223, 226, 240
    management, 34
Water
    borne causes of diarrhoea, 133
    drinking water, 3, 16, 20, 121, 124, 127, 128, 188, 189, 203, 206, 208, 209, 211, 216, 218–222, 224–226
    quality, 4, 20, 187, 189, 203, 204, 206–209, 217, 219, 220, 222, 223, 225, 234
    sampling, 205, 208, 217, 218, 221, 222, 225
    and sanitation, 199, 217

Water (*cont.*)
  storage, 141, 145, 148, 150, 166, 167, 178, 181, 206, 209, 219, 222, 225, 256
  supply, 70, 73, 169, 175, 177, 182, 220, 224, 260
  supply policies, 73
  water management, 7, 34, 147, 148, 180
Water buffalo, 191, 196
Water filters, ôLe Seauö, 127
Wenger, E., 243
Wiese, M., 133–136
Wilcox, B., 244
Wild edible plants, 69, 73–77
Wildlife, 115, 120, 241
Witchcraft, 147
Women
  gender equity, 260–261
  maternal health, 260, 261
  maternal roles and responsibilities, 260, 261
  pregnancy, 139
  pregnant women, 139
  violence against women, 16, 123, 188
  women and malaria, 16, 135, 139, 141
  women empowerment, 122
  women farmers, 56
  women head of household, 16
  women in research, 16, 35, 37, 39, 51, 65, 71, 73–75, 114, 181, 205,
  women's health, 16, 102, 260
Woollard, R.F., 243
World Health Organization (WHO), 4, 61, 72

Y

Yaounde. *See* Cameroon
Yemen, 35, 69–78, 256, 261
Yogyakarta. *See* Indonesia

Z

Zinc, 120
Zoonosis, zoonoses, 134, 136, 191
Zoophilic mosquito, 141, 145
Zooprophylaxis, 145